The Fast Solution of Boundary Integral Equations

MATHEMATICAL AND ANALYTICAL TECHNIQUES WITH APPLICATIONS TO ENGINEERING

Series Editor
Alan Jeffrey

The importance of mathematics in the study of problems arising from the real world, and the increasing success with which it has been used to model situations ranging from the purely deterministic to the stochastic, in all areas of today's Physical Sciences and Engineering, is well established. The progress in applicable mathematics has been brought about by the extension and development of many important analytical approaches and techniques, in areas both old and new, frequently aided by the use of computers without which the solution of realistic problems in modern Physical Sciences and Engineering would otherwise have been impossible. The purpose of the series is to make available authoritative, up to date, and self-contained accounts of some of the most important and useful of these analytical approaches and techniques. Each volume in the series will provide a detailed introduction to a specific subject area of current importance, and then will go beyond this by reviewing recent contributions, thereby serving as a valuable reference source.

Series Titles:
THE FAST SOLUTION OF BOUNDARY INTEGRAL EQUATIONS
  Sergej Rjasanow & Olaf Steinbach, ISBN 978-0-387-34041-8
THEORY OF STOCHASTIC DIFFERENTIAL EQUATIONS WITH JUMPS AND APPLICATIONS
  Rong Situ, ISBN 978-0-387-25083-0
METHODS FOR CONSTRUCTING EXACT SOLUTIONS OF PARTIAL DIFFERENTIAL EQUATIONS
  S.V. Meleshko, ISBN 978-0-387-25060-1
INVERSE PROBLEMS
  Alexander G. Ramm, ISBN 978-0-387-23195-2
SINGULAR PERTURBATION THEORY
  Robin S. Johnson, ISBN 978-0-387-23200-3
INVERSE PROBLEMS IN ELECTRIC CIRCUITS AND ELECTROMAGNETICS
  N.V. Korovkin, ISBN 978-0-387-33524-7

The Fast Solution of Boundary Integral Equations

Sergej Rjasanow
Universität des Saarlandes

Olaf Steinbach
Technische Universität Graz

Sergej Rjasanow
Fachrichtung 6.1 – Mathematik
Universität des Saarlandes
Postfach 151150
D-66041 Saarbrücken
GERMANY

Olaf Steinbach
Institut für Numerische Mathematik
Technische Universität Graz
Steyrergasse 30
A-8010 Graz
AUSTRIA

ISBN 978-1-4419-4160-2
e-ISBN 978-0-387-34042-5

Printed on acid-free paper.

9 8 7 6 5 4 3 2 1

springer.com

Preface

Boundary Element Methods (BEM) play an important role in modern numerical computations in the applied and engineering sciences. Such algorithms are often more convenient than the traditional Finite Element Method (FEM), since the corresponding equations are formulated on the boundary, and, therefore, a significant reduction of dimensionality takes place. Especially when the physical description of the problem leads to an unbounded domain, traditional methods like FEM become unalluring.

A numerical procedure, called Boundary Element Methods (BEM), has been developed in the physics and engineering community since the 1950s. This method turns out to be a powerful tool for numerical studies of various physical phenomena. The most prominent examples of such phenomena are the potential equation (Laplace equation) in electromagnetism, gravitation theory, and in perfect fluids. A further example leading to the Laplace equation is the steady state heat flow. One of the most popular applications of the BEM is, however, the system of linear elastostatics which can be considered in both bounded and unbounded domains. A simple model for a fluid flow, the Stokes system, can also be solved by the use of the BEM. The most important examples for the Helmholtz equation are the acoustic scattering and the sound radiation.

It has been known for a long time that boundary value problems for elliptic partial differential equations can be reformulated in terms of boundary integral equations. The trace of the solution on the boundary and its conormal derivative (Cauchy data) can be found by solving these equations numerically. The solution of the problem as well as its gradients or even high order derivatives are then given by the application of Green's third formula (representation formula); this method based on Green's formula is called the direct BEM approach. Another possibility is to use the property that single or double layer potentials solve the partial differential equation exactly for any given density function. Thus, this function can be used in order to fulfill the boundary conditions. The density function obtained this way has, in general, no physical meaning. Therefore, these boundary element methods are called indirect.

When boundary integral equations are approximated and solved numerically, the study of stability and convergence is the most important issue. The most popular numerical methods are the Galerkin methods which perfectly fit to the variational formulation of the boundary integral equations. The theoretical study of the Galerkin methods is now completed and provides a powerful theoretical background for BEM. Traditionally, however, the collocation methods were widely used, especially in the engineering community. These methods provide an easier practical implementation compared with the Galerkin methods. However, the stability and convergence theory for collocation methods is available only for two-dimensional problems. Furthermore, the error analysis of the collocation methods for three-dimensional problems, when assuming their stability, shows that the rate of convergence of the Galerkin methods is better, when assuming that the solution is smooth enough.

In any case, a numerical procedure applied to the boundary integral equation leads to a linear system of algebraic equations. The matrix of this system is in general dense, i.e. almost all its entries are different from zero, and, therefore, have to be stored in computer memory. It is clear that this is the main disadvantage of the BEM compared with FEM which leads to sparse matrices. This quadratic amount of computer memory sets very strong, unattractive bounds for the discretisation parameters and, often, force the user to switch to the out–of–core programming. However, so called fast BEM have been developed in the last two decades. The original methods are the Fast Multipole Method and the Panel Clustering; another example is the use of wavelets. Furthermore, the Adaptive Cross Approximation (ACA) was introduced and successfully applied to many practical problems in the last years.

The purpose of this book is twofold. The first goal is to give an exact mathematical description of various mathematical formulations and numerical methods for boundary integral equations in the three-dimensional case in an uniform and possibly compact form. The second goal is a systematic numerical treatment of a variety of boundary value problems for the Laplace equation, for the linear elastostatics system, and for the Helmholtz equation. This study will illustrate both the convergence of the Galerkin methods corresponding to the theory and the fast realisation of BEM based on the ACA method. We restrict our numerical tests to some more or less artificial surface examples. The simplest one is the surface of the unit sphere. Furthermore, two TEAM examples (Testing Electromagnetic Analysis Methods) will be considered besides some other non-trivial surfaces.

This book is subdivided into four parts. Chapter 1 provides an overview of the direct and indirect reformulations of second order boundary value problems by using boundary integral equations, and it discusses the mapping properties of all boundary integral operators involved. From this, the unique solvability of the resulting boundary integral equations and the continuous dependence of the solution on the given boundary data can be deduced. Chapter 2 is concerned with boundary element methods, especially with the Galerkin method. The discrete version of the boundary integral equations from Chapter 1 and their variational formulations lead to systems of linear equations with different matrices. The entries of these matrices are explicitly derived for all integral operators involved. Chapter 3 describes the Adaptive Cross Approximation of dense matrices and provides, in addition to the theory, some first numerical examples. The largest part of the book, Chapter 4, contains some results of numerical experiments. First, the Laplace equation is considered, where we study Dirichlet, Neumann, and mixed boundary value problems as well as an inhomogeneous interface problem. Then, two mixed boundary value problems of linear elastostatics will be presented, and, finally, many examples for the Helmholtz equation are described. We consider again Dirichlet and Neumann, interior and exterior boundary value problems as well as multifrequency analysis. Many auxiliary results are collected in three appendices.

The chapters are relatively independent of one another. Necessary notations and formulas are not only cross-referred to other chapters but usually repeated at the appropriate places.

In 2003, Prof. Allan Jeffrey approached us with the idea to write a book about fast solutions of boundary integral equations. It has been delightful to write this book and we are also very thankful for his providing the opportunity to get this book published.

We would like to thank our colleagues from the BEM community for many useful discussions and suggestions. We are grateful to our home institutions, the University of Saarland in Saarbrücken and the Technical University in Graz, for providing an excellent scientific environment and financial funding to our research.

We appreciate the help of Jürgen Rachor, who read the manuscript and made valuable comments and corrections. Furthermore, the authors would very much like to express their appreciation to Richard Grzibovski for his help in performing numerical tests.

Saarbrücken and Graz
March 2007

Sergej Rjasanow
Olaf Steinbach

Contents

1

Boundary Integral Equations

The solutions of second order partial differential equations can be described by certain surface and volume potentials when a fundamental solution of the underlying partial differential equation is known. Although the existence of such a fundamental solution can be guaranteed for a wide class of partial differential operators, see for example [53], the explicit construction of fundamental solutions is a more difficult task in the general case. Hence, we consider here partial differential operators with constant coefficients only. In particular, we restrict ourselves to the Laplace operator, the Helmholtz operator, and the systems of linear elastostatics and of Stokes, which include the most important applications of boundary integral equations and boundary element methods.

When using either a representation formula stemming from Green's second formula or when considering indirect surface potential methods, one has to find unknown density functions from the given boundary conditions. This is done by applying the corresponding trace operators to the surface and volume potentials yielding appropriate boundary integral equations to be solved. Depending on the given boundary conditions one can derive different formulations of first or second kind boundary integral equations. Although on the continuous level all boundary integral equations are equivalent to the original boundary value problem, and, therefore, to each other, they admit quite different properties when applying a numerical scheme to obtain an approximate solution.

In this chapter we give an overview of direct and indirect reformulations of second order boundary value problems by using boundary integral equations and discuss the mapping properties of all boundary integral operators involved. From this we can deduce the unique solvability of the resulting boundary integral equations and the continuous dependence of the solution on the given boundary data.

1.1 Laplace Equation

The simplest example for a second order partial differential equation is the Laplace equation for a scalar function $u : \mathbb{R}^3 \to \mathbb{R}$ satisfying

$$-\Delta u(x) = -\sum_{i=1}^{3} \frac{\partial^2}{\partial x_i^2} u(x) = 0 \quad \text{for } x \in \Omega \subset \mathbb{R}^3. \tag{1.1}$$

This equation is used for the modelling of, for example, the stationary heat transfer, of electrostatic potentials, and of ideal fluids.

In (1.1), $\Omega \subset \mathbb{R}^3$ is a bounded, multiply or simply connected domain with a Lipschitz boundary $\Gamma = \partial\Omega$.

Multiplying the partial differential equation (1.1) with a test function v, integrating over Ω, and applying integration by parts, this gives Green's first formula

$$\int_{\Omega} (-\Delta u(y)) \, v(y) dy = a(u,v) - \int_{\Gamma} \gamma_1^{\text{int}} u(y) \gamma_0^{\text{int}} v(y) ds_y \tag{1.2}$$

with the symmetric bilinear form

$$a(u,v) = \int_{\Omega} (\nabla u(y), \nabla v(y)) dy,$$

with the interior trace operator

$$\gamma_0^{\text{int}} v(y) = \lim_{\widetilde{y} \in \Omega, \, \widetilde{y} \to y \in \Gamma} v(\widetilde{y}),$$

and with the interior conormal derivative of u on Γ,

$$\gamma_1^{\text{int}} u(y) = \lim_{\widetilde{y} \in \Omega, \, \widetilde{y} \to y \in \Gamma} \left(\underline{n}(y), \nabla_{\widetilde{y}} u(\widetilde{y}) \right).$$

Here, $\underline{n}(y)$ is the outer normal vector defined for almost all $y \in \Gamma$.

From Green's first formula (1.2) and by the use of the symmetry of the bilinear form $a(\cdot, \cdot)$, we deduce Green's second formula

$$\int_{\Omega} (-\Delta v(y)) u(y) dy + \int_{\Gamma} \gamma_1^{\text{int}} v(y) \gamma_0^{\text{int}} u(y) ds_y \tag{1.3}$$

$$= \int_{\Omega} (-\Delta u(y)) v(y) dy + \int_{\Gamma} \gamma_1^{\text{int}} u(y) \gamma_0^{\text{int}} v(y) ds_y.$$

Inserting $v = v_0 \equiv 1$, we then obtain the compatibility condition

$$\int_{\Omega} (-\Delta u(y)) dy + \int_{\Gamma} \gamma_1^{\text{int}} u(y) ds_y = 0. \tag{1.4}$$

Now, choosing in Green's second formula (1.3) as a test function v a fundamental solution $u^* : \mathbb{R}^3 \times \mathbb{R}^3 \to \mathbb{R}$ satisfying

$$\int_\Omega \big(-\Delta_y u^*(x,y) \big) u(y)\, dy = u(x) \quad \text{for } x \in \Omega, \tag{1.5}$$

the solution of the Laplace equation (1.1) is given by the representation formula

$$u(x) = \int_\Gamma u^*(x,y)\gamma_1^{\mathrm{int}} u(y)\, ds_y - \int_\Gamma \gamma_{1,y}^{\mathrm{int}} u^*(x,y)\gamma_0^{\mathrm{int}} u(y)\, ds_y \tag{1.6}$$

for $x \in \Omega$. The fundamental solution of the Laplace equation is

$$u^*(x,y) = \frac{1}{4\pi}\frac{1}{|x-y|} \quad \text{for } x,y \in \mathbb{R}^3. \tag{1.7}$$

For a domain Ω with Lipschitz boundary $\Gamma = \partial\Omega$, the solution (1.6) of the partial differential equation (1.1) has to be understood in a weak or distributional sense. For this, appropriate Sobolev spaces $H^\alpha(\Omega)$ and $H^\beta(\Gamma)$ have to be introduced; see Appendix A.1.

To derive suitable boundary integral equations from the representation formula (1.6), we first have to investigate the surface potentials in (1.6) as well as their interior trace and conormal derivative.

Single Layer Potential

First we consider the single layer potential

$$(\widetilde{V}w)(x) = \int_\Gamma u^*(x,y)w(y)\, ds_y = \frac{1}{4\pi}\int_\Gamma \frac{w(y)}{|x-y|}\, ds_y \quad \text{for } x \in \Omega,$$

which defines a continuous map from a given density function w on the boundary Γ to a harmonic function $\widetilde{V}w$ in the domain Ω. In particular,

$$\widetilde{V} : H^{-1/2}(\Gamma) \to H^1(\Omega)$$

is continuous and $\widetilde{V}w \in H^1(\Omega)$ is a weak solution of the Laplace equation (1.1) for any $w \in H^{-1/2}(\Gamma)$. Using the mapping property of the interior trace operator

$$\gamma_0^{\mathrm{int}} : H^1(\Omega) \to H^{1/2}(\Gamma),$$

we can define the corresponding boundary integral operator

$$V = \gamma_0^{\mathrm{int}}\widetilde{V}$$

with the following mapping properties, see for example [24, 71, 105].

Lemma 1.1. *The single layer potential operator*

$$V = \gamma_0^{int}\widetilde{V} : H^{-1/2}(\Gamma) \to H^{1/2}(\Gamma)$$

is bounded with

$$\|Vw\|_{H^{1/2}(\Gamma)} \leq c_2^V \|w\|_{H^{-1/2}(\Gamma)} \quad \text{for all } w \in H^{-1/2}(\Gamma),$$

and $H^{-1/2}(\Gamma)$*–elliptic,*

$$\langle Vw, w\rangle_\Gamma \geq c_1^V \|w\|_{H^{-1/2}(\Gamma)}^2 \quad \text{for all } w \in H^{-1/2}(\Gamma).$$

Moreover, for $w \in L_\infty(\Gamma)$ *there holds the representation*

$$(Vw)(x) = \int_\Gamma u^*(x,y)w(y)ds_y = \frac{1}{4\pi}\int_\Gamma \frac{w(y)}{|x-y|}ds_y \quad \text{for } x \in \Gamma$$

as a weakly singular surface integral.

Double Layer Potential

Next we consider the double layer potential

$$(Wv)(x) = \int_\Gamma \gamma_{1,y}^{int}u^*(x,y)v(y)ds_y = \frac{1}{4\pi}\int_\Gamma \frac{(x-y,\underline{n}(y))}{|x-y|^3}v(y)ds_y$$

for $x \in \Omega$, which again defines a continuous map from a given density function v on the boundary Γ to a harmonic function Wv in the domain Ω. In particular,

$$W : H^{1/2}(\Gamma) \to H^1(\Omega)$$

is continuous and $Wv \in H^1(\Omega)$ is a weak solution of the Laplace equation (1.1) for any $v \in H^{1/2}(\Gamma)$. Applying the interior trace operator

$$\gamma_0^{int} : H^1(\Omega) \to H^{1/2}(\Gamma),$$

this defines an associated boundary integral operator [24, 71, 105].

Lemma 1.2. *The boundary integral operator*

$$\gamma_0^{int}W : H^{1/2}(\Gamma) \to H^{1/2}(\Gamma)$$

is bounded with

$$\|\gamma_0^{int}Wv\|_{H^{1/2}(\Gamma)} \leq c_2^W \|v\|_{H^{1/2}(\Gamma)} \quad \text{for all } v \in H^{1/2}(\Gamma).$$

For $v \in H^{1/2}(\Gamma)$ *there holds the representation*

$$\gamma_0^{\mathrm{int}}(Wv)(x) = (-1 + \sigma(x))v(x) + (Kv)(x) \quad for\ x \in \Gamma,$$

with the double layer potential operator

$$(Kv)(x) = \lim_{\varepsilon \to 0} \int_{y \in \Gamma: |y-x| \geq \varepsilon} \gamma_{1,y}^{\mathrm{int}} u^*(x,y) v(y) ds_y$$

$$= \frac{1}{4\pi} \lim_{\varepsilon \to 0} \int_{y \in \Gamma: |y-x| \geq \varepsilon} \frac{(x-y, \underline{n}(y))}{|x-y|^3} v(y) ds_y$$

and

$$\sigma(x) = \lim_{\varepsilon \to 0} \frac{1}{4\pi} \frac{1}{\varepsilon^2} \int_{y \in \Omega: |y-x| = \varepsilon} ds_y \quad for\ x \in \Gamma.$$

Moreover, for $v = v_0 \equiv 1$, we have

$$\sigma(x)v_0(x) + (Kv_0)(x) = 0 \quad for\ x \in \Gamma.$$

If $x \in \Gamma$ is on a smooth part of the boundary $\Gamma = \partial\Omega$, then we obtain

$$\sigma(x) = \frac{1}{2}.$$

Otherwise, if $x \in \Gamma$ is on an edge or in a corner point of the boundary $\Gamma = \partial\Omega$, $\sigma(x)$ is related to the interior angle of Ω in $x \in \Gamma$. However, without loss of generality, we assume $\sigma(x) = 1/2$ for almost all $x \in \Gamma$.

By applying the interior trace operator γ_0^{int} to the representation formula (1.6),

$$u(x) = (\widetilde{V}\gamma_1^{\mathrm{int}} u)(x) - (W\gamma_0^{\mathrm{int}} u)(x) \quad for\ x \in \Omega,$$

we obtain the boundary integral equation

$$\gamma_0^{\mathrm{int}} u(x) = (V\gamma_1^{\mathrm{int}} u)(x) + \frac{1}{2}\gamma_0^{\mathrm{int}} u(x) - (K\gamma_0^{\mathrm{int}} u)(x) \tag{1.8}$$

for almost all $x \in \Gamma$. In particular, this is a weakly singular boundary integral equation,

$$\int_\Gamma u^*(x,y)\gamma_1^{\mathrm{int}} u(y) ds_y - \frac{1}{2}\gamma_0^{\mathrm{int}} u(x) + \int_\Gamma \gamma_{1,y}^{\mathrm{int}} u^*(x,y)\gamma_0^{\mathrm{int}} u(y) ds_y$$

for $x \in \Gamma$, or,

$$\frac{1}{4\pi}\int_\Gamma \frac{1}{|x-y|}\gamma_1^{\mathrm{int}} u(y) ds_y = \frac{1}{2}\gamma_0^{\mathrm{int}} u(x) + \frac{1}{4\pi}\int_\Gamma \frac{(x-y, \underline{n}(y))}{|x-y|^3}\gamma_0^{\mathrm{int}} u(y) ds_y.$$

Instead of the interior trace operator γ_0^{int}, we may also apply the interior conormal derivative operator γ_1^{int} to the representation formula (1.6). To do so, we first need to investigate the interior conormal derivatives of the single and double layer potentials $\widetilde{V}w$ and Wv, which are both harmonic in Ω. Then,

$$\gamma_1^{\text{int}} : H^1(\Omega, \Delta) \to H^{-1/2}(\Gamma),$$

where

$$H^1(\Omega, \Delta) = \left\{ v \in H^1(\Omega) : \Delta v \in \widetilde{H}^{-1}(\Omega) \right\}.$$

Adjoint Double Layer Potential

Lemma 1.3. *The boundary integral operator*

$$\gamma_1^{\text{int}}\widetilde{V} : H^{-1/2}(\Gamma) \to H^{-1/2}(\Gamma)$$

is bounded with

$$\|\gamma_1^{\text{int}}\widetilde{V}w\|_{H^{-1/2}(\Gamma)} \le c_2^{\gamma_1^{\text{int}}\widetilde{V}} \|w\|_{H^{-1/2}(\Gamma)} \quad \text{for all } w \in H^{-1/2}(\Gamma).$$

For $w \in H^{-1/2}(\Gamma)$ there holds the representation

$$\gamma_1^{\text{int}}(\widetilde{V}w)(x) = \frac{1}{2}w(x) + (K'w)(x)$$

in the sense of $H^{-1/2}(\Gamma)$, with the adjoint double layer potential operator

$$(K'w)(x) = \lim_{\varepsilon \to 0} \int_{y \in \Gamma : |y-x| \ge \varepsilon} \gamma_{1,x}^{\text{int}} u^*(x,y) w(y) ds_y$$

$$= \frac{1}{4\pi} \lim_{\varepsilon \to 0} \int_{y \in \Gamma : |y-x| \ge \varepsilon} \frac{(y-x, \underline{n}(x))}{|x-y|^3} w(y) ds_y.$$

In particular, we have

$$\langle \gamma_1^{\text{int}}\widetilde{V}w, v \rangle_\Gamma = \frac{1}{2}\langle w, v \rangle_\Gamma + \langle K'w, v \rangle_\Gamma = \frac{1}{2}\langle w, v \rangle_\Gamma + \langle w, Kv \rangle_\Gamma$$

for all $v \in H^{1/2}(\Gamma)$.

Hypersingular Integral Operator

In the same way as for the single layer potential $\widetilde{V}w$, we now consider the interior conormal derivate of the double layer potential Wv.

Lemma 1.4. *The operator*

$$D = -\gamma_1^{\text{int}}W : H^{1/2}(\Gamma) \to H^{-1/2}(\Gamma)$$

is bounded with

$$\|Dv\|_{H^{-1/2}(\Gamma)} \leq c_2^D \|v\|_{H^{1/2}(\Gamma)} \quad \text{for all } v \in H^{1/2}(\Gamma),$$

and $H^{1/2}(\Gamma)$*–semi–elliptic,*

$$\langle Dv, v \rangle_\Gamma \geq c_1^D |v|^2_{H^{1/2}(\Gamma)} \quad \text{for all } v \in H^{1/2}(\Gamma).$$

In particular, for $v = v_0 \equiv 1$, *we have*

$$(Dv_0)(x) = 0 \quad \text{for } x \in \Gamma.$$

Moreover, for continuous functions $u, v \in H^{1/2}(\Gamma) \cap C(\Gamma)$ *there holds the representation*

$$\langle Du, v \rangle_\Gamma = \frac{1}{4\pi} \int_\Gamma \int_\Gamma \frac{(\underline{\mathrm{curl}}_\Gamma u(y), \underline{\mathrm{curl}}_\Gamma v(x))}{|x-y|} ds_y ds_x \qquad (1.9)$$

where

$$\underline{\mathrm{curl}}_\Gamma u(x) = \underline{n}(x) \times \nabla_x \tilde{u}(x) \quad \text{for } x \in \Gamma$$

is the surface curl operator and $\tilde{u}$ *is some (locally defined) extension of* u *into the neighbourhood of* Γ.

The boundary integral operator $D = -\gamma_1^{\mathrm{int}} W$ does not exhibit an explicit representation as a Cauchy singular surface integral, in particular,

$$(Dv)(x) = -\gamma_1^{\mathrm{int}}(Wv)(x) = - \lim_{\tilde{x} \in \Omega, \tilde{x} \to x \in \Gamma} \left(\underline{n}(x), \nabla_{\tilde{x}}(Wv)(\tilde{x}) \right) =$$

$$\frac{1}{4\pi} \lim_{\varepsilon \to 0} \int_{y \in \Gamma: |y-x| \geq \varepsilon} \left(3\frac{(x-y, \underline{n}(y))(x-y, \underline{n}(x))}{|x-y|^5} - \frac{(\underline{n}(x), \underline{n}(y))}{|x-y|^3} \right) v(y) ds_y$$

does not exist. Therefore the boundary integral operator D is called hypersingular operator and it requires some appropriate regularisation procedure.

Since $u(x) = 1$ for $x \in \Omega$ is a solution of the Laplace equation $-\Delta u(x) = 0$, the representation formula (1.6) reads for this special choice

$$-\int_\Gamma \gamma_{1,y}^{\mathrm{int}} u^*(\tilde{x}, y) ds_y = 1 \quad \text{for } \tilde{x} \in \Omega.$$

Thus, we have

$$-\nabla_{\tilde{x}} \int_\Gamma \gamma_{1,y}^{\mathrm{int}} u^*(\tilde{x}, y) ds_y = 0 \quad \text{for } \tilde{x} \in \Omega.$$

Then,

$$(Dv)(x) = -\gamma_1^{\text{int}}(Wv)(x) = - \lim_{\tilde{x} \in \Omega, \tilde{x} \to x \in \Gamma} \left(\underline{n}(x), \nabla_{\tilde{x}}(Wv)(\tilde{x}) \right)$$

$$= - \lim_{\tilde{x} \in \Omega, \tilde{x} \to x \in \Gamma} \left(\underline{n}(x), \nabla_{\tilde{x}} \int_\Gamma \gamma_{1,y}^{\text{int}} u^*(\tilde{x}, y) v(y) ds_y \right)$$

$$= - \lim_{\tilde{x} \in \Omega, \tilde{x} \to x \in \Gamma} \left(\underline{n}(x), \nabla_{\tilde{x}} \int_\Gamma \gamma_{1,y}^{\text{int}} u^*(\tilde{x}, y) \Big(v(y) - v(x) \Big) ds_y \right)$$

exists as a Cauchy singular surface integral,

$$(Dv)(x) =$$
$$\frac{1}{4\pi} \int_\Gamma \left(3 \frac{(x - y, \underline{n}(y))(x - y, \underline{n}(x))}{|x - y|^5} - \frac{(\underline{n}(x), \underline{n}(y))}{|x - y|^3} \right) \Big(v(y) - v(x) \Big) ds_y .$$

However, using integration by parts as in the derivation of formula (1.9) of Lemma 1.4, the induced bilinear form of the hypersingular boundary integral operator can be transformed to a weakly singular bilinear form including some surface curl operators.

Boundary Integral Equations

Applying now the interior conormal derivative operator γ_1^{int} to the representation formula (1.6),

$$u(x) = (\tilde{V} \gamma_1^{\text{int}} u)(x) - (W \gamma_0^{\text{int}} u)(x) \quad \text{for } x \in \Omega,$$

this gives the boundary integral equation

$$\gamma_1^{\text{int}} u(x) = \frac{1}{2} \gamma_1^{\text{int}} u(x) + (K' \gamma_1^{\text{int}} u)(x) + (D \gamma_0^{\text{int}} u)(x) \qquad (1.10)$$

in the sense of $H^{-1/2}(\Gamma)$. In particular, this is a hypersingular boundary integral equation,

$$-\gamma_{1,x}^{\text{int}} \int_\Gamma \gamma_{1,y}^{\text{int}} u^*(x,y) \gamma_0^{\text{int}} u(y) ds_y = \frac{1}{2} \gamma_1^{\text{int}} u(x) - \int_\Gamma \gamma_{1,x}^{\text{int}} u^*(x,y) \gamma_1^{\text{int}} u(y) ds_y,$$

or,

$$\frac{1}{4\pi} \int_\Gamma \left(3 \frac{(y - x, \underline{n}(y))(y - x, \underline{n}(x))}{|x - y|^5} - \frac{(\underline{n}(x), \underline{n}(y))}{|x - y|^3} \right) \times$$

$$\Big(\gamma_0^{\text{int}} u(y) - \gamma_0^{\text{int}} u(x) \Big) ds_y = \frac{1}{2} \gamma_1^{\text{int}} u(x) - \int_\Gamma \frac{(y - x, \underline{n}(x))}{|x - y|^3} \gamma_1^{\text{int}} u(y) ds_y .$$

Combining (1.8) and (1.10), we can write both boundary integral equations by the use of the Calderon projector as

$$\begin{pmatrix} \gamma_0^{\text{int}} u \\ \gamma_1^{\text{int}} u \end{pmatrix} = \begin{pmatrix} \frac{1}{2}I - K & V \\ D & \frac{1}{2}I + K' \end{pmatrix} \begin{pmatrix} \gamma_0^{\text{int}} u \\ \gamma_1^{\text{int}} u \end{pmatrix}. \tag{1.11}$$

From the boundary integral equations of the Calderon projector (1.11) one can derive some important properties of boundary integral operators, i.e.

$$VK' = KV, \quad VD = \left(\frac{1}{2}I - K\right)\left(\frac{1}{2}I + K\right).$$

Steklov–Poincaré Operator

Since the single layer potential V is $H^{-1/2}(\Gamma)$–elliptic and therefore invertible, we obtain from the first equation in (1.11) the Dirichlet to Neumann map

$$\gamma_1^{\text{int}} u(x) = V^{-1}\left(\frac{1}{2}I + K\right)\gamma_0^{\text{int}} u(x) = (S^{\text{int}}\gamma_0^{\text{int}} u)(x) \quad \text{for } x \in \Gamma \tag{1.12}$$

which defines the Steklov–Poincaré operator

$$S^{\text{int}} : H^{1/2}(\Gamma) \to H^{-1/2}(\Gamma)$$

associated to the partial differential equation (1.1). Inserting the Dirichlet to Neumann map (1.12) into the second equation of the Calderon projector (1.11), this gives

$$\gamma_1^{\text{int}} u(x) =$$
$$\left(D + \left(\frac{1}{2}I + K'\right)V^{-1}\left(\frac{1}{2}I + K\right)\right)\gamma_0^{\text{int}} u(x) = (S^{\text{int}}\gamma_0^{\text{int}} u)(x)$$

with the symmetric representation of the Steklov–Poincaré operator

$$S^{\text{int}} = D + \left(\frac{1}{2}I + K'\right)V^{-1}\left(\frac{1}{2}I + K\right). \tag{1.13}$$

Note that also the representation (1.12) of the Steklov–Poincaré operator is symmetric. However, due to Lemma 1.1 and Lemma 1.4, we conclude from the symmetric representation (1.13)

$$\left\langle S^{\text{int}}v, v \right\rangle_\Gamma = \left\langle Dv, v \right\rangle_\Gamma + \left\langle V^{-1}\left(\frac{1}{2}I + K\right)v, \left(\frac{1}{2}I + K\right)v \right\rangle_\Gamma$$
$$\geq \left\langle Dv, v \right\rangle_\Gamma \geq c_1^D |v|_{H^{1/2}(\Gamma)}^2$$

for all $v \in H^{1/2}(\Gamma)$, and, therefore, S^{int} is $H^{1/2}(\Gamma)$–semi–elliptic. In particular, for $v = v_0 \equiv 1$, we have $(S^{\text{int}}v_0)(x) = 0$ for $x \in \Gamma$.

The Steklov–Poincaré operator S^{int} defines the Dirichlet to Neumann map, $\gamma_1^{\text{int}}u = S^{\text{int}}\gamma_0^{\text{int}}u$, which is a relation of the Cauchy data associated to a solution of the homogeneous partial differential equation. This map will be used to handle more general, e.g. nonlinear, boundary conditions. Moreover, Steklov–Poincaré operators play an important role in domain decomposition methods, e.g. when considering boundary value problems with piecewise constant coefficients, or when considering the coupling of finite and boundary element methods, see, for example, [60, 107].

1.1.1 Interior Dirichlet Boundary Value Problem

We first consider the interior Dirichlet boundary value problem for the Laplace equation, i.e.,

$$-\Delta u(x) = 0 \quad \text{for } x \in \Omega, \quad \gamma_0^{\text{int}}u(x) = g(x) \quad \text{for } x \in \Gamma. \tag{1.14}$$

Using the representation formula (1.6), the solution of the Dirichlet boundary value problem (1.14) is given by

$$u(x) = \int_\Gamma u^*(x,y)t(y)ds_y - \int_\Gamma \gamma_{1,y}^{\text{int}}u^*(x,y)g(y)ds_y \quad \text{for } x \in \Omega,$$

where $t = \gamma_1^{\text{int}}u$ is the unknown conormal derivative of u on Γ which has to be determined from some appropriate boundary integral equations.

Direct Single Layer Potential Formulation

Using the first equation in the Calderon projector (1.11), we have to solve a first kind boundary integral equation to find $t \in H^{-1/2}(\Gamma)$, such that

$$(Vt)(x) = \frac{1}{2}g(x) + (Kg)(x) \quad \text{for } x \in \Gamma, \tag{1.15}$$

or,

$$\frac{1}{4\pi}\int_\Gamma \frac{t(y)}{|x-y|}ds_y = \frac{1}{2}g(x) + \frac{1}{4\pi}\int_\Gamma \frac{(x-y,\underline{n}(y))}{|x-y|^3}g(y)ds_y.$$

Using duality arguments, the boundary integral equation

$$Vt = \frac{1}{2}g + Kg \in H^{1/2}(\Gamma)$$

corresponds to

$$0 = \left\|Vt - \frac{1}{2}g - Kg\right\|_{H^{1/2}(\Gamma)} = \sup_{0\neq w \in H^{-1/2}(\Gamma)} \frac{\langle Vt - \frac{1}{2}g - Kg, w\rangle_\Gamma}{\|w\|_{H^{-1/2}(\Gamma)}}.$$

Therefore, $t \in H^{-1/2}(\Gamma)$ is the solution of the variational problem

$$\left\langle Vt, w \right\rangle_\Gamma = \left\langle \left(\frac{1}{2}I + K\right)g, w \right\rangle_\Gamma \quad \text{for all } w \in H^{-1/2}(\Gamma), \tag{1.16}$$

or,

$$\frac{1}{4\pi} \int_\Gamma w(x) \int_\Gamma \frac{t(y)}{|x-y|} ds_y ds_x =$$

$$= \frac{1}{2} \int_\Gamma w(x)g(x)ds_x + \frac{1}{4\pi} \int_\Gamma w(x) \int_\Gamma \frac{(x-y, \underline{n}(y))}{|x-y|^3} g(y)ds_y ds_x.$$

Theorem 1.5. *Let $g \in H^{1/2}(\Gamma)$ be given. Then there exists a unique solution $t \in H^{-1/2}(\Gamma)$ of the variational problem* (1.16). *Moreover,*

$$\|t\|_{H^{-1/2}(\Gamma)} \le \frac{1}{c_1^V}\left(1 + c_2^W\right)\|g\|_{H^{1/2}(\Gamma)}.$$

Because the boundary integral equation (1.15) results from the representation formula (1.6) this approach is called direct.

Since both the single and the double layer potentials are harmonic in Ω, the solution of the Dirichlet boundary value problem (1.14) can be represented also either by a single or by a double layer potential alone. Then the unknown density functions have no physical meaning in general. The resulting methods are called indirect and have a long history when solving boundary value problems for second order partial differential equations, see, e.g., [33].

Indirect Single Layer Potential Formulation

Let us consider the indirect single layer potential approach

$$u(x) = (\tilde{V}w)(x) = \frac{1}{4\pi} \int_\Gamma \frac{w(y)}{|x-y|} ds_y \quad \text{for } x \in \Omega,$$

where we have to find the unknown density function $w \in H^{-1/2}(\Gamma)$. Applying the interior trace operator γ_0^{int}, from the given Dirichlet boundary conditions, we then obtain the first kind boundary integral equation

$$(Vw)(x) = \int_\Gamma u^*(x,y)w(y)ds_y = g(x) \quad \text{for } x \in \Gamma, \tag{1.17}$$

which is equivalent to the variational problem

$$\langle Vw, z \rangle_\Gamma = \langle g, z \rangle_\Gamma \quad \text{for all } z \in H^{-1/2}(\Gamma), \tag{1.18}$$

or,

$$\frac{1}{4\pi} \int_\Gamma z(x) \int_\Gamma \frac{w(y)}{|x-y|} ds_y ds_x = \int_\Gamma z(x)g(x)ds_x.$$

Theorem 1.6. *Let $g \in H^{1/2}(\Gamma)$ be given. Then there exists a unique solution $w \in H^{-1/2}(\Gamma)$ of the variational problem (1.18). Moreover,*

$$\|w\|_{H^{-1/2}(\Gamma)} \leq \frac{1}{c_1^V} \|g\|_{H^{1/2}(\Gamma)}.$$

Note that both boundary integral equations (1.15) and (1.17) are of the same structure, while they are different in the definition of the right hand side. In fact, the boundary integral equation (1.15) of the direct approach involves the application of the double layer potential K to the given Dirichlet datum g, while the right hand side of the boundary integral equation (1.17) of the indirect approach is just the given Dirichlet datum g itself.

Indirect Double Layer Potential Formulation

Instead of the indirect single layer potential $u = \widetilde{V}w$ we now consider the indirect double layer potential approach

$$u(x) = -(Wv)(x) = -\frac{1}{4\pi} \int_{\Gamma} \frac{(x - y, \underline{n}(y))}{|x - y|^3} v(y) ds_y \quad \text{for } x \in \Omega$$

which leads, by applying the interior trace operator γ_0^{int} and by the use of Lemma 1.2, to a second kind boundary integral equation to find $v \in H^{1/2}(\Gamma)$ such that

$$\frac{1}{2}v(x) - (Kv)(x) = g(x) \quad \text{for } x \in \Gamma, \tag{1.19}$$

or,

$$\frac{1}{2}v(x) - \frac{1}{4\pi} \int_{\Gamma} \frac{(x - y, \underline{n}(y))}{|x - y|^3} v(y) ds_y = g(x) \quad \text{for } x \in \Gamma.$$

Since this boundary integral equation is formulated in $H^{1/2}(\Gamma)$, the equivalent variational problem is to find $v \in H^{1/2}(\Gamma)$ such that

$$\left\langle \left(\frac{1}{2}I - K\right)v, w \right\rangle_{\Gamma} = \left\langle g, w \right\rangle_{\Gamma} \quad \text{for all } w \in H^{-1/2}(\Gamma).$$

The solution of the second kind boundary integral equation (1.19) is given by the Neumann series

$$v(x) = \sum_{\ell=0}^{\infty} \left(\frac{1}{2}I + K\right)^{\ell} g(x) \quad \text{for } x \in \Gamma. \tag{1.20}$$

The convergence of the Neumann series (1.20) and therefore the unique solvability of the boundary integral equation (1.19) can be established when using an appropriate norm in the Sobolev space $H^{1/2}(\Gamma)$, see [108].

Theorem 1.7. *Let $g \in H^{1/2}(\Gamma)$ be given. Then there exists a unique solution $v \in H^{1/2}(\Gamma)$ of the boundary integral equation (1.19). Moreover,*

$$\|v\|_{V^{-1}} \leq \frac{1}{1 - c_K} \|g\|_{V^{-1}}$$

where $c_K < 1$ is the contraction rate,

$$\left\|\left(\frac{1}{2}I + K\right)z\right\|_{V^{-1}} \leq c_K \|z\|_{V^{-1}} \quad \text{for all } z \in H^{1/2}(\Gamma)$$

with respect to the norm induced by the inverse single layer potential,

$$\|z\|_{V^{-1}}^2 = \langle V^{-1}z, z \rangle_\Gamma \quad \text{for all } z \in H^{1/2}(\Gamma).$$

It seems to be a natural setting to consider the second kind boundary integral equation (1.19) in the trace space $H^{1/2}(\Gamma)$, where Theorem 1.7 ensures the unique solvability. However, for practical reasons, the boundary integral equation (1.19) is often considered in $L_2(\Gamma)$. While it is known that the shifted double layer potential operator

$$\frac{1}{2}I - K \; : \; L_2(\Gamma) \to L_2(\Gamma)$$

is bounded, see [112], it is an open problem whether this operator is invertible in $L_2(\Gamma)$ or not for general Lipschitz boundaries $\Gamma = \partial\Omega$.

1.1.2 Interior Neumann Boundary Value Problem

For a simply connected domain $\Omega \subset \mathbb{R}^3$, we now consider the interior Neumann boundary value problem for the Laplace equation,

$$-\Delta u(x) = 0 \quad \text{for } x \in \Omega, \quad \gamma_1^{\text{int}}u(x) = g(x) \quad \text{for } x \in \Gamma. \tag{1.21}$$

From (1.4), we have to assume the solvability condition

$$\int_\Gamma g(y)ds_y = 0. \tag{1.22}$$

Note that the solution of the Neumann boundary value problem (1.21) is only unique up to an additive constant.

Using the representation formula (1.6), a solution of the Neumann boundary value problem (1.21) is given by

$$u(x) = \int_\Gamma u^*(x,y)g(y)ds_y - \int_\Gamma \gamma_{1,y}^{\text{int}}u^*(x,y)\gamma_0^{\text{int}}u(y)ds_y, \quad x \in \Omega.$$

Hence, we have to find the yet unknown Dirichlet datum $\bar{u} = \gamma_0^{\text{int}}u$ on Γ.

Direct Double Layer Potential Formulation

Using the first equation in (1.11), we have to solve a second kind boundary integral equation to find $\bar{u} = \gamma_0^{int} u \in H^{1/2}(\Gamma)$ such that

$$\frac{1}{2}\bar{u}(x) + (K\bar{u})(x) = (Vg)(x) \quad \text{for } x \in \Gamma, \tag{1.23}$$

or,

$$\frac{1}{2}\bar{u}(x) + \frac{1}{4\pi}\int_\Gamma \frac{(x-y,\underline{n}(y))}{|x-y|^3}\bar{u}(y)ds_y = \frac{1}{4\pi}\int_\Gamma \frac{g(y)}{|x-y|}ds_y \quad \text{for } x \in \Gamma.$$

As for the second kind boundary integral equation (1.19) for the Dirichlet boundary value problem (1.14), a solution of the second kind boundary integral equation (1.23) is given by the Neumann series

$$\bar{u}(x) = \sum_{\ell=0}^\infty \left(\frac{1}{2}I - K\right)^\ell (Vg)(x) \quad \text{for } x \in \Gamma. \tag{1.24}$$

Since the given Neumann datum $g \in H^{-1/2}(\Gamma)$ has to satisfy the solvability condition (1.22), and since $v_0 \equiv 1$ is the eigenfunction corresponding to the zero eigenvalue of $1/2\,I + K$, all members of the Neumann series (1.24), and, therefore, $\bar{u}$ are in the subspace $H_*^{1/2}(\Gamma) \subset H^{1/2}(\Gamma)$ defined as follows

$$H_*^{1/2}(\Gamma) = \left\{ v \in H^{1/2}(\Gamma) : \langle V^{-1}v, 1\rangle_\Gamma = 0 \right\}.$$

The general solution of the second kind boundary integral equation (1.23) is then given by $\bar{u}_\alpha = \bar{u} + c$ where $c \in \mathbb{R}$ is an arbitrary constant. To fix the constant, we may require the scaling condition

$$\langle \bar{u}_\alpha, 1\rangle_\Gamma = \int_\Gamma \bar{u}_\alpha(y)ds_y = \alpha, \tag{1.25}$$

where $\alpha \in \mathbb{R}$ can be arbitrary, but prescribed. This finally leads to a variational problem to find $\bar{u}_\alpha \in H^{1/2}(\Gamma)$ such that

$$\left\langle \left(\frac{1}{2}I + K\right)\bar{u}_\alpha, w\right\rangle_\Gamma + \left\langle \bar{u}_\alpha, 1\right\rangle_\Gamma \left\langle w, 1\right\rangle_\Gamma = \left\langle Vg, w\right\rangle_\Gamma + \alpha\left\langle w, 1\right\rangle_\Gamma \tag{1.26}$$

is satisfied for all $w \in H^{-1/2}(\Gamma)$. Note that the bilinear form of the extended variational problem (1.26) is regular due to the additional term $\langle \bar{u}_\alpha, 1\rangle_\Gamma \langle w, 1\rangle_\Gamma$, which regularises the singular operator $1/2\,I + K$. Summarising the above, we obtain the following result:

Theorem 1.8. *Let $g \in H^{-1/2}(\Gamma)$ and $\alpha \in \mathbb{R}$ be given. Then there exists a unique solution $\bar{u}_\alpha \in H^{1/2}(\Gamma)$ of the extended variational problem (1.26) satisfying*

$$\|\bar{u}_\alpha\|_{H^{1/2}(\Gamma)} \leq c \left(\|g\|_{H^{-1/2}(\Gamma)} + |\alpha| \right).$$

If $g \in H^{-1/2}(\Gamma)$ satisfies the solvability condition (1.22), then $\bar{u}_\alpha \in H^{1/2}(\Gamma)$ is the unique solution of the boundary integral equation (1.23) satisfying the scaling condition (1.25).

Direct Hypersingular Integral Operator Formulation

When using the second equation in (1.11), we have to solve a first kind boundary integral equation to find $\bar{u} = \gamma_0^{\text{int}} u \in H^{1/2}(\Gamma)$ such that

$$(D\bar{u})(x) = \frac{1}{2}g(x) - (K'g)(x) \quad \text{for } x \in \Gamma \tag{1.27}$$

is satisfied in a weak sense, in particular, in the sense of $H^{-1/2}(\Gamma)$. Since the hypersingular boundary integral operator D has a non–trivial kernel, we have to consider the equation (1.27) in suitable subspaces. For this we define

$$H_{**}^{1/2}(\Gamma) = \left\{ v \in H^{1/2}(\Gamma) : \langle v, 1 \rangle_\Gamma = 0 \right\}.$$

Then the variational problem of the boundary integral equation (1.27) reads to find $\bar{u} \in H_{**}^{1/2}(\Gamma)$ such that

$$\left\langle D\bar{u}, v \right\rangle_\Gamma = \left\langle \left(\frac{1}{2}I - K' \right)g, v \right\rangle_\Gamma \tag{1.28}$$

is satisfied for all $v \in H_{**}^{1/2}(\Gamma)$. The general solution of the first kind boundary integral equation (1.27) is then given by $\bar{u}_\alpha = \bar{u} + c$ where $c \in \mathbb{R}$ is a constant which can be determined by the scaling condition (1.25) afterwards.

Instead of solving the variational problem (1.28) in the subspace $H_{**}^{1/2}(\Gamma)$ and finding the unique solution afterwards from the scaling condition (1.25), we can formulate an extended variational problem to find $\bar{u}_\alpha \in H^{1/2}(\Gamma)$ such that

$$\left\langle D\bar{u}_\alpha, v \right\rangle_\Gamma + \left\langle \bar{u}_\alpha, 1 \right\rangle_\Gamma \left\langle v, 1 \right\rangle_\Gamma = \left\langle \left(\frac{1}{2}I - K' \right)g, v \right\rangle_\Gamma + \alpha \left\langle v, 1 \right\rangle_\Gamma \tag{1.29}$$

is satisfied for all $v \in H^{1/2}(\Gamma)$.

Theorem 1.9. *Let $g \in H^{-1/2}(\Gamma)$ and $\alpha \in \mathbb{R}$ be given. Then there exists a unique solution $\bar{u}_\alpha \in H^{1/2}(\Gamma)$ of the extended variational problem (1.29) satisfying*

$$\|\bar{u}_\alpha\|_{H^{1/2}(\Gamma)} \leq c \left(\|g\|_{H^{-1/2}(\Gamma)} + |\alpha| \right).$$

If $g \in H^{-1/2}(\Gamma)$ satisfies the solvability condition (1.22), then $\bar{u}_\alpha$ is the unique solution of the hypersingular boundary integral equation (1.27) satisfying the scaling condition (1.25).

Steklov–Poincaré Operator Formulation

Instead of the hypersingular boundary integral equation (1.27) we may also consider a Steklov–Poincaré operator equation to find $\bar{u} = \gamma_0^{\mathrm{int}} u \in H^{1/2}(\Gamma)$ such that

$$(S^{\mathrm{int}}\bar{u})(x) = g(x) \quad \text{for } x \in \Gamma,\tag{1.30}$$

where the Steklov–Poincaré operator S^{int} is given either by the Dirichlet to Neumann map (1.12) or in the symmetric form (1.13). As for the hypersingular boundary integral equation (1.27), one can formulate an extended variational formulation to find $\bar{u}_\alpha \in H^{1/2}(\Gamma)$ such that

$$\langle S^{\mathrm{int}}\bar{u}_\alpha, v\rangle_\Gamma + \langle \bar{u}_\alpha, 1\rangle_\Gamma \langle v, 1\rangle_\Gamma = \langle g, v\rangle_\Gamma + \alpha\langle v, 1\rangle_\Gamma\tag{1.31}$$

is satisfied for all $v \in H^{1/2}(\Gamma)$, and where $\alpha \in \mathbb{R}$ is given by the scaling condition (1.25).

Theorem 1.10. *Let $g \in H^{-1/2}(\Gamma)$ and $\alpha \in \mathbb{R}$ be given. Then there exists a unique solution $\bar{u}_\alpha \in H^{1/2}(\Gamma)$ of the extended variational problem (1.31) satisfying*

$$\|\bar{u}_\alpha\|_{H^{1/2}(\Gamma)} \leq c\left(\|g\|_{H^{-1/2}(\Gamma)} + |\alpha|\right).$$

If $g \in H^{-1/2}(\Gamma)$ satisfies the solvability condition (1.22), then $\bar{u}_\alpha$ is the unique solution of the Steklov–Poincaré operator equation (1.30) satisfying the scaling condition (1.25).

Indirect Single and Double Layer Potential Formulations

When using the indirect single layer potential ansatz $u = \widetilde{V}w$ in Ω, the application of the interior conormal derivative operator γ_1^{int} gives the second kind boundary integral equation

$$\frac{1}{2}w(x) + (K'w)(x) = g(x) \quad \text{for } x \in \Gamma.\tag{1.32}$$

As for the second kind boundary integral equation (1.23), the solution of the boundary integral equation (1.32) is given by the Neumann series

$$w(x) = \sum_{\ell=0}^{\infty} \left(\frac{1}{2}I - K'\right)^\ell g(x) \quad \text{for } x \in \Gamma.\tag{1.33}$$

The convergence of the series (1.33) follows as in Theorem 1.7 due to the contraction estimate, see [108],

$$\left\|\left(\frac{1}{2}I - K'\right)w\right\|_V \leq c_K \|w\|_V \quad \text{for all } w \in H^{-1/2}(\Gamma) : \langle w, 1\rangle_\Gamma = 0$$

with $c_K < 1$.

The indirect double layer potential approach $u = -Wv$ in Ω leads, finally, to the hypersingular boundary integral equation

$$(Dv)(x) = g(x) \quad \text{for } x \in \Gamma,$$

which is of the same structure and hence can be handled like the hypersingular boundary integral equation (1.27); we skip the details.

1.1.3 Mixed Boundary Value Problem

In most applications we have to deal with boundary value problems with boundary conditions of mixed type, e.g. with Dirichlet or Neumann boundary conditions on different non–overlapping parts Γ_D and Γ_N of the boundary $\Gamma = \overline{\Gamma}_D \cup \overline{\Gamma}_N$, respectively. Therefore, we now consider the mixed boundary value problem

$$\begin{aligned}
-\Delta u(x) &= 0 && \text{for } x \in \Omega, \\
\gamma_0^{\text{int}} u(x) &= g(x) && \text{for } x \in \Gamma_D, \\
\gamma_1^{\text{int}} u(x) &= f(x) && \text{for } x \in \Gamma_N.
\end{aligned} \tag{1.34}$$

Note that for simplicity the domain Ω is supposed to be simply connected. The solution of the mixed boundary value problem (1.34) is then given by the representation formula

$$\begin{aligned}
u(x) = &\int_{\Gamma_D} u^*(x,y)\gamma_1^{\text{int}} u(y) ds_y + \int_{\Gamma_N} u^*(x,y) f(y) ds_y \\
&- \int_{\Gamma_D} \gamma_{1,y}^{\text{int}} u^*(x,y) g(y) ds_y - \int_{\Gamma_N} \gamma_{1,y}^{\text{int}} u^*(x,y) \gamma_0^{\text{int}} u(y) ds_y \quad \text{for } x \in \Omega,
\end{aligned}$$

where we have to find the yet unknown Cauchy data $\gamma_0^{\text{int}} u$ on Γ_N and $\gamma_1^{\text{int}} u$ on Γ_D. As we have seen in the two previous subsections on the Dirichlet and on the Neumann problem, there exist different approaches leading to different boundary integral equations to find the unknown Cauchy data. However, we consider here only two direct methods, which seem to be the most convenient approaches to solve mixed boundary value problems by boundary element methods. The definition of the Sobolev spaces $\widetilde{H}^{1/2}(\Gamma_N)$ and $\widetilde{H}^{-1/2}(\Gamma_D)$ can be seen in Appendix A.1.

Symmetric Formulation of Boundary Integral Equations

The symmetric formulation (cf. [103]) is based on the use of the first kind boundary integral equation (1.15) to find the unknown Neumann datum $\gamma_1^{\text{int}} u$ on the Dirichlet part Γ_D, while the hypersingular boundary integral equation (1.27) is used to find the unknown Dirichlet datum $\gamma_0^{\text{int}} u$ on the Neumann part Γ_N:

$$(V\gamma_1^{int}u)(x) = \frac{1}{2}g(x) + (K\gamma_0^{int}u)(x) \text{ for } x \in \Gamma_D,$$

$$(D\gamma_0^{int}u)(x) = \frac{1}{2}f(x) - (K'\gamma_1^{int}u)(x) \text{ for } x \in \Gamma_N.$$

Let $\widetilde{g} \in H^{1/2}(\Gamma)$ and $\widetilde{f} \in H^{-1/2}(\Gamma)$ be some arbitrary, but fixed extensions of the given boundary data $g \in H^{1/2}(\Gamma_D)$ and $f \in H^{-1/2}(\Gamma_N)$, respectively. Then, we have to find

$$\widetilde{u} = \gamma_0^{int}u - \widetilde{g} \in \widetilde{H}^{1/2}(\Gamma_N), \quad \widetilde{t} = \gamma_1^{int}u - \widetilde{f} \in \widetilde{H}^{-1/2}(\Gamma_D)$$

satisfying the system of boundary integral equations

$$(V\widetilde{t})(x) - (K\widetilde{u})(x) = \frac{1}{2}g(x) + (K\widetilde{g})(x) - (V\widetilde{f})(x) \quad \text{for } x \in \Gamma_D,$$

$$(K'\widetilde{t})(x) + (D\widetilde{u})(x) = \frac{1}{2}f(x) - (K'\widetilde{f})(x) - (D\widetilde{g})(x) \quad \text{for } x \in \Gamma_N.$$

The associated variational problem is to find

$$(\widetilde{t}, \widetilde{u}) \in \widetilde{H}^{-1/2}(\Gamma_D) \times \widetilde{H}^{1/2}(\Gamma_N)$$

such that

$$a(\widetilde{t}, \widetilde{u}; w, v) = F(w, v) \tag{1.35}$$

is satisfied for all $(w, v) \in \widetilde{H}^{-1/2}(\Gamma_D) \times \widetilde{H}^{1/2}(\Gamma_N)$ with

$$a(\widetilde{t}, \widetilde{u}; w, v) = \langle V\widetilde{t}, w \rangle_{\Gamma_D} - \langle K\widetilde{u}, w \rangle_{\Gamma_D} + \langle K'\widetilde{t}, v \rangle_{\Gamma_N} + \langle D\widetilde{u}, v \rangle_{\Gamma_N},$$

$$F(w, v) = \left\langle \frac{1}{2}g + K\widetilde{g} - V\widetilde{f}, w \right\rangle_{\Gamma_D} + \left\langle \frac{1}{2}f - K'\widetilde{f} - D\widetilde{g}, v \right\rangle_{\Gamma_N}.$$

Since the bilinear form $a(\cdot, \cdot; \cdot, \cdot)$ is skew–symmetric, i.e.

$$a(w, v; w, v) = \langle Vw, w \rangle_{\Gamma_D} + \langle Dv, v \rangle_{\Gamma_N},$$

the unique solvability of the variational problem (1.35) follows from the mapping properties of the single layer potential V and of the hypersingular integral operator D.

Theorem 1.11. *Let $g \in H^{1/2}(\Gamma_D)$ and $f \in H^{-1/2}(\Gamma_N)$ be given. Then there exists a unique solution $(\widetilde{t}, \widetilde{u}) \in \widetilde{H}^{-1/2}(\Gamma_D) \times \widetilde{H}^{1/2}(\Gamma_N)$ of the variational problem (1.35) satisfying*

$$\|\widetilde{t}\|^2_{\widetilde{H}^{-1/2}(\Gamma_D)} + \|\widetilde{u}\|^2_{\widetilde{H}^{1/2}(\Gamma_N)} \leq c \left(\|g\|^2_{H^{1/2}(\Gamma_D)} + \|f\|^2_{H^{-1/2}(\Gamma_N)} \right).$$

Steklov–Poincaré Operator Formulation

Instead of using the weakly singular boundary integral equation (1.15) on Γ_D and the hypersingular boundary integral equation (1.27) on Γ_N, we may also use the Dirichlet to Neumann map (1.12) to derive a second boundary integral approach to find the yet unknown Cauchy data. Then we have to solve an operator equation to find $\widetilde{u} \in \widetilde{H}^{1/2}(\Gamma_N)$ such that

$$(S^{\text{int}}\widetilde{u})(x) = f(x) - (S^{\text{int}}\widetilde{g})(x) \quad \text{for } x \in \Gamma_N , \qquad (1.36)$$

where the Steklov–Poincaré operator $S^{\text{int}} : H^{1/2}(\Gamma) \to H^{-1/2}(\Gamma)$ is either given by the representation (1.12) or by the symmetric version (1.13). Although both representations are equivalent in the continuous case, they exhibit different stability properties when applying some numerical approximation schemes.

Theorem 1.12. *Let $g \in H^{1/2}(\Gamma_D)$ and $f \in H^{-1/2}(\Gamma_N)$ be given. Then there exists a unique solution $\widetilde{u} \in \widetilde{H}^{1/2}(\Gamma_N)$ of the Steklov–Poincaré operator equation (1.36) satisfying*

$$\|\widetilde{u}\|_{\widetilde{H}^{1/2}(\Gamma_N)} \leq c \left(\|g\|_{H^{1/2}(\Gamma_D)} + \|f\|_{H^{-1/2}(\Gamma_N)} \right) .$$

When the Dirichlet datum $\gamma_0^{\text{int}}u = \widetilde{u} + \widetilde{g}$ is known on the whole boundary Γ, we can find the complete Neumann datum $\gamma_1^{\text{int}}u$ by solving the corresponding Dirichlet boundary value problem afterwards.

1.1.4 Robin Boundary Value Problem

Besides of standard Dirichlet or Neumann boundary conditions also linear or nonlinear boundary conditions of Robin type have to be included, as for example in radiosity transfer problems.

Linear Robin Boundary Conditions

Hence we now consider the Robin boundary value problem

$$-\Delta u(x) - 0 \text{ for } x \in \Omega, \ \gamma_1^{\text{int}}u(x) + \kappa(x)\gamma_0^{\text{int}}u(x) = g(x) \text{ for } x \in \Gamma, \quad (1.37)$$

where $\kappa \in L_\infty(\Gamma)$ is strictly positive with $\kappa(x) \geq \kappa_0 > 0$ for $x \in \Gamma$. Using the Dirichlet to Neumann map $\gamma_1^{\text{int}}u = S^{\text{int}}\gamma_0^{\text{int}}u$ on Γ with the Steklov–Poincaré operator $S^{\text{int}} : H^{1/2}(\Gamma) \to H^{-1/2}(\Gamma)$ either defined by (1.12) or by (1.13), we can find the unknown Dirichlet datum $\gamma_0^{\text{int}}u \in H^{1/2}(\Gamma)$ by solving the boundary integral equation

$$(S^{\text{int}}\gamma_0^{\text{int}}u)(x) + \kappa(x)\gamma_0^{\text{int}}u(x) = g(x) \quad \text{for } x \in \Gamma .$$

Since κ is assumed to be strictly positive, the additive term regularises the $H^{1/2}(\Gamma)$–semi–elliptic Steklov–Poincaré operator S^{int} yielding the unique solvability of the equivalent variational problem to find $\bar{u} \in H^{1/2}(\Gamma)$ such that

$$\langle S^{\mathrm{int}} \bar{u}, v \rangle_\Gamma + \langle \kappa \bar{u}, v \rangle_\Gamma = \langle g, v \rangle_\Gamma \quad \text{for all } v \in H^{1/2}(\Gamma). \tag{1.38}$$

Theorem 1.13. *Let $g \in H^{-1/2}(\Gamma)$ and $\kappa \in L_\infty(\Gamma)$ with $\kappa(x) \geq \kappa_0 > 0$ for $x \in \Gamma$ be given. Then there exists a unique solution of the variational problem (1.38). Moreover,*

$$\|\bar{u}\|_{H^{1/2}(\Gamma)} \leq c \|g\|_{H^{-1/2}(\Gamma)}.$$

When the complete Dirichlet datum $\bar{u} = \gamma_0^{\mathrm{int}} u \in H^{1/2}(\Gamma)$ is known we can find the Neumann datum $\gamma_1^{\mathrm{int}} u$ by solving the corresponding Dirichlet boundary value problem.

Nonlinear Robin Boundary Conditions

Instead of linear Robin boundary conditions in the boundary value problem (1.37), we may also consider a boundary value problem with nonlinear Robin boundary conditions,

$$-\Delta u(x) = 0 \text{ for } x \in \Omega, \; \gamma_1^{\mathrm{int}} u(x) + f(\gamma_0^{\mathrm{int}} u, x) = g(x) \text{ for } x \in \Gamma,$$

where $f(\cdot, \cdot)$ is nonlinear in the first argument, for example $f(u, x) = \big(u(x)\big)^m$ with $m \in \mathbb{N}$, typical choices are $m = 3$ or $m = 4$. Using again the Dirichlet to Neumann map $\gamma_1^{\mathrm{int}} u = S^{\mathrm{int}} \gamma_0^{\mathrm{int}} u$ on Γ, we can find the unknown Dirichlet datum $\bar{u} = \gamma_0^{\mathrm{int}} u \in H^{1/2}(\Gamma)$ by solving the nonlinear boundary integral equation

$$(S^{\mathrm{int}} \bar{u})(x) + f(\bar{u}, x) = g(x) \quad \text{for } x \in \Gamma.$$

The equivalent variational problem is to find $\bar{u} \in H^{1/2}(\Gamma)$ such that

$$\langle S^{\mathrm{int}} \bar{u}, v \rangle_\Gamma + \langle f(\bar{u}, \cdot), v \rangle_\Gamma = \langle g, v \rangle_\Gamma \quad \text{for all } v \in H^{1/2}(\Gamma). \tag{1.39}$$

The unique solvability of the nonlinear variational problem (1.39) follows from appropriate assumptions on the nonlinear function f, see, e.g., [32, 95].

Theorem 1.14. *Let $g \in H^{-1/2}(\Gamma)$ be given and let f be strongly monotone satisfying*

$$\Big\langle f(u, \cdot) - f(v, \cdot), u - v \Big\rangle_\Gamma \geq c \|u - v\|_{L_2(\Gamma)}^2 \quad \text{for all } u, v \in L_2(\Gamma).$$

Then there exists a unique solution $\bar{u} \in H^{1/2}(\Gamma)$ of the nonlinear variational problem (1.39) satisfying

$$\|\bar{u}\|_{H^{1/2}(\Gamma)} \leq c \|g\|_{H^{-1/2}(\Gamma)}.$$

1.1.5 Exterior Dirichlet Boundary Value Problem

One of the main advantages in using boundary element methods for the approximate solution of boundary value problems is their applicability to problems in exterior unbounded domains. As a first model problem we consider the exterior Dirichlet boundary value problem

$$-\Delta u(x) = 0 \quad \text{for } x \in \Omega^e = \mathbb{R}^3 \backslash \overline{\Omega}, \quad \gamma_0^{\text{ext}} u(x) = g(x) \quad \text{for } x \in \Gamma \quad (1.40)$$

with the radiation condition

$$|u(x) - u_0| = \mathcal{O}\left(\frac{1}{|x|}\right) \quad \text{as } |x| \to \infty, \quad (1.41)$$

where $u_0 \in \mathbb{R}$ is given. We denote by

$$\gamma_0^{\text{ext}} u(x) = \lim_{\widetilde{x} \in \Omega^e,\, \widetilde{x} \to x \in \Gamma} u(\widetilde{x})$$

the exterior trace of u on Γ and by

$$\gamma_1^{\text{ext}} u(x) = \lim_{\widetilde{x} \in \Omega^e,\, \widetilde{x} \to x \in \Gamma} \left(\underline{n}(x), \nabla_{\widetilde{x}} u(\widetilde{x})\right)$$

the exterior conormal derivative of u on Γ. Note that the outer normal vector $\underline{n}(x)$ is still defined with respect to the interior domain Ω.

For a fixed $y_0 \in \Omega$ and $R > 2 \operatorname{diam} \Omega$, let $B_R(y_0)$ be a ball of radius R with centre in y_0 and including Ω. The solution of the boundary value problem (1.40) is then given by the representation formula, see (1.6), for $x \in B_R(y_0)\backslash \overline{\Omega}$

$$u(x) = -\int_{\Gamma} u^*(x,y) \gamma_1^{\text{ext}} u(y) ds_y + \int_{\Gamma} \gamma_{1,y}^{\text{ext}} u^*(x,y) g(y) ds_y$$

$$+ \int_{\partial B_R(y_0)} u^*(x,y) \gamma_1^{\text{int}} u(y) ds_y - \int_{\partial B_R(y_0)} \gamma_{1,y}^{\text{int}} u^*(x,y) \gamma_0^{\text{int}} u(y) ds_y.$$

Taking the limit $R \to \infty$ and incorporating the radiation condition (1.41), this gives the representation formula in the exterior domain Ω^e

$$u(x) = u_0 - \int_{\Gamma} u^*(x,y) \gamma_1^{\text{ext}} u(y) ds_y + \int_{\Gamma} \gamma_{1,y}^{\text{ext}} u^*(x,y) g(y) ds_y \quad (1.42)$$

for $x \in \Omega^e$. To find the yet unknown Neumann datum $t = \gamma_1^{\text{ext}} u \in H^{-1/2}(\Gamma)$, we apply the exterior trace operator γ_0^{ext} to obtain the boundary integral equation

$$(Vt)(x) = -\frac{1}{2} g(x) + (Kg)(x) + u_0 \quad \text{for } x \in \Gamma. \quad (1.43)$$

As for the direct and the indirect approach for the interior Dirichlet boundary value problem, we can conclude the unique solvability of the first kind boundary integral equation (1.43) from Lemma 1.1. We then obtain the Dirichlet to Neumann map

$$\gamma_1^{\text{ext}} u(x) = V^{-1}\left(-\frac{1}{2}I + K\right)\gamma_0^{\text{ext}} u(x) + (V^{-1}u_0)(x) \quad \text{for } x \in \Gamma \quad (1.44)$$

associated to the exterior Dirichlet boundary value problem (1.40).

Applying the exterior conormal derivative to the representation formula (1.42), and inserting the Dirichlet to Neumann map (1.44), this gives

$$\gamma_1^{\text{ext}} u(x) = \left(\frac{1}{2}I - K'\right)\gamma_1^{\text{ext}} u(x) - (D\gamma_0^{\text{ext}} u)(x)$$

$$= \left(\frac{1}{2}I - K'\right)\left(V^{-1}\left(-\frac{1}{2}I + K\right)\gamma_0^{\text{ext}} u(x) + (V^{-1}u_0)(x)\right) - (D\gamma_0^{\text{ext}} u)(x)$$

$$= -(S^{\text{ext}}\gamma_0^{\text{ext}} u)(x) + \left(\frac{1}{2}I - K'\right)(V^{-1}u_0)(x) \quad (1.45)$$

with the Steklov–Poincaré operator (cf. (1.13))

$$S^{\text{ext}} = D + \left(-\frac{1}{2}I + K'\right)V^{-1}\left(-\frac{1}{2}I + K\right) : H^{1/2}(\Gamma) \to H^{-1/2}(\Gamma) \quad (1.46)$$

associated to the exterior boundary value problem (1.40).

1.1.6 Exterior Neumann Boundary Value Problem

Instead of the exterior Dirichlet boundary value problem (1.40), we now consider the exterior Neumann boundary value problem

$$-\Delta u(x) = 0 \quad \text{for } x \in \Omega^e, \quad \gamma_1^{\text{ext}} u(x) = g(x) \quad \text{for } x \in \Gamma \quad (1.47)$$

with the radiation condition (1.41)

$$|u(x) - u_0| = \mathcal{O}\left(\frac{1}{|x|}\right) \quad \text{as } |x| \to \infty,$$

where $u_0 \in \mathbb{R}$ is given. Note that, due to the radiation condition, we have unique solvability of the exterior Neumann boundary value problem (1.47). As for the exterior Dirichlet boundary value problem, the solution of the exterior Neumann boundary value problem is given by the representation formula (1.42)

$$u(x) = u_0 - \int_\Gamma u^*(x,y)g(y)ds_y + \int_\Gamma \gamma_{1,y}^{\text{ext}} u^*(x,y)\gamma_0^{\text{ext}} u(y)ds_y. \quad (1.48)$$

for $x \in \Omega^e$. To find the yet unknown Dirichlet datum $\bar{u} = \gamma_0^{\text{ext}} \in H^{1/2}(\Gamma)$, we apply the exterior trace operator γ_0^{ext} to obtain the boundary integral equation

$$\frac{1}{2}\bar{u}(x) - (K\bar{u})(x) = u_0 - (Vg)(x) \quad \text{for } x \in \Gamma. \tag{1.49}$$

As for the indirect double layer potential formulation for the interior Dirichlet boundary value problem, the solution of the boundary integral equation (1.49) is given by the Neumann series

$$\bar{u}(x) = u_0 + \sum_{\ell=0}^{\infty} \left(\frac{1}{2}I + K\right)^{\ell} (Vg)(x) \quad \text{for } x \in \Gamma, \tag{1.50}$$

where we have used $(1/2\, I + K)u_0 = 0$. The convergence of the Neumann series (1.50) in $H^{1/2}(\Gamma)$ follows from Theorem 1.7. Note that the boundary integral equation (1.49), and, therefore, the exterior Neumann boundary value problem (1.47) with the radiation condition (1.41) is uniquely solvable for any given $g \in H^{-1/2}(\Gamma)$.

When applying the exterior conormal derivative γ_1^{ext} to the representation formula (1.48), this gives the hypersingular boundary integral equation to find $\bar{u} = \gamma_0^{\text{ext}}u \in H^{1/2}(\Gamma)$ satisfying

$$(D\bar{u})(x) = -\frac{1}{2}g(x) - (K'g)(x) \quad \text{for } x \in \Gamma. \tag{1.51}$$

The boundary integral equation (1.51) is equivalent to the variational problem to find $\bar{u} \in H^{1/2}(\Gamma)$ such that

$$\left\langle D\bar{u}, v \right\rangle_{\Gamma} = -\left\langle g, \left(\frac{1}{2}I + K\right)v \right\rangle_{\Gamma} \tag{1.52}$$

is satisfied for all $v \in H^{1/2}(\Gamma)$. Using the test function $v = v_0 \equiv 1$, this gives the trivial equality

$$\left\langle D\bar{u}, v_0 \right\rangle_{\Gamma} = \left\langle \bar{u}, Dv_0 \right\rangle_{\Gamma} = -\left\langle g, \left(\frac{1}{2}I + K\right)v_0 \right\rangle_{\Gamma} = 0.$$

This shows that the variational problem (1.52) has to be considered in a subspace of $H^{1/2}(\Gamma)$ which is orthogonal to constants. In particular, the solution of the variational problem (1.52) is only unique up to a constant. Since the hypersingular boundary integral operator $D : H^{1/2}(\Gamma) \to H^{-1/2}(\Gamma)$ is only $H^{1/2}(\Gamma)$–semi elliptic, see Lemma 1.4, a suitable regularisation of the hypersingular boundary integral operator has to be introduced. As in (1.29), we obtain an extended variational problem to find $\bar{u} \in H^{1/2}(\Gamma)$ such that

$$\left\langle D\bar{u}, v \right\rangle_{\Gamma} + \left\langle \bar{u}, 1 \right\rangle_{\Gamma}\left\langle v, 1 \right\rangle_{\Gamma} = -\left\langle g, \left(\frac{1}{2}I + K\right)v \right\rangle_{\Gamma} \tag{1.53}$$

is satisfied for all $v \in H^{1/2}(\Gamma)$. The extended variational problem (1.53) is uniquely solvable yielding a solution $\bar{u} \in H^{1/2}(\Gamma)$ satisfying the orthogonality

$\langle \bar{u}, 1 \rangle_\Gamma = 0$. Since $u(x) = 1$ for $x \in \Omega^e$ is a solution of the Laplace equation $-\Delta u(x) = 0$ with the radiation condition (1.41) for $u_0 = 1$, the representation formula (1.48) reads

$$u(x) = u_0 + \int\limits_\Gamma \gamma_{1,y}^{\text{ext}} u^*(x,y) ds_y$$

implying

$$\int\limits_\Gamma \gamma_{1,y}^{\text{ext}} u^*(x,y) ds_y = 0 \quad \text{for } x \in \Omega^e .$$

This shows that the scaling condition for the solution $\bar{u}$ of the extended variational problem (1.53) can be chosen in an arbitrary way, the representation formula (1.48) describes the correct solution for any scaling parameter.

1.1.7 Poisson Problem

Instead of the homogeneous Laplace equation (1.1), we now consider an inhomogeneous Poisson equation with some given right hand side. The Dirichlet boundary value problem for the Poisson equation reads

$$-\Delta u(x) = f(x) \quad \text{for } x \in \Omega, \quad \gamma_0^{\text{int}} u(x) = g(x) \quad \text{for } x \in \Gamma. \tag{1.54}$$

From Green's second formula (1.3), we then obtain the representation formula

$$u(x) = \int\limits_\Gamma u^*(x,y) t(y) ds_y - \int\limits_\Gamma \gamma_{1,y}^{\text{int}} u^*(x,y) g(y) ds_y + \int\limits_\Omega u^*(x,y) f(y) dy$$

for $x \in \Omega$, where $t = \gamma_1^{\text{int}} u$ is the yet unknown Neumann datum. As for the interior Dirichlet boundary value problem (1.14), we have to solve a first kind boundary integral equation to find $t \in H^{-1/2}(\Gamma)$ such that

$$(Vt)(x) = \frac{1}{2} g(x) + (Kg)(x) - (N_0 f)(x) \quad \text{for } x \in \Gamma, \tag{1.55}$$

where

$$(N_0 f)(x) = \int\limits_\Omega u^*(x,y) f(y) dy \quad \text{for } x \in \Gamma$$

is the Newton potential entering the right hand side. Hence, the unique solvability of the boundary integral equation (1.55) follows, as in Theorem 1.5, for the first kind boundary integral equation (1.15), which is associated to the Dirichlet boundary value problem (1.14).

The drawback in considering the boundary integral equation (1.55) is the evaluation of the Newton potential $N_0 f$. Besides a direct computation there exist several approaches leading to more efficient methods.

Particular Solution Approach

Let u_p be a particular solution of the Poisson equation in (1.54) satisfying

$$-\Delta u_p(x) = f(x) \quad \text{for } x \in \Omega.$$

Then, instead of (1.54), we consider a Dirichlet boundary value problem for the Laplace operator,

$$-\Delta u_0(x) = 0 \quad \text{for } x \in \Omega, \quad \gamma_0^{int} u_0(x) = g(x) - \gamma_0^{int} u_p(x) \quad \text{for } x \in \Gamma.$$

The solution u of (1.54) is then given by $u_0 + u_p$. The unknown Neumann datum $t_0 = \gamma_1^{int} u_0$ is the unique solution of the boundary integral equation

$$(V t_0)(x) = \frac{1}{2}\left(g(x) - \gamma_0^{int} u_p(x)\right) + \left(K\left(g - \gamma_0^{int} u_p\right)\right)(x) \quad \text{for } x \in \Gamma.$$

On the other hand we have

$$t_0 = \gamma_1^{int} u_0 = \gamma_1^{int}\left(u - u_p\right) = t - \gamma_1^{int} u_p(x).$$

Hence, we obtain

$$(V t)(x) = \frac{1}{2} g(x) + (Kg)(x) - \frac{1}{2}\gamma_0^{int} u_p(x) - (K\gamma_0^{int} u_p)(x) + (V\gamma_1^{int} u_p)(x)$$

for $x \in \Gamma$, and, therefore,

$$(N_0 f)(x) = \frac{1}{2}\gamma_0^{int} u_p(x) + (K\gamma_0^{int} u_p)(x) - (V\gamma_1^{int} u_p(x)) \quad \text{for } x \in \Gamma.$$

Thus, we can evaluate a Newton potential $N_0 f$ by the use of the surface potentials, when a particular solution u_p of the Poisson equation is known.

Integration by Parts

In several applications the given function f in (1.54) satisfies a certain homogeneous partial differential equation. For simplicity, we assume that

$$-\Delta f(x) = 0 \quad \text{for } x \in \Omega.$$

Using

$$u^*(x, y) = \frac{1}{4\pi}\frac{1}{|x-y|} - \Delta_y\left(\frac{1}{8\pi}|x-y|\right)$$

we obtain from the Green's second formula (1.3)

$$\int_\Omega u^*(x, y) f(y) dy = \frac{1}{8\pi}\int_\Omega f(y)\Delta_y |x-y| dy$$

$$= \frac{1}{8\pi}\int_\Gamma \gamma_{1,y}^{int}|x-y|\gamma_0^{int} f(y) ds_y - \frac{1}{8\pi}\int_\Gamma \gamma_{0,y}^{int}|x-y|\gamma_1^{int} f(y) dy.$$

1.1.8 Interface Problem

In addition to interior and exterior boundary value problems, we may also consider an interface problem, i.e.,

$$-\alpha_i \Delta u_i(x) = f(x) \quad \text{for } x \in \Omega, \quad -\alpha_e \Delta u_e(x) = 0 \quad \text{for } x \in \Omega^e, \qquad (1.56)$$

with transmission conditions describing the continuity of the potential and of the flux, respectively,

$$\gamma_0^{\text{int}} u_i(x) = \gamma_0^{\text{ext}} u_e(x), \quad \alpha_i \gamma_1^{\text{int}} u_i(x) = \alpha_e \gamma_1^{\text{ext}} u_e(x) \quad \text{for } x \in \Gamma, \qquad (1.57)$$

and with the radiation condition for a given $u_0 \in \mathbb{R}$,

$$\left| u_e(x) - u_0 \right| = \mathcal{O}\left(\frac{1}{|x|}\right) \quad \text{as } |x| \to \infty. \qquad (1.58)$$

The solution of the above interface problem is given by the representation formula

$$u_i(x) = \int_\Gamma u^*(x,y)\gamma_1^{\text{int}} u_i(x)ds_x - \int_\Gamma \gamma_{1,y}^{\text{int}} u^*(x,y)\gamma_0^{\text{int}} u_i(y)ds_y$$

$$+\frac{1}{\alpha_i}\int_\Omega u^*(x,y)f(y)dy$$

for $x \in \Omega$ and

$$u_e(x) = u_0 - \int_\Gamma u^*(x,y)\gamma_1^{\text{ext}} u_e(x)ds_x + \int_\Gamma \gamma_{1,y}^{\text{ext}} u^*(x,y)\gamma_0^{\text{ext}} u_e(y)ds_y$$

for $x \in \Omega^e$. To find the unknown Cauchy data $\gamma_0^{\text{int/ext}} u$ and $\gamma_1^{\text{int/ext}} u$, which are linked via the transmission conditions (1.57), we have to solve appropriate boundary integral equations on the interface boundary Γ. Using the Dirichlet to Neumann map associated to the interior Dirichlet boundary value problem (1.54), in particular, solving the boundary integral equation (1.55),

$$(V\gamma_1^{\text{int}} u_i)(x) = \frac{1}{2}\gamma_0^{\text{int}} u_i(x) + (K\gamma_0^{\text{int}} u_i)(x) - \frac{1}{\alpha_i}(N_0 f)(x) \quad \text{for } x \in \Gamma,$$

we obtain

$$\gamma_1^{\text{int}} u_i(x) = V^{-1}\left(\frac{1}{2}I + K\right)\gamma_0^{\text{int}} u_i(x) - \frac{1}{\alpha_i}V^{-1}(N_0 f)(x) \quad \text{for } x \in \Gamma.$$

Let us assume that there is given a particular solution u_p satisfying

$$-\Delta u_p(x) = f(x) \quad \text{for } x \in \Omega.$$

Hence, we obtain

$$(N_0 f)(x) = \frac{1}{2}\gamma_0^{\text{int}} u_p(x) + (K\gamma_0^{\text{int}} u_p)(x) - (V\gamma_1^{\text{int}} u_p)(x) \quad \text{for } x \in \Gamma,$$

and, therefore,

$$\gamma_1^{\text{int}} u_i(x) =$$
$$V^{-1}\left(\frac{1}{2}I + K\right)\gamma_0^{\text{int}} u_i(x) + \frac{1}{\alpha_i}\gamma_1^{\text{int}} u_p(x) - \frac{1}{\alpha_i}V^{-1}\left(\frac{1}{2}I + K\right)\gamma_0^{\text{int}} u_p(x) =$$
$$(S^{\text{int}}\gamma_0^{\text{int}} u_i)(x) + \frac{1}{\alpha_i}\gamma_1^{\text{int}} u_p(x) - \frac{1}{\alpha_i}(S^{\text{int}}\gamma_0^{\text{int}} u_p)(x)$$

for $x \in \Gamma$ with the Steklov–Poincaré operator S^{int}. Correspondingly, the Dirichlet to Neumann map (1.45) associated to the exterior Dirichlet boundary value problem (1.40) gives

$$\gamma_1^{\text{ext}} u_e(x) = -(S^{\text{ext}}\gamma_0^{\text{ext}} u_e)(x) + \left(\frac{1}{2}I - K'\right)(V^{-1}u_0)(x) \quad \text{for } x \in \Gamma.$$

Inserting the transmission conditions (1.57),

$$\bar{u} = \gamma_0^{\text{ext}} u_e(x) = \gamma_0^{\text{int}} u_i(x), \quad \alpha_i\gamma_1^{\text{int}} u_i(x) = \alpha_e\gamma_1^{\text{ext}} u_e(x) \quad \text{for } x \in \Gamma,$$

we obtain a coupled Steklov–Poincaré operator equation to find $\bar{u} \in H^{1/2}(\Gamma)$ such that

$$\alpha_i(S^{\text{int}}\bar{u})(x) + \alpha_e(S^{\text{ext}}\bar{u})(x) =$$
$$(S^{\text{int}}\gamma_0^{\text{int}} u_p)(x) - \gamma_1^{\text{int}} u_p(x) + \alpha_e\left(\frac{1}{2}I - K'\right)(V^{-1}u_0)(x)$$

is satisfied for $x \in \Gamma$. This is equivalent to a variational problem to find $\bar{u} \in H^{1/2}(\Gamma)$ such that

$$\left\langle (\alpha_i S^{\text{int}} + \alpha_e S^{\text{ext}})\bar{u}, v \right\rangle_\Gamma = \tag{1.59}$$
$$\left\langle S^{\text{int}}\gamma_0^{\text{int}} u_p - \gamma_1^{\text{int}} u_p + \alpha_e\left(\frac{1}{2}I - K'\right)V^{-1}u_0, v \right\rangle_\Gamma$$

is satisfied for all $v \in H^{1/2}(\Gamma)$. The unique solvability of (1.59) finally follows from the ellipticity estimates for the interior and exterior Steklov–Poincaré operators S^{int} and S^{ext}.

1.2 Lamé Equations

In linear isotropic elastostatics the displacement field $\underline{u}$ of an elastic body occupying some reference configuration $\Omega \subset \mathbb{R}^3$ satisfies the equilibrium equations

$$-\sum_{j=1}^{3}\frac{\partial}{\partial x_j}\sigma_{ij}(\underline{u},x) = 0 \quad \text{for } x \in \Omega, \ i = 1,2,3 \,, \tag{1.60}$$

where $\sigma \in \mathbb{R}^{3\times 3}$ denotes the stress tensor. For a homogeneous isotropic material, the linear stress–strain relation is given by Hooke's law

$$\sigma_{ij}(\underline{u},x) = \frac{E\,\nu}{(1+\nu)(1-2\nu)}\delta_{ij}\sum_{k=1}^{3}e_{kk}(\underline{u},x) + \frac{E}{1+\nu}e_{ij}(\underline{u},x)$$

for $i,j = 1,2,3$. Here, $E > 0$ is the Young modulus, and $\nu \in (0,1/2)$ denotes the Poisson ratio. The strain tensor e is defined as follows,

$$e_{ij}(\underline{u},x) = \frac{1}{2}\left(\frac{\partial}{\partial x_i}u_j(x) + \frac{\partial}{\partial x_j}u_i(x)\right) \quad \text{for } i,j = 1,2,3 \,.$$

Inserting the strain and stress tensors, we obtain from (1.60) the Navier system

$$-\mu\Delta\underline{u}(x) - (\lambda+\mu)\text{grad div}\,\underline{u}(x) = \underline{0} \quad \text{for } x \in \Omega$$

with the Lamé constants

$$\lambda = \frac{E\,\nu}{(1+\nu)(1-2\nu)}, \quad \mu = \frac{E}{2(1+\nu)}\,.$$

Multiplying the equilibrium equations (1.60) with some test function v_i, integrating over Ω, applying integration by parts, and taking the sum over $i = 1,2,3$, this gives the first Betti formula

$$-\int_{\Omega}\sum_{i,j=1}^{3}\frac{\partial}{\partial y_j}\sigma_{ij}(\underline{u},y)v_i(y)dy = a(\underline{u},\underline{v}) - \int_{\Gamma}\left(\gamma_1^{\text{int}}\underline{u}(y),\gamma_0^{\text{int}}\underline{v}(y)\right)ds_y \tag{1.61}$$

with the symmetric bilinear form

$$a(\underline{u},\underline{v}) = \int_{\Omega}\sum_{i,j=1}^{3}\sigma_{ij}(\underline{u},y)e_{ij}(\underline{v},y)dy$$

$$= 2\mu\int_{\Omega}\sum_{i,j=1}^{3}e_{ij}(\underline{u},y)e_{ij}(\underline{v},y)dy + \lambda\int_{\Omega}\text{div}\,\underline{u}(y)\,\text{div}\,\underline{v}(y)dy$$

and with the boundary stress operator

$$(\gamma_1^{\text{int}}\underline{u})_i(y) = \sum_{j=1}^{3}\sigma_{ij}(\underline{u},y)n_j(y) \quad \text{for } y \in \Gamma \,,$$

and for $i = 1,2,3$, which can be written as

$$(\gamma_1^{\text{int}}\underline{u})(y) = \lambda \operatorname{div}\underline{u}(y)\,\underline{n}(y) + 2\mu\frac{\partial}{\partial\underline{n}(y)}\underline{u}(y) + \mu\,\underline{n}(y) \times \operatorname{curl}\underline{u}(y) \quad \text{for } y \in \Gamma.$$

From (1.61) and using the symmetry of the bilinear form $a(\cdot,\cdot)$, we can deduce the second Betti formula

$$-\int_\Omega \sum_{i,j=1}^3 \frac{\partial}{\partial y_j}\sigma_{ij}(\underline{v},y)u_i(y)dy + \int_\Gamma \left(\gamma_1^{\text{int}}\underline{v}(y), \gamma_0^{\text{int}}\underline{u}(y)\right)ds_y \qquad (1.62)$$

$$= -\int_\Omega \sum_{i,j=1}^3 \frac{\partial}{\partial y_j}\sigma_{ij}(\underline{u},x)v_i(x)dy + \int_\Gamma \left(\gamma_1^{\text{int}}\underline{u}(y), \gamma_0^{\text{int}}\underline{v}(y)\right)ds_y.$$

Let

$$\mathcal{R} = \operatorname{span}\left\{\begin{pmatrix}1\\0\\0\end{pmatrix}, \begin{pmatrix}0\\1\\0\end{pmatrix}, \begin{pmatrix}0\\0\\1\end{pmatrix}, \begin{pmatrix}-x_2\\x_1\\0\end{pmatrix}, \begin{pmatrix}0\\-x_3\\x_2\end{pmatrix}, \begin{pmatrix}x_3\\0\\-x_1\end{pmatrix}\right\} \qquad (1.63)$$

be the space of the rigid body motions which are solutions of the homogeneous Neumann boundary value problem

$$-\sum_{j=1}^3 \frac{\partial}{\partial x_j}\sigma_{ij}(\underline{v},x) = 0 \quad \text{for } x \in \Omega, \quad (\gamma_1^{\text{int}}\underline{v})_i(x) = 0 \quad \text{for } x \in \Gamma,$$

for $i = 1,2,3$, and $\underline{v} \in \mathcal{R}$. Then there holds

$$-\int_\Omega \sum_{i,j=1}^3 \frac{\partial}{\partial y_j}\sigma_{ij}(\underline{u},y)v_i(y)dy + \int_\Gamma \left(\gamma_1^{\text{int}}\underline{u}(y), \gamma_0^{\text{int}}\underline{v}(y)\right)ds_y = 0$$

for $i = 1,2,3$, and for all $\underline{v} \in \mathcal{R}$.

Choosing in (1.62) as a test function $\underline{v}$ a fundamental solution $\underline{U}_\ell^*(x,y)$ having the property

$$-\int_\Omega \sum_{i,j=1}^3 \frac{\partial}{\partial y_j}\sigma_{ij}(\underline{U}_\ell^*(x,y),y)u_i(y)dy = u_\ell(x), \qquad (1.64)$$

the displacement field $\underline{u}$ satisfying the equilibrium equations (1.60) is given by the Somigliana identity

$$u_\ell(x) = \int_\Gamma \gamma_1^{\text{int}}\left(\underline{U}_\ell^*(x,y), \underline{u}(y)\right)ds_y - \int_\Gamma \left(\gamma_{1,y}^{\text{int}}\underline{U}_\ell^*(x,y), \gamma_0^{\text{int}}\underline{u}(y)\right)ds_y \quad (1.65)$$

for $x \in \Omega$ and $\ell = 1,2,3$. The fundamental solution of linear elastostatics is given by the Kelvin tensor

$$U_{k\ell}^*(x,y) = \frac{1}{8\pi}\frac{1}{E}\frac{1+\nu}{1-\nu}\left((3-4\nu)\frac{\delta_{k\ell}}{|x-y|} + \frac{(x_k-y_k)(x_\ell-y_\ell)}{|x-y|^3}\right) \qquad (1.66)$$

for $x,y \in \mathbb{R}^3$ and $k,\ell = 1,2,3$. Note that the fundamental solution is defined even in the incompressible case $\nu = 1/2$.

The mapping properties of all boundary potentials and the related boundary integral operators follow as in the case of the Laplace operator.

Single Layer Potential

The single layer potential of linear elastostatics is given as

$$(\widetilde{V}^{\mathrm{Lame}}\underline{w})_k(x) = \frac{1}{2}\frac{1}{E}\frac{1+\nu}{1-\nu}\left((3-4\nu)(\widetilde{V}w_k)(x) + \sum_{\ell=1}^{3}(\widetilde{V}_{k\ell}w_\ell)(x)\right),$$

where

$$(\widetilde{V}w_k)(x) = \frac{1}{4\pi}\int_{\Gamma}\frac{w_k(y)}{|x-y|}ds_y$$

is the single layer potential of the Laplace operator, and

$$(\widetilde{V}_{k\ell}w_\ell)(x) = \frac{1}{4\pi}\int_{\Gamma}\frac{(x_k-y_k)(x_\ell-y_\ell)}{|x-y|^3}w_\ell(y)ds_y$$

$$= \frac{1}{4\pi}\int_{\Gamma}w_\ell(y)(x_k-y_k)\frac{\partial}{\partial y_\ell}\frac{1}{|x-y|}ds_y$$

for $k,\ell = 1,2,3$. The single layer potential $\widetilde{V}^{\mathrm{Lame}}$ defines a continuous map from a given vector function $\underline{w}$ on the boundary Γ to a vector field $\widetilde{V}^{\mathrm{Lame}}\underline{w}$ which is a solution of the homogeneous equilibrium equations (1.60). In particular,

$$\widetilde{V}^{\mathrm{Lame}} : [H^{-1/2}(\Gamma)]^3 \to [H^1(\Omega)]^3$$

is continuous. Using the mapping property of the interior trace operator

$$\gamma_0^{\mathrm{int}} : H^1(\Omega) \to H^{1/2}(\Gamma)$$

for $(\widetilde{V}^{\mathrm{Lame}}\underline{w})_\ell$, $\ell = 1,2,3$, this defines a continuous boundary integral operator $V^{\mathrm{Lame}} = \gamma_0^{\mathrm{int}}\widetilde{V}^{\mathrm{Lame}}$.

Lemma 1.15. *The single layer potential operator*

$$V^{\mathrm{Lame}} : [H^{-1/2}(\Gamma)]^3 \to [H^{1/2}(\Gamma)]^3$$

is bounded with

$$\|V^{\mathrm{Lame}}\underline{w}\|_{[H^{1/2}(\Gamma)]^3} \leq c_2^V\|\underline{w}\|_{[H^{-1/2}(\Gamma)]^3} \quad \textit{for all } \underline{w} \in [H^{-1/2}(\Gamma)]^3$$

and, if $\nu \in (0, 1/2)$, $[H^{-1/2}(\Gamma)]^3$–elliptic,

$$\langle V^{\text{Lame}} \underline{w}, \underline{w} \rangle_\Gamma \geq c_1^V \|\underline{w}\|^2_{[H^{-1/2}(\Gamma)]^3} \quad \text{for all } \underline{w} \in [H^{-1/2}(\Gamma)]^3,$$

where the duality pairing $\langle \cdot, \cdot \rangle$ is now defined as follows

$$\langle \underline{u}, \underline{v} \rangle = \int_\Gamma (\underline{u}(y), \underline{v}(y)) ds_y.$$

Moreover, for $\underline{w} \in [L_\infty(\Gamma)]^3$ there holds the representation

$$(V^{\text{Lame}} \underline{w})_k(x) = \frac{1}{2} \frac{1}{E} \frac{1+\nu}{1-\nu} \left((3 - 4\nu)(V w_k)(x) + \sum_{\ell=1}^{3} (V_{k\ell} w_\ell)(x) \right),$$

where

$$(V w_k)(x) = \frac{1}{4\pi} \int_\Gamma \frac{w_k(y)}{|x-y|} ds_y$$

is the single layer potential of the Laplace operator, and

$$(V_{k\ell} w_\ell)(x) = \frac{1}{4\pi} \int_\Gamma w_\ell(y)(x_k - y_k) \frac{\partial}{\partial y_\ell} \frac{1}{|x-y|} ds_y$$

for $k, \ell = 1, 2, 3$, all defined as weakly singular surface integrals.

Note that the single layer potential V^{Lame} of linear elastostatics can be written as

$$(V^{\text{Lame}} \underline{w})_k(x) =$$
$$\frac{1}{2} \frac{1}{E} \frac{1+\nu}{1-\nu} \left((V w_k)(x) + \sum_{\ell=1}^{3} (V_{k\ell} w_\ell)(x) \right) + \frac{1}{E} \frac{1+\nu}{1-\nu}(1 - 2\nu)(V w_k)(x),$$

where the first part corresponds to the single layer potential V^{Stokes} of the Stokes problem (see Section 1.3). From $V^{\text{Stokes}} \underline{n} = \underline{0}$, we then obtain (cf. [106])

$$(V^{\text{Lame}} \underline{n}, \underline{n}) = \frac{1}{E} \frac{1+\nu}{1-\nu}(1 - 2\nu) \sum_{k=1}^{3} (V n_k, n_k),$$

showing that the ellipticity constant c_1^V behaves like $\mathcal{O}(1 - 2\nu)$ for $\nu \to 1/2$. In particular, we have

$$\lim_{\nu \to 1/2} c_1^V(\nu) = 0.$$

Double Layer Potential

The double layer potential of linear elastostatics is

$$(W^{\text{Lame}}\underline{v})_\ell(x) \;=\; \int_\Gamma (\gamma^{\text{int}}_{1,y}\underline{U}^*_\ell(x,y), \underline{v}(y))\, ds_y$$

for $\ell = 1, 2, 3$. The double layer potential W^{Lame} defines a continuous map from a given vector function $\underline{v}$ on the boundary Γ to a vector field $W^{\text{Lame}}\underline{v}$ which is a solution of the homogeneous equilibrium equations (1.60). In particular,

$$W \,:\, [H^{1/2}(\Gamma)]^3 \to [H^1(\Omega)]^3$$

is continuous. Using the mapping property of the interior trace operator

$$\gamma^{\text{int}}_0 \,:\, H^1(\Omega) \to H^{1/2}(\Gamma)$$

applied to the components $(W\underline{v})_\ell$, $\ell = 1, 2, 3$, this defines an associated boundary integral operator.

Lemma 1.16. *The boundary integral operator*

$$\gamma^{\text{int}}_0 W^{\text{Lame}}\underline{v} : [H^{1/2}(\Gamma)]^3 \to [H^{1/2}(\Gamma)]^3$$

is bounded with

$$\|\gamma^{\text{int}}_0 W^{\text{Lame}}\underline{v}\|_{[H^{1/2}(\Gamma)]^3} \leq c^W_2 \, \|\underline{v}\|_{[H^{1/2}(\Gamma)]^3} \quad \text{for all } \underline{v} \in [H^{1/2}(\Gamma)]^3.$$

For continuous $\underline{v}$ there holds the representation

$$\gamma^{\text{int}}_0 (W^{\text{Lame}}\underline{v})(x) \;=\; \frac{1}{2}\underline{v} + (K^{\text{Lame}}\underline{v})(x)$$

for $x \in \Gamma$ with the double layer potential operator

$$(K^{\text{Lame}}\underline{v})(x) \;=\; (K\underline{v})(x) - (VM(\partial, \underline{n})\underline{v})(x) + \frac{E}{1+\nu}(V^{\text{Lame}}M(\partial, \underline{n})\underline{v})(x)\,,$$

where K and V are the double and single layer potential for the Laplace operator, and V^{Lame} is the single layer potential of linear elasticity, respectively. In addition, we have used the matrix surface curl operator given by

$$M_{ij}(\partial_y, \underline{n}(y)) \;=\; n_j(y)\frac{\partial}{\partial y_i} - n_i(y)\frac{\partial}{\partial y_j} \qquad (1.67)$$

for $i, j = 1, 2, 3$. Moreover, we have

$$\left(\frac{1}{2}I + K^{\text{Lame}}\right)\underline{v}(x) \;=\; \underline{0} \quad \text{for all } \underline{v} \in \mathcal{R}\,,$$

where $\mathcal{R}$ is the space of the rigid body motions.

By applying the interior trace operator γ_0^{int} to the representation formula (1.65), we obtain the first boundary integral equation

$$\gamma_0^{\text{int}}\underline{u}(x) = (V^{\text{Lame}}\gamma_1^{\text{int}}\underline{u})(x) + \frac{1}{2}\gamma_0^{\text{int}}\underline{u}(x) - (K^{\text{Lame}}\gamma_0^{\text{int}}\underline{u})(x) \qquad (1.68)$$

for $x \in \Gamma$. Instead of the interior trace operator γ_0^{int}, we may also apply the interior boundary stress operator γ_1^{int} to the representation formula (1.65). To do so, we first need to investigate the application of the boundary stress operator to the single and double layer potentials $\widetilde{V}^{\text{Lame}}\underline{w}$ and $W^{\text{Lame}}\underline{v}$ which are both solutions of the homogeneous equilibrium equations (1.60).

Adjoint Double Layer Potential

Lemma 1.17. *The boundary integral operator*

$$\gamma_1^{\text{int}}\widetilde{V}^{\text{Lame}} : [H^{-1/2}(\Gamma)]^3 \rightarrow [H^{-1/2}(\Gamma)]^3$$

is bounded with

$$\|\gamma_1^{\text{int}}\widetilde{V}^{\text{Lame}}\underline{w}\|_{[H^{-1/2}(\Gamma)]^3} \leq c_2^{\gamma_1^{\text{int}}\widetilde{V}} \|\underline{w}\|_{[H^{-1/2}(\Gamma)]^3} \quad \text{for all } \underline{w} \in [H^{-1/2}(\Gamma)]^3.$$

For $\underline{w} \in [H^{-1/2}(\Gamma)]^3$ there holds the representation

$$(\gamma_1^{\text{int}}\widetilde{V}^{\text{Lame}}\underline{w})(x) = \frac{1}{2}\underline{w}(x) + \left((K^{\text{Lame}})'\underline{w}\right)(x)$$

in the sense of $[H^{-1/2}(\Gamma)]^3$. In particular, for $\underline{v} \in [H^{1/2}(\Gamma)]^3$ we have

$$\langle \gamma_1^{\text{int}}\widetilde{V}^{\text{Lame}}\underline{w}, \underline{v}\rangle_\Gamma = \frac{1}{2}\langle\underline{w}, \underline{v}\rangle_\Gamma + \langle\underline{w}, K^{\text{Lame}}\underline{v}\rangle_\Gamma.$$

Hypersingular Integral Operator

In the same way as for the single layer potential $\widetilde{V}^{\text{Lame}}\underline{w}$, we now consider the application of the boundary stress operator γ_1^{int} to the double layer potential $W^{\text{Lame}}\underline{v}$.

Lemma 1.18. *The boundary integral operator*

$$D^{\text{Lame}} = -\gamma_1^{\text{int}}W^{\text{Lame}} : [H^{1/2}(\Gamma)]^3 \rightarrow [H^{-1/2}(\Gamma)]^3$$

is bounded with

$$\|D^{\text{Lame}}\underline{v}\|_{[H^{-1/2}(\Gamma)]^3} \leq c_2^D \|\underline{v}\|_{[H^{1/2}(\Gamma)]^3} \quad \text{for all } \underline{v} \in [H^{1/2}(\Gamma)]^3$$

and $H_{\mathcal{R}}^{1/2}(\Gamma)$–elliptic,

$$\langle D^{\mathrm{Lame}}\underline{v},\underline{v}\rangle_\Gamma \geq c_1^D \,\|\underline{v}\|^2_{[H^{1/2}(\Gamma)]^3} \quad \textit{for all } \underline{v} \in H_{\mathcal{R}}^{1/2}(\Gamma),$$

where $H_{\mathcal{R}}^{1/2}(\Gamma)$ is the space of all vector functions which are orthogonal to the space $\mathcal{R}$ of rigid body motions. In particular, there holds

$$(D^{\mathrm{Lame}}\underline{v})(x) = \underline{0} \quad \textit{for all } \underline{v} \in \mathcal{R}.$$

Moreover, for continuous vector functions $\underline{u},\underline{v} \in [H^{1/2}(\Gamma) \cap C(\Gamma)]^3$, there holds the representation

$$\langle D^{\mathrm{Lame}}\underline{u},\underline{v}\rangle_\Gamma =$$

$$\frac{\mu}{4\pi} \int_\Gamma \int_\Gamma \frac{1}{|x-y|} \sum_{k=1}^3 \Big(\frac{\partial}{\partial S_k(y)}\underline{u}(y), \frac{\partial}{\partial S_k(x)}\underline{v}(x)\Big) ds_y ds_x +$$

$$\int_\Gamma \int_\Gamma (M(\partial_x, \underline{n}(x))\underline{v}(x))^\top \times$$

$$\Big(\frac{\mu}{2\pi}\frac{I}{|x-y|} - 4\mu^2 U^*(x,y)\Big) M(\partial_y, \underline{n}(y))\underline{u}(y) ds_y ds_x +$$

$$\frac{\mu}{4\pi} \int_\Gamma \int_\Gamma \sum_{i,j,k=1}^3 M_{kj}(\partial_x, \underline{n}(x))v_i(x) \frac{1}{|x-y|} M_{ki}(\partial_y, \underline{n}(y))v_j(y) ds_y ds_x$$

with the surface curl operator $M(\partial, \underline{n})$ as defined in (1.67) and

$$\frac{\partial}{\partial S_1(x)} = M_{32}(\partial_x, \underline{n}(x)),$$

$$\frac{\partial}{\partial S_2(x)} = M_{13}(\partial_x, \underline{n}(x)),$$

$$\frac{\partial}{\partial S_3(x)} = M_{21}(\partial_x, \underline{n}(x)).$$

Boundary Integral Equations

Applying the interior boundary stress operator γ_1^{int} to the Somigliana identity (1.65),

$$\underline{u}(x) = (\widetilde{V}^{\mathrm{Lame}}\gamma_1^{\mathrm{int}}\underline{u})(x) - (W^{\mathrm{Lame}}\gamma_0^{\mathrm{int}}\underline{u})(x) \quad \text{for } x \in \Omega,$$

this gives a second boundary integral equation

$$\gamma_1^{\mathrm{int}}\underline{u}(x) = \frac{1}{2}\gamma_1^{\mathrm{int}}\underline{u}(x) + \Big((K^{\mathrm{Lame}})'\gamma_1^{\mathrm{int}}\underline{u}\Big)(x) + (D^{\mathrm{Lame}}\gamma_0^{\mathrm{int}}\underline{u})(x) \quad (1.69)$$

for $x \in \Gamma$. As in (1.11), we can write the boundary integral equations (1.68) and (1.69) by the use of the Calderon projector as

$$\begin{pmatrix} \gamma_0^{\text{int}} \underline{u} \\ \gamma_1^{\text{int}} \underline{u} \end{pmatrix} \begin{pmatrix} \frac{1}{2} I - K^{\text{Lame}} & V^{\text{Lame}} \\ D^{\text{Lame}} & \frac{1}{2} I + (K^{\text{Lame}})' \end{pmatrix} \begin{pmatrix} \gamma_0^{\text{int}} \underline{u} \\ \gamma_1^{\text{int}} \underline{u} \end{pmatrix}. \tag{1.70}$$

Since the single layer potential V^{Lame} is $[H^{-1/2}(\Gamma)]^3$–elliptic and therefore invertible, we obtain from the first equation in (1.70) the Dirichlet to Neumann map

$$\gamma_1^{\text{int}} \underline{u}(x) = (S^{\text{Lame}} \gamma_0^{\text{int}} \underline{u})(x) \quad \text{for } x \in \Gamma \tag{1.71}$$

with the Steklov–Poincaré operator

$$S^{\text{Lame}} = \left(V^{\text{Lame}}\right)^{-1} \left(\frac{1}{2} I + K^{\text{Lame}}\right)$$

$$= D^{\text{Lame}} + \left(\frac{1}{2} I + \left(K^{\text{Lame}}\right)'\right) \left(V^{\text{Lame}}\right)^{-1} \left(\frac{1}{2} I + K^{\text{Lame}}\right).$$

Note that it holds

$$(S^{\text{Lame}} \gamma_0^{\text{int}} \underline{v})(x) = \underline{0} \quad \text{for all } \underline{v} \in \mathcal{R}.$$

1.2.1 Dirichlet Boundary Value Problem

When considering the Dirichlet boundary value problem of linear elastostatics,

$$-\sum_{j=1}^{3} \frac{\partial}{\partial x_j} \sigma_{ij}(\underline{u}, x) = 0 \qquad \text{for } x \in \Omega, \, i = 1, 2, 3,$$

$$\gamma_0^{\text{int}} \underline{u}(x) = \underline{g}(x) \qquad \text{for } x \in \Gamma,$$

the displacement field $\underline{u}$ can be described by the Somigliana identity

$$u_k(x) = \int_{\Gamma} \left(\underline{U}_k^*(x, y), \underline{t}(y)\right) ds_y - \int_{\Gamma} \left(\gamma_{1,y}^{\text{int}} \underline{U}_k^*(x, y), \underline{g}(y)\right) ds_y$$

for $x \in \Omega$ and $k = 1, 2, 3$, where the boundary stress $\underline{t} = \gamma_1^{\text{int}} \underline{u}$ has to be determined from some appropriate boundary integral equation.

Using the first equation in the Calderon projector (1.70), we have to solve a first kind boundary integral equation to find $\underline{t} \in [H^{-1/2}(\Gamma)]^3$ such that

$$(V^{\text{Lame}} \underline{t})(x) = \frac{1}{2} \underline{g}(x) + (K^{\text{Lame}} \underline{g})(x) \quad \text{for } x \in \Gamma.$$

This boundary integral equation corresponds to finding the solution $\underline{t}$ of the variational problem

$$\left\langle V^{\text{Lame}} \underline{t}, \underline{w} \right\rangle_{\Gamma} = \left\langle \left(\frac{1}{2} I + K^{\text{Lame}}\right) \underline{g}, \underline{w} \right\rangle_{\Gamma} \tag{1.72}$$

in $[H^{-1/2}(\Gamma)]^3$ for all test functions $\underline{w} \in [H^{-1/2}(\Gamma)]^3$. Since the single layer potential V^{Lame} is $[H^{-1/2}(\Gamma)]^3$–elliptic, the unique solvability of the variational problem (1.72) follows due to the Lax–Milgram theorem.

1.2.2 Neumann Boundary Value Problem

For a simply connected domain $\Omega \subset \mathbb{R}^3$, we now consider the Neumann boundary value problem

$$-\sum_{j=1}^{3} \frac{\partial}{\partial x_j} \sigma_{ij}(\underline{u}, x) = 0 \qquad \text{for } x \in \Omega,\ i = 1, 2, 3,$$

$$\gamma_1^{\text{int}} \underline{u}(x) = \underline{g}(x) \qquad \text{for } x \in \Gamma \tag{1.73}$$

where we have to assume the solvability conditions

$$\int_{\Gamma} \left(\underline{g}(y), \gamma_0^{\text{int}} \underline{v}(y) \right) ds_y = 0 \quad \text{for all } \underline{v} \in \mathcal{R}. \tag{1.74}$$

Note that the solution of the Neumann boundary value problem (1.73) is only unique up to the rigid body motions $\underline{v} \in \mathcal{R}$.

Using the Somigliana identity (1.65), a solution of the Neumann boundary value problem (1.73) is given by the representation formula

$$u_\ell(x) = \int_{\Gamma} \left(\underline{U}_\ell^*(x, y), \underline{g}(y) \right) ds_y - \int_{\Gamma} \left(\gamma_{1,y}^{\text{int}} \underline{U}_\ell^*(x, y), \gamma_0^{\text{int}} \underline{u}(y) \right) ds_y$$

for $x \in \Omega$ and $\ell = 1, 2, 3$. Hence, we have to find the yet unknown Dirichlet datum $\bar{\underline{u}} = \gamma_0^{\text{int}} \underline{u}$ on Γ.

When using the second equation in the Calderon projector (1.70), we have to solve a first kind boundary integral equation to find $\bar{\underline{u}} \in [H^{1/2}(\Gamma)]^3$ such that

$$(D^{\text{Lame}} \bar{\underline{u}})(x) = \frac{1}{2} \underline{g}(x) - \left(\left(K^{\text{Lame}} \right)' \underline{g} \right)(x) \quad \text{for } x \in \Gamma \tag{1.75}$$

is satisfied in a weak sense, in particular, in the sense of $[H^{-1/2}(\Gamma)]^3$. Since the hypersingular boundary integral operator D^{Lame} has the non–trivial kernel of the rigid body motions, we have to consider the boundary integral equation (1.75) in suitable subspaces. To this end, we define

$$H_{\mathcal{R}}^{1/2}(\Gamma) = \left\{ \underline{u} \in [H^{1/2}(\Gamma)]^3 \,:\, \langle \underline{u}, \underline{v} \rangle_\Gamma = 0 \quad \text{for all } \underline{v} \in \mathcal{R} \right\}.$$

Then the variational problem of the boundary integral equation (1.75) is to find $\bar{\underline{u}} \in H_{\mathcal{R}}^{1/2}(\Gamma)$ such that

$$\left\langle D^{\text{Lame}} \bar{\underline{u}}, \underline{v} \right\rangle_\Gamma = \left\langle \left(\frac{1}{2} I - \left(K^{\text{Lame}} \right)' \right) \underline{g}, \underline{v} \right\rangle_\Gamma \tag{1.76}$$

is satisfied for all $\underline{v} \in H_{\mathcal{R}}^{1/2}(\Gamma)$. The general solution of the hypersingular boundary integral equation (1.75) is then given by

$$\bar{u}_\alpha(x) = \bar{u}(x) + \sum_{k=1}^{6} c_k \, \underline{v}_k(x),$$

where the vectors $\underline{v}_k$, $k = 1, \ldots, 6$ build a basis in the space of rigid body motions (cf. (1.63)). To fix the constants c_k, we may require the scaling conditions

$$\int_\Gamma \left(\bar{u}(y), \gamma_0^{\text{int}} \underline{v}_k(y) \right) ds_y = \alpha_k \qquad (1.77)$$

for $k = 1, \ldots, 6$, where the $\alpha_k \in \mathbb{R}$ can be arbitrary, but prescribed.

Instead of solving the variational problem (1.76) in the subspace $H_{\mathcal{R}}^{1/2}(\Gamma)$ and finding the unique solution afterwards from the scaling conditions (1.77), one can formulate an extended variational problem to find $\bar{u}_\alpha \in [H^{1/2}(\Gamma)]^3$ such that

$$\left\langle D^{\text{Lame}} \bar{u}_\alpha, \underline{v} \right\rangle_\Gamma + \sum_{k=1}^{6} \left\langle \bar{u}_\alpha, \underline{v}_k \right\rangle_\Gamma \left\langle \underline{v}, \underline{v}_k \right\rangle_\Gamma = \qquad (1.78)$$

$$\left\langle \left(\frac{1}{2} I - \left(K^{\text{Lame}} \right)' \right) \underline{g}, \underline{v} \right\rangle_\Gamma + \sum_{k=1}^{6} \alpha_k \left\langle \underline{v}, \underline{v}_k \right\rangle_\Gamma$$

is satisfied for all $\underline{v} \in [H^{1/2}(\Gamma)]^3$. The extended variational problem (1.78) is uniquely solvable for any given $\underline{g} \in [H^{-1/2}(\Gamma)]^3$. If $\underline{g}$ satisfies the solvability conditions (1.74), then $\bar{u}_\alpha$ is the unique solution of the hypersingular boundary integral equation (1.75) satisfying the scaling conditions (1.77).

1.2.3 Mixed Boundary Value Problem

Let $\Omega \subset \mathbb{R}^3$ be simply connected. Then we consider the mixed boundary value problem

$$-\sum_{j=1}^{3} \frac{\partial}{\partial x_j} \sigma_{ij}(\underline{u}, x) = 0 \qquad \text{for } x \in \Omega,$$

$$\gamma_0^{\text{int}} u_i(x) = g_i(x) \qquad \text{for } x \in \Gamma_{D,i}, \qquad (1.79)$$

$$\sum_{j=1}^{3} \sigma_{ij}(\underline{u}, x) n_j(x) = f_i(x) \qquad \text{for } x \in \Gamma_{N,i},$$

and for $i = 1, 2, 3$. We assume that

$$\Gamma = \overline{\Gamma}_{N,i} \cup \overline{\Gamma}_{D,i}, \quad \Gamma_{N,i} \cap \Gamma_{D,i} = \varnothing, \quad \text{meas } \Gamma_{D,i} > 0$$

for $i = 1, 2, 3$ is satisfied.

Using the Somigliana identity (1.65), the solution of the mixed boundary value problem (1.79) is given by the representation formula

$$u_\ell(x) = \sum_{i=1}^{3} \int_{\Gamma_{N,i}} f_i(x) U_{\ell i}^*(x,y) ds_y - \sum_{i=1}^{3} \int_{\Gamma_{D,i}} g_i(y) (\gamma_{1,y}^{\mathrm{int}} \underline{U}_\ell^*)_i(x,y) ds_y +$$

$$\sum_{i=1}^{3} \int_{\Gamma_{D,i}} (\gamma_1^{\mathrm{int}} \underline{u})_i(x) U_{\ell i}^*(x,y) ds_y - \sum_{i=1}^{3} \int_{\Gamma_{N,i}} \gamma_0^{\mathrm{int}} u_i(y) (\gamma_{1,y}^{\mathrm{int}} \underline{U}_\ell^*)_i(x,y) ds_y$$

for $x \in \Omega$ and for $\ell = 1,2,3$. Hence, we have to find the yet unknown Cauchy data $(\gamma_1^{\mathrm{int}} \underline{u})_i$ on $\Gamma_{D,i}$ and $\gamma_0^{\mathrm{int}} u_i$ on $\Gamma_{N,i}$.

The symmetric formulation of boundary integral equations is based on the use of the first kind boundary integral equation (1.68) for those components, where the boundary displacement $\gamma_0^{\mathrm{int}} u_i = g_i$ is given, while the hypersingular boundary integral equation (1.69) is used when the boundary stress $(\gamma_1^{\mathrm{int}} \underline{u})_i = f_i$ is prescribed.

Let $\widetilde{g}_i \in H^{1/2}(\Gamma)$ and $\widetilde{f}_i \in H^{-1/2}(\Gamma)$ be some arbitrary but fixed extensions of the given boundary data $g_i \in H^{1/2}(\Gamma_{D,i})$ and $f_i \in H^{-1/2}(\Gamma_{N,i})$, respectively. Then, we have to find

$$\widetilde{u}_i = \gamma_0^{\mathrm{int}} u_i - \widetilde{g}_i \in \widetilde{H}^{1/2}(\Gamma_{N,i}), \quad \widetilde{t}_i = (\gamma_1^{\mathrm{int}} \underline{u})_i - \widetilde{f}_i \in \widetilde{H}^{-1/2}(\Gamma_{D,i})$$

satisfying a system of boundary integral equations,

$$(V^{\mathrm{Lame}} \widetilde{\underline{t}})_i(x) - (K^{\mathrm{Lame}} \widetilde{\underline{u}})_i(x) =$$
$$\frac{1}{2} g_i(x) + (K^{\mathrm{Lame}} \widetilde{\underline{g}})_i(x) - (V^{\mathrm{Lame}} \widetilde{\underline{f}})_i(x)$$

for $x \in \Gamma_{D,i}$, and

$$(D^{\mathrm{Lame}} \widetilde{\underline{u}})_i(x) + \left((K^{\mathrm{Lame}})' \widetilde{\underline{t}} \right)_i(x) =$$
$$\frac{1}{2} f_i(x) - \left((K^{\mathrm{Lame}})' \widetilde{\underline{f}} \right)_i(x) - (D^{\mathrm{Lame}} \widetilde{\underline{g}})_i(x)$$

for $x \in \Gamma_{N,i}$, and for $i = 1,2,3$. The associated variational problem is to find

$$(\widetilde{\underline{t}}, \widetilde{\underline{u}}) \in \prod_{i=1}^{3} \widetilde{H}^{-1/2}(\Gamma_{D,i}) \times \prod_{i=1}^{3} \widetilde{H}^{1/2}(\Gamma_{N,i})$$

such that

$$a(\widetilde{\underline{t}}, \widetilde{\underline{u}}; \underline{w}, \underline{v}) = F(\underline{w}, \underline{v}) \tag{1.80}$$

is satisfied for all

$$(\underline{w}, \underline{v}) \in \prod_{i=1}^{3} \widetilde{H}^{-1/2}(\Gamma_{D,i}) \times \prod_{i=1}^{3} \widetilde{H}^{1/2}(\Gamma_{N,i})$$

with the bilinear form

$$a(\widetilde{\underline{t}}, \widetilde{\underline{u}}; \underline{w}, \underline{v}) =$$

$$\sum_{i=1}^{3} \left\langle (V^{\text{Lame}}\widetilde{\underline{t}})_i, w_i \right\rangle_{\Gamma_{D,i}} - \sum_{i=1}^{3} \left\langle (K^{\text{Lame}}\widetilde{\underline{u}})_i, w_i \right\rangle_{\Gamma_{D,i}} +$$

$$\sum_{i=1}^{3} \left\langle \widetilde{t}_i, (K'^{\text{Lame}}\underline{v})_i \right\rangle_{\Gamma_{N,i}} + \sum_{i=1}^{3} \left\langle (D^{\text{Lame}}\widetilde{\underline{u}})_i, v_i \right\rangle_{\Gamma_{N,i}}$$

and with the linear form

$$F(\underline{w}, \underline{v}) =$$

$$\sum_{i=1}^{3} \left(\frac{1}{2} \left\langle g_i, w_i \right\rangle_{\Gamma_{D,i}} + \left\langle (K^{\text{Lame}}\widetilde{\underline{g}})_i, w_i \right\rangle_{\Gamma_{D,i}} - \left\langle (V^{\text{Lame}}\widetilde{\underline{f}})_i, w_i \right\rangle_{\Gamma_{D,i}} \right) +$$

$$\sum_{i=1}^{3} \left(\frac{1}{2} \left\langle f_i, v_i \right\rangle_{\Gamma_{N,i}} - \left\langle \widetilde{f}_i, (K'^{\text{Lame}}\underline{v})_i \right\rangle_{\Gamma_{N,i}} - \left\langle (D^{\text{Lame}}\widetilde{\underline{g}})_i, v_i \right\rangle_{\Gamma_{N,i}} \right).$$

Since the bilinear form $a(\cdot, \cdot ; \cdot, \cdot)$ is skew–symmetric, the unique solvability of the variational problem (1.80) follows from the mapping properties of all boundary integral operators involved.

In the mixed boundary value problem (1.79), different boundary conditions in the cartesian coordinate system are prescribed. In many practical applications, however, boundary conditions are given with respect to some different orthogonal coordinate system. As an example, we consider the mixed boundary value problem

$$-\sum_{j=1}^{3} \frac{\partial}{\partial x_j} \sigma_{ij}(\underline{u}, x) = 0 \qquad \text{for } x \in \Omega, \quad i = 1, 2, 3$$

$$(\gamma_0^{\text{int}}\underline{u}(x), \underline{n}(x)) = g(x) \text{ for } x \in \Gamma, \tag{1.81}$$

$$\gamma_1^{\text{int}}\underline{u}(x) - (\gamma_1^{\text{int}}\underline{u}(x), \underline{n}(x))\underline{n}(x) = \underline{0} \qquad \text{for } x \in \Gamma.$$

An elastic body, which is modelled by the mixed boundary value problem (1.81), can slide in tangential direction while in the normal direction a displacement is given. Note that the boundary value problem (1.81) may arise when considering a linearisation of nonlinear contact (Signorini) boundary conditions.

Using the Dirichlet to Neumann map (1.71), it remains to find the boundary displacements $\gamma_0^{\text{int}}\underline{u}$ and the boundary stresses $\gamma_1^{\text{int}}\underline{u}$ satisfying

$$\gamma_1^{\text{int}}\underline{u}(x) = (S^{\text{Lame}}\gamma_0^{\text{int}}\underline{u}(x) \quad \text{for } x \in \Gamma,$$

as well as the boundary conditions

$$(\gamma_0^{\text{int}}\underline{u}(x), \underline{n}(x)) = g(x), \quad \gamma_1^{\text{int}}\underline{u}(x) - (\gamma_1^{\text{int}}\underline{u}(x), \underline{n}(x))\underline{n}(x) = \underline{0} \quad \text{for } x \in \Gamma.$$

Using

$$\gamma_0^{\text{int}} \underline{u}(x) = g(x)\underline{n}(x) + \underline{u}_T(x) \,,$$

we have to find a tangential displacement field

$$\underline{u}_T \in H_T^{1/2}(\Gamma) = \left\{ \underline{v} \in [H^{1/2}(\Gamma)]^3 : (\underline{v}(x), \underline{n}(x)) = 0 \quad \text{for } x \in \Gamma \right\}$$

as a solution of the boundary integral equation

$$(S^{\text{Lame}} g\underline{n})(x) + (S^{\text{Lame}} \underline{u}_T)(x) - (\gamma_1^{\text{int}} \underline{u}(x), \underline{n}(x))\underline{n}(x) = \underline{0} \quad \text{for } x \in \Gamma.$$

For a test function $\underline{v}_T \in H_T^{1/2}(\Gamma)$, we then obtain the variational problem

$$\langle S^{\text{Lame}} \underline{u}_T, \underline{v}_T \rangle_\Gamma = -\langle S^{\text{Lame}} g\underline{n}, \underline{v}_T \rangle_\Gamma \,,$$

which is uniquely solvable due to the mapping properties of the Steklov–Poincaré operator. Note that one may also consider mixed boundary value problems with sliding boundary conditions only on a part Γ_S, but standard Dirichlet or Neumann boundary conditions elsewhere. However, to ensure uniqueness, one needs to assume Dirichlet boundary conditions somewhere for each component.

1.3 Stokes System

The Stokes problem is to find a velocity field $\underline{u} \in \mathbb{R}^3$ and a pressure p such that

$$-\varrho \Delta \underline{u}(x) + \nabla p(x) = 0, \quad \text{div } \underline{u}(x) = 0 \quad \text{for } x \in \Omega \qquad (1.82)$$

is satisfied, where ϱ is the viscosity of the fluid. Note that the Stokes system (1.82) also arises in the limiting case when considering the Navier system

$$-\mu \Delta \underline{u}(x) - (\lambda + \mu)\text{grad div } \underline{u}(x) = \underline{0} \quad \text{for } x \in \Omega$$

for incompressible materials. Introducing the pressure

$$p(x) = -(\lambda + \mu)\text{div } \underline{u}(x) \quad \text{for } x \in \Omega \,,$$

we get

$$-\mu \Delta \underline{u}(x) + \nabla p(x) = \underline{0} \quad \text{for } x \in \Omega \,,$$

as well as

$$\text{div } \underline{u}(x) = -\frac{1}{\lambda + \mu} p(x) = -\frac{2}{E}(1 + \nu)(1 - 2\nu)p(x) = 0$$

in the incompressible case $\nu = 1/2$.

Using integration by parts, we obtain from the second equation in (1.82) the compatibility condition

$$0 = \int_\Omega \operatorname{div} \underline{u}(y)\, dy = \int_\Gamma (\underline{u}(y), \underline{n}(y))\, ds_y . \tag{1.83}$$

Green's first formula for the Stokes system (1.82) reads

$$a(\underline{u}, \underline{v}) = \int_\Omega \sum_{i=1}^{3} \left(-\varrho \Delta u_i(y) + \frac{\partial}{\partial y_i} p(y) \right) v_i(y) dy \tag{1.84}$$

$$+ \int_\Omega p(y) \operatorname{div} \underline{v}(y) dy + \int_\Gamma \sum_{i=1}^{3} t_i(\underline{u}(y), p(y)) v_i(y) ds_y$$

with the symmetric bilinear form

$$a(\underline{u}, \underline{v}) = 2\varrho \int_\Omega \sum_{i,j=1}^{3} e_{ij}(\underline{u}, y) e_{ij}(\underline{v}, y) dy - \int_\Omega \operatorname{div} \underline{u}(y) \operatorname{div} \underline{v}(y)\, dy$$

and with the associated boundary stress

$$t_i(\underline{u}(y), p(y)) = -p(y)n_i(y) + 2\varrho \sum_{j=1}^{3} e_{ij}(\underline{u}, y) n_j(y), \quad y \in \Gamma, \ i = 1, 2, 3.$$

From Green's first formula (1.84), we now derive Green's second formula which reads for the solution $(\underline{u}, p)$ of (1.82) as

$$\int_\Omega \sum_{i=1}^{3} \left(-\varrho \Delta v_i(y) + \frac{\partial}{\partial y_i} q(y) \right) u_i(y) dy - \int_\Omega p(y) \operatorname{div} \underline{v}(y) dy$$

$$= \int_\Gamma \sum_{i=1}^{3} t_i(\underline{u}(y), p(y)) v_i(y) ds_y - \int_\Gamma \sum_{i=1}^{3} t_i(\underline{v}(y), q(y)) u_i(y) ds_y.$$

Choosing as test functions a pair of fundamental solutions $\underline{U}_\ell^*(x, y)$ and $q_\ell^*(x, y)$, i.e. satisfying

$$\int_\Omega \sum_{i=1}^{3} \left(-\varrho \Delta U_{\ell i}^*(x, y) + \frac{\partial}{\partial y_i} q_\ell^*(x, y) \right) u_i(y) dy = u_\ell(x), \quad \operatorname{div} \underline{U}_\ell^*(x, y) = 0,$$

we obtain a representation formula for $x \in \Omega$

$$u_\ell(x) = \int_\Gamma \sum_{k=1}^{3} t_k(\underline{u}, p) U_{\ell k}^*(x, y) ds_y - \int_\Gamma \sum_{k=1}^{3} t_k(\underline{U}_\ell^*(x, y), q_\ell^*) u_k(y) ds_y.$$

for $\ell = 1, 2, 3$. The fundamental solution of the Stokes system is given by

$$U_{k\ell}^*(x, y) = \frac{1}{8\pi} \frac{1}{\varrho} \left(\frac{\delta_{k\ell}}{|x-y|} + \frac{(x_k - y_k)(x_\ell - y_\ell)}{|x-y|^3} \right) \tag{1.85}$$

for $k, \ell = 1, 2, 3$, and

$$q_\ell^*(x, y) = \frac{1}{4\pi} \frac{y_\ell - x_\ell}{|x-y|^3}$$

for $\ell = 1, 2, 3$. Note that the fundamental solution (1.85) coincides with the Kelvin tensor (1.66) for

$$\nu = \frac{1}{2}, \quad \varrho = \frac{E}{3}.$$

Hence, we can define and analyse all the boundary integral operators and related boundary integral equations as for the system of linear elastostatics. The only exception is the Dirichlet boundary value problem of the Stokes system which requires a special treatment of the associated single layer potential.

As in linear elastostatics the single layer potential of the Stokes system is given by

$$(\widetilde{V}^{\text{Stokes}}\underline{w})_k = \frac{1}{2\varrho} \left((\widetilde{V}w_k)(x) + \sum_{\ell=1}^{3} (\widetilde{V}_{k\ell}w_\ell)(x) \right)$$

for $k = 1, 2, 3$. As before,

$$\widetilde{V}^{\text{Stokes}} : [H^{-1/2}(\Gamma)]^3 \to [H^1(\Omega)]^3$$

defines a continuous map. Combining this with the mapping properties of the interior trace operator

$$\gamma_0^{\text{int}} : H^1(\Omega) \to H^{1/2}(\Gamma),$$

we can define the continuous boundary integral operator

$$V^{\text{Stokes}} = \gamma_0^{\text{int}}\widetilde{V}^{\text{Stokes}} : [H^{-1/2}(\Gamma)]^3 \to [H^{1/2}(\Gamma)]^3$$

allowing the representation

$$(V^{\text{Stokes}}\underline{w})_k = \frac{1}{2} \frac{1}{\varrho} \left((Vw_k)(x) + \sum_{\ell=1}^{3} (V_{k\ell}w_\ell)(x) \right) \quad \text{for } x \in \Gamma,$$

and for $k = 1, 2, 3$ as a weakly singular surface integral; see also Lemma 1.15.

When considering the interior Dirichlet boundary value problem for the Stokes system

$$\begin{aligned}
-\varrho \Delta \underline{u}(x) + \nabla p(x) &= 0 && \text{for } x \in \Omega, \\
\operatorname{div} \underline{u}(x) &= 0 && \text{for } x \in \Omega, \\
\gamma_0^{\text{int}}\underline{u}(x) &= \underline{g}(x) && \text{for } x \in \Gamma,
\end{aligned} \tag{1.86}$$

and using (1.83), we first have to assume the solvability condition

$$\int\limits_{\Gamma} (g(y), n(y)) \, ds_y = 0. \tag{1.87}$$

On the other hand, it is obvious, that the pressure p satisfying the first equation in (1.86) is only unique up to an additive constant. In particular, the homogeneous Dirichlet boundary value problem

$$-\varrho\Delta \underline{u}(x) + \nabla p(x) = 0, \quad \operatorname{div} \underline{u}(x) = 0 \text{ for } x \in \Omega, \ \underline{u}(x) = \underline{0} \text{ for } x \in \Gamma$$

has the non–trivial pair of solutions $\underline{u}^*(x) = \underline{0}$ and $p^*(x) = -1$ for $x \in \Omega$.

The first kind boundary integral equation of the direct approach for the Dirichlet boundary value problem (1.86) is

$$(V^{\text{Stokes}}\underline{t})(x) = \frac{1}{2}\underline{g}(x) + (K^{\text{Stokes}}\underline{g})(x) \quad \text{for } x \in \Gamma. \tag{1.88}$$

For the homogeneous Dirichlet boundary value problem with $\underline{g} = \underline{0}$, we therefore obtain

$$(V^{\text{Stokes}}\underline{t}^*)(x) = \underline{0} \quad \text{for } x \in \Gamma$$

with

$$\underline{t}^*(\underline{u}^*(x), p^*(x)) = -p^*(x)\underline{n}(x) = \underline{n}(x) \quad \text{for } x \in \Gamma.$$

Thus, $\underline{t}^* = \underline{n}$ is an eigenfunction of the single layer potential V^{Stokes} yielding a zero eigenvalue. Therefore, we conclude that the boundary integral equation (1.88) is only solvable modulo $\underline{t}^*$, and we have to consider the boundary integral equation (1.88) in an appropriate factor space [90]. Hence, we define

$$H_*^{-1/2}(\Gamma) = \left\{ \underline{w} \in [H^{-1/2}(\Gamma)]^3 \ : \ \langle \underline{w}, \underline{n} \rangle_V = \sum_{k=1}^{3} \langle Vw_k, n_k \rangle_{\Gamma} = 0 \right\},$$

where

$$V : H^{-1/2}(\Gamma) \to H^{1/2}(\Gamma)$$

is the single layer potential of the Laplace operator. Considering the boundary integral equation (1.88) in $H_*^{-1/2}(\Gamma)$, this can be rewritten as an extended variational problem to find $\underline{t} \in [H^{-1/2}(\Gamma)]^3$ such that

$$\left\langle V^{\text{Stokes}}\underline{t}, \underline{w} \right\rangle_{\Gamma} + \left\langle \underline{t}, \underline{n} \right\rangle_V \left\langle \underline{w}, \underline{n} \right\rangle_V = \left\langle \left(\frac{1}{2}I + K^{\text{Stokes}}\right)\underline{g}, \underline{w} \right\rangle_{\Gamma} \tag{1.89}$$

is satisfied for all $\underline{w} \in [H^{-1/2}(\Gamma)]^3$. Note that there exists a unique solution $\underline{t} \in [H^{-1/2}(\Gamma)]^3$ of the extended variational problem (1.89) for any given Dirichlet datum $\underline{g} \in [H^{1/2}(\Gamma)]^3$. If $\underline{g}$ satisfies the solvability condition (1.87), we then obtain $\underline{t} \in H_*^{-1/2}(\Gamma)$.

1.4 Helmholtz Equation

Let $U : \mathbb{R}_+ \times \Omega \to \mathbb{R}$ be a scalar function which satisfies the wave equation

$$\frac{1}{c^2} \frac{\partial^2}{\partial t^2} U(t,x) = \Delta U(t,x) \quad \text{for } t > 0, \, x \in \Omega. \tag{1.90}$$

The equation (1.90) is valid for the wave propagation in a homogeneous, isotrop, friction-free medium having the constant speed of sound c. The most important examples are the acoustic scattering and the sound radiation.

The time harmonic acoustic waves are of the form

$$U(t,x) = \operatorname{Re}\left(u(x)e^{-\imath \omega t}\right), \tag{1.91}$$

where $\imath$ is the imaginary unit. In (1.91), $u : \Omega \to \mathbb{C}$ is a scalar, complex valued function and $\omega > 0$ denotes the frequency. Inserting (1.91) into the wave equation (1.90), we obtain the reduced wave equation or the Helmholtz equation

$$-\Delta u(x) - \kappa^2 u(x) = 0 \quad \text{for } x \in \Omega, \tag{1.92}$$

where $\kappa = \omega/c > 0$ is the wave number.

First we consider the Helmholtz equation (1.92) in a bounded domain $\Omega \subset \mathbb{R}^3$. Multiplying this equation (1.92) with a test function v, integrating over Ω, and applying integration by parts, this gives Green's first formula

$$\int_{\Omega} (-\Delta u(y) - \kappa^2 u(y)) v(y) dy = a(u,v) - \int_{\Gamma} \gamma_1^{\text{int}} u(y) \gamma_0^{\text{int}} v(y) ds_y \tag{1.93}$$

with the symmetric bilinear form

$$a(u,v) = \int_{\Omega} \left(\nabla u(y), \nabla v(y)\right) dy - \kappa^2 \int_{\Omega} u(y) v(y) dy.$$

From Green's formula (1.93) and by the use of the symmetry of the bilinear form $a(\cdot, \cdot)$, we deduce Green's second formula,

$$\int_{\Omega} (-\Delta u(y) - \kappa^2 u(y)) v(y) dy + \int_{\Gamma} \gamma_1^{\text{int}} u(y) \gamma_0^{\text{int}} v(y) ds_y =$$

$$\int_{\Omega} (-\Delta v(y) - \kappa^2 v(y)) u(y) dy + \int_{\Gamma} \gamma_1^{\text{int}} v(y) \gamma_0^{\text{int}} u(y) ds_y.$$

Now, choosing as a test function v a fundamental solution $u_\kappa^* : \mathbb{R}^3 \times \mathbb{R}^3 \to \mathbb{C}$ satisfying

$$\int_{\Omega} \left(-\Delta u_\kappa^*(x,y) - \kappa^2 u_\kappa^*(x,y)\right) u(y) dy = u(x) \quad \text{for } x \in \Omega, \tag{1.94}$$

the solution of the Helmholtz equation (1.92) is given by the representation formula

$$u(x) = \int_\Gamma u_\kappa^*(x,y)\gamma_1^{\text{int}}u(y)ds_y - \int_\Gamma \gamma_{1,y}^{\text{int}}u_\kappa^*(x,y)\gamma_0^{\text{int}}u(y)ds_y \qquad (1.95)$$

for $x \in \Omega$. The fundamental solution of the Helmholtz equation (1.92) is

$$u_\kappa^*(x,y) = \frac{1}{4\pi}\frac{e^{\iota\kappa|x-y|}}{|x-y|} \qquad \text{for } x,y \in \mathbb{R}^3. \qquad (1.96)$$

As for the Laplace operator, we consider the single layer potential

$$(\widetilde{V}_\kappa w)(x) = \int_\Gamma u_\kappa^*(x,y)w(y)ds_y = \frac{1}{4\pi}\int_\Gamma \frac{e^{\iota\kappa|x-y|}}{|x-y|}w(y)ds_y \quad \text{for } x \in \Omega$$

which defines a continuous map from a given density function w on the boundary Γ to a function $\widetilde{V}_\kappa w$, which satisfies the partial differential equation (1.92) in Ω. In particular,

$$\widetilde{V}_\kappa : H^{-1/2}(\Gamma) \to H^1(\Omega)$$

is continuous and $\widetilde{V}_\kappa w \in H^1(\Omega)$ is a weak solution of the Helmholtz equation (1.92) for any $w \in H^{-1/2}(\Gamma)$. Using the mapping properties of the interior trace operators

$$\gamma_0^{\text{int}} : H^1(\Omega) \to H^{1/2}(\Gamma)$$

and

$$\gamma_1^{\text{int}} : H^1(\Omega, \Delta + \kappa^2) \to H^{-1/2}(\Gamma),$$

we can define corresponding boundary integral operators, e.g. the single layer potential operator

$$V_\kappa = \gamma_0 \widetilde{V}_\kappa : H^{-1/2}(\Gamma) \to H^{1/2}(\Gamma),$$

as follows:

$$(V_\kappa w)(x) = \int_\Gamma u_\kappa^*(x,y)w(y)ds_y = \frac{1}{4\pi}\int_\Gamma \frac{e^{\iota\kappa|x-y|}}{|x-y|}w(y)ds_y \quad \text{for } x \in \Gamma.$$

Its conormal derivative is

$$\gamma_1^{\text{int}}\widetilde{V}_\kappa = \frac{1}{2}I + K_\kappa' : H^{-1/2}(\Gamma) \to H^{-1/2}(\Gamma)$$

with the adjoint double layer potential operator

$$(K_\kappa' w)(x) = \lim_{\varepsilon \to 0} \int_{y\in\Gamma:|y-x|\geq\varepsilon} \gamma_{1,x}^{\text{int}}u_\kappa^*(x,y)w(y)ds_y$$

$$= \lim_{\varepsilon \to 0}\frac{1}{4\pi} \int_{y\in\Gamma:|y-x|\geq\varepsilon} \left(\nabla_x\frac{e^{\iota\kappa|x-y|}}{|x-y|}, \underline{n}(x)\right)w(y)ds_y.$$

Note that

$$H^1(\Omega, \Delta + \kappa^2) = \left\{ v \in H^1(\Omega) : \Delta v + \kappa^2 v \in \tilde{H}^{-1}(\Omega) \right\}.$$

Since the density functions of the boundary integral operators introduced above may be complex valued, we consider

$$\langle v, w \rangle_\Gamma = \int_\Gamma v(x)\overline{w(x)}ds_x$$

as an appropriate duality pairing for $v \in H^{1/2}(\Gamma)$ and $w \in H^{-1/2}(\Gamma)$. Then the single layer potential operator is complex symmetric, i.e. the following property holds for $w, z \in H^{-1/2}(\Gamma)$

$$\langle V_\kappa w, z \rangle_\Gamma = \frac{1}{4\pi} \int_\Gamma \int_\Gamma \frac{e^{i\kappa|x-y|}}{|x-y|} w(y)ds_y \overline{z(x)}ds_x$$

$$= \frac{1}{4\pi} \int_\Gamma w(y) \overline{\int_\Gamma \frac{e^{-i\kappa|x-y|}}{|x-y|} z(x)ds_x} \, ds_y = \langle w, V_{-\kappa}z \rangle_\Gamma.$$

If Γ is a Lipschitz boundary, the operator

$$V_\kappa - V_0 : H^{-1/2}(\Gamma) \to H^{1/2}(\Gamma)$$

is compact. Since the single layer potential V_0 of the Laplace operator is $H^{-1/2}(\Gamma)$–elliptic (see Lemma 1.1), the single layer potential V_κ is coercive, i.e. with the compact operator $C = V_0 - V_\kappa$, the Gårding's inequality

$$\langle (V_k + C)w, w \rangle_\Gamma = \langle V_0 w, w \rangle_\Gamma \geq c_1^{V_0} \|w\|_{H^{-1/2}(\Gamma)}^2 \qquad (1.97)$$

is satisfied for all $w \in H^{-1/2}(\Gamma)$.

Next we consider the double layer potential

$$(W_\kappa v)(x) = \int_\Gamma \gamma_{1,y}^{\mathrm{int}} u_\kappa^*(x,y)v(y)ds_y = \frac{1}{4\pi} \int_\Gamma \left(\nabla_y \frac{e^{i\kappa|x-y|}}{|x-y|}, \underline{n}(y) \right) v(y)ds_y$$

for $x \in \Omega$, which again defines a continuous map from a given density function v on the boundary Γ to a function $W_\kappa v$ satisfying the Helmholtz equation (1.92). In particular,

$$W_\kappa : H^{1/2}(\Gamma) \to H^1(\Omega)$$

is continuous and $W_\kappa v \in H^1(\Omega)$ is a weak solution of the Helmholtz equation (1.92) for any $v \in H^{1/2}(\Gamma)$. Using the mapping properties of the interior trace operator

$$\gamma_0^{\mathrm{int}} : H^1(\Omega) \to H^{1/2}(\Gamma)$$

and

$$\gamma_1^{\text{int}} \,:\, H^1(\Omega, \Delta + \kappa^2) \rightarrow H^{-1/2}(\Gamma)\,,$$

we can define corresponding boundary integral operators, i.e. the trace

$$\gamma_0^{\text{int}} W = -\frac{1}{2}I + K_\kappa$$

with the double layer potential operator

$$(K_\kappa v)(x) \;=\; \lim_{\varepsilon \to 0} \frac{1}{4\pi} \int\limits_{y \in \Gamma : |y - x| \geq \varepsilon} \left(\nabla_y \frac{e^{i\,\kappa|x-y|}}{|x-y|}, \underline{n}(y) \right) v(y) ds_y \quad \text{for } x \in \Gamma.$$

As for the single layer potential, we have

$$\langle K_\kappa v, w \rangle_\Gamma \;=\; \langle v, K'_{-\kappa} w \rangle_\Gamma$$

for all $v \in H^{1/2}(\Gamma)$ and $w \in H^{-1/2}(\Gamma)$.

The conormal derivative of the double layer potential defines the hypersingular boundary integral operator

$$D_\kappa = -\gamma_1^{\text{int}} W_\kappa \,:\, H^{1/2}(\Gamma) \rightarrow H^{-1/2}(\Gamma)\,.$$

For a Lipschitz boundary Γ, the operator

$$D_\kappa - D_0 \,:\, H^{1/2}(\Gamma) \rightarrow H^{-1/2}(\Gamma)$$

is compact. Since the regularised hypersingular boundary integral operator $D_0 + I$ of the Laplace operator is $H^{1/2}(\Gamma)$–elliptic, and since the embedding $H^{1/2}(\Gamma) \rightarrow H^{-1/2}(\Gamma)$ is compact, the hypersingular boundary integral operator D_κ is coercive, i.e. with the compact operator $C = D_\kappa - D_0 - I$, the Gårding's inequality

$$\langle (D_\kappa + C)v, v \rangle_\Gamma \;=\; \langle (D_0 + I)v, v \rangle_\Gamma \;\geq\; c_1^{\tilde{D}_0} \|v\|_{H^{1/2}(\Gamma)}^2 \tag{1.98}$$

is satisfied for all $v \in H^{1/2}(\Gamma)$.

As for the bilinear form for the hypersingular boundary integral operator for the Laplace equation (see (1.9)), there holds an analogue result for the Helmholtz equation, see [78]:

$$\int\limits_\Gamma (D_\kappa u)(x) v(x) ds_x = \frac{1}{4\pi} \int\limits_\Gamma \int\limits_\Gamma \frac{e^{i\,\kappa|x-y|}}{|x-y|} \big(\underline{\operatorname{curl}}_\Gamma u(y), \underline{\operatorname{curl}}_\Gamma v(x)\big) ds_y ds_x$$

$$- \frac{\kappa^2}{4\pi} \int\limits_\Gamma \int\limits_\Gamma \frac{e^{i\,\kappa|x-y|}}{|x-y|} u(y) v(x) \big(\underline{n}(x), \underline{n}(y)\big) ds_y ds_x\,. \tag{1.99}$$

In addition to the interior boundary value problem for the Helmholtz equation (1.92), we also consider the exterior boundary value problem

$$-\Delta u(x) - \kappa^2 u(x) = 0 \quad \text{for } x \in \Omega^e = \mathbb{R}^3 \backslash \overline{\Omega}, \tag{1.100}$$

where we have to add the Sommerfeld radiation condition

$$\left| \left(\frac{x}{|x|}, \nabla u(x) \right) - \imath \kappa u(x) \right| = \mathcal{O}\left(\frac{1}{|x|^2} \right) \quad \text{as } |x| \to \infty. \tag{1.101}$$

For a fixed $y_0 \in \Omega$ and $R > 2 \operatorname{diam} \Omega$, let $B_R(y_0)$ be a ball of radius R with centre y_0 and including Ω. Let u be a solution of the exterior boundary value problem for the Helmholtz equation (1.100) satisfying the radiation condition (1.101). Considering Green's first formula (1.93) with respect to the bounded domain $\Omega_R = B_R(y_0) \backslash \overline{\Omega}$ and choosing $v = \overline{u}$ as test function, we obtain

$$\int_{\Omega_R} |\nabla u(y)|^2 dy - k^2 \int_{\Omega_R} |u(y)|^2 = \int_{\partial \Omega_R} \gamma_1^{\mathrm{int}} u(y) \overline{\gamma_0^{\mathrm{int}} u(y)} ds_y$$

$$= \int_{\partial B_R(y_0)} \gamma_1^{\mathrm{int}} u(y) \overline{\gamma_0^{\mathrm{int}} u(y)} ds_y - \int_{\Gamma} \gamma_1^{\mathrm{ext}} u(y) \overline{\gamma_0^{\mathrm{ext}} u(y)} ds_y,$$

when taking into account the opposite direction of the normal vector $\underline{n}(x)$ for $x \in \Gamma$. Since the left hand side of the above equation is real, we conclude

$$\operatorname{Im} \int_{\partial B_R(y_0)} \gamma_1^{\mathrm{int}} u(y) \overline{\gamma_0^{\mathrm{int}} u(y)} ds_y = \operatorname{Im} \int_{\Gamma} \gamma_1^{\mathrm{ext}} u(y) \overline{\gamma_0^{\mathrm{ext}} u(y)} ds_y.$$

By the use of this property, the Sommerfeld radiation condition (1.101) implies

$$0 = \lim_{R \to \infty} \int_{\partial B_R(y_0)} \left| \gamma_1^{\mathrm{int}} u(y) - \imath \kappa \gamma_0^{\mathrm{int}} u(y) \right|^2 ds_y$$

$$= \lim_{R \to \infty} \left[\int_{\partial B_R(y_0)} |\gamma_1^{\mathrm{int}} u(y)|^2 ds_y + \kappa^2 \int_{\partial B_R(y_0)} |\gamma_0^{\mathrm{int}} u(y)|^2 ds_y \right.$$

$$\left. -2\kappa \operatorname{Im} \int_{\partial B_R(y_0)} \gamma_1^{\mathrm{int}} u(y) \overline{\gamma_0^{\mathrm{int}} u(y)} ds_y \right]$$

$$= \lim_{R \to \infty} \left[\int_{\partial B_R(y_0)} |\gamma_1^{\mathrm{int}} u(y)|^2 ds_y + \kappa^2 \int_{\partial B_R(y_0)} |\gamma_0^{\mathrm{int}} u(y)|^2 ds_y \right.$$

$$\left. -2\kappa \operatorname{Im} \int_{\Gamma} \gamma_1^{\mathrm{ext}} u(y) \overline{\gamma_0^{\mathrm{ext}} u(y)} ds_y \right]$$

and, therefore,

$$2\kappa \operatorname{Im} \int_\Gamma \gamma_1^{\text{ext}} u(y)\overline{\gamma_0^{\text{ext}} u(y)}ds_y$$

$$= \lim_{R\to\infty}\left[\int_{\partial B_R(y_0)}|\gamma_1^{\text{int}} u(y)|^2 ds_y + \kappa^2\int_{\partial B_R(y_0)}|\gamma_0^{\text{int}} u(y)|^2 ds_y\right] \geq 0.$$

In particular, this gives

$$\lim_{R\to\infty}\int_{\partial B_R(y_0)}|u(y)|^2 ds_y = \mathcal{O}(1),$$

and, therefore,

$$|u(x)| = \mathcal{O}\left(\frac{1}{|x|}\right) \quad \text{as } |x|\to\infty. \tag{1.102}$$

For the bounded domain Ω_R, we can apply the representation formula (1.95) to obtain

$$u(x) = -\int_\Gamma u_\kappa^*(x,y)\gamma_1^{\text{int}} u(y)ds_y + \int_\Gamma \gamma_{1,y}^{\text{int}} u_\kappa^*(x,y)\gamma_0^{\text{int}} u(y)ds_y$$

$$+ \int_{\partial B_R(y_0)} u_\kappa^*(x,y)\gamma_1^{\text{int}} u(y)ds_y - \int_{\partial B_R(y_0)} \gamma_{1,y}^{\text{int}} u_\kappa^*(x,y)\gamma_0^{\text{int}} u(y)ds_y$$

for $x \in \Omega_R$. Taking the limit $R\to\infty$ and incorporating the radiation conditions (1.101) and (1.102), this gives the representation formula in the exterior domain Ω^e, i.e. for $x\in\Omega^e$

$$u(x) = -\int_\Gamma u_\kappa^*(x,y)\gamma_1^{\text{int}} u(y)ds_y + \int_\Gamma \gamma_{1,y}^{\text{int}} u_\kappa^*(x,y)\gamma_0^{\text{int}} u(y)ds_y. \tag{1.103}$$

1.4.1 Interior Dirichlet Boundary Value Problem

We first consider the interior Dirichlet boundary value problem for the Helmholtz equation, i.e.

$$-\Delta u(x) - \kappa^2 u(x) = 0 \quad \text{for } x\in\Omega, \quad \gamma_0^{\text{int}} u(x) = g(x) \quad \text{for } x\in\Gamma. \tag{1.104}$$

Using the representation formula (1.95), the solution of the above Dirichlet boundary value problem is given by

$$u(x) = \int_\Gamma u_\kappa^*(x,y)t(y)ds_y - \int_\Gamma \gamma_{1,y}^{\text{int}} u^*(x,y)g(y)ds_y \quad \text{for } x\in\Omega,$$

where $t = \gamma_1^{\text{int}} u$ is the unknown conormal derivative of u on Γ which has to be determined from some appropriate boundary integral equation.

Applying the interior trace operator γ_0^{int} to the representation formula, this gives a boundary integral equation to find $t \in H^{-1/2}(\Gamma)$ such that

$$(V_\kappa t)(x) = \frac{1}{2}g(x) + (K_\kappa g)(x) \quad \text{for } x \in \Gamma. \qquad (1.105)$$

Note that $t \in H^{-1/2}(\Gamma)$ is the solution of the variational problem

$$\left\langle V_\kappa t, w \right\rangle_\Gamma = \left\langle \left(\frac{1}{2}I + K_\kappa\right)g, w \right\rangle_\Gamma \quad \text{for all } w \in H^{-1/2}(\Gamma). \qquad (1.106)$$

When applying the interior normal derivative γ_1^{int} to the representation formula, this gives a second kind boundary integral equation to find $t \in H^{-1/2}(\Gamma)$ such that

$$\frac{1}{2}t(x) - (K_\kappa' t)(x) = (D_\kappa g)(x) \quad \text{for } x \in \Gamma. \qquad (1.107)$$

To investigate the unique solvability of the variational problem (1.106), and, therefore, of the boundary integral equation (1.105) as well as of the boundary integral equation (1.107), we first consider the Dirichlet eigenvalue problem for the Laplace operator,

$$-\Delta u(x) = \lambda u(x) \quad \text{for } x \in \Omega, \quad \gamma_0^{int} u(x) = 0 \quad \text{for } x \in \Gamma. \qquad (1.108)$$

Let $\lambda \in \mathbb{R}_+$ be a certain eigenvalue, and let u_λ be the corresponding eigenfunction. Since the eigenvalue problem (1.108) can be seen as the Helmholtz equation with the wave number κ satisfying $\kappa^2 = \lambda$, we obtain for the conormal derivative $t_\lambda = \gamma_1^{int} u_\lambda$ the boundary integral equations

$$(V_\kappa t_\lambda)(x) = \frac{1}{2}\gamma_0^{int} u_\lambda(x) + (K_\kappa \gamma_0^{int} u_\lambda)(x) = 0 \quad \text{for } x \in \Gamma$$

and

$$\frac{1}{2}t_\lambda(x) - (K_\kappa' t_\lambda)(x) = (D_\kappa \gamma_0^{int} u_\lambda)(x) = 0 \quad \text{for } x \in \Gamma.$$

Thus, the boundary integral operators V_κ and $1/2\,I - K_\kappa'$ are singular, and, therefore, not invertible, if $\kappa^2 = \lambda$ is an eigenvalue of the Dirichlet eigenvalue problem (1.108). On the other hand, if κ^2 is not an eigenvalue of the Dirichlet eigenvalue problem (1.108), the single layer potential V_κ is injective and hence, since V_κ is coercive, also invertible. This shows the unique solvability of the variational problem (1.106) and of the boundary integral equation (1.105) in this case. Note that also in this case the second kind boundary integral equation (1.107) is uniquely solvable.

1.4.2 Interior Neumann Boundary Value Problem

Next we consider the interior Neumann boundary value problem for the Helmholtz equation, i.e.

$$-\Delta u(x) - \kappa^2 u(x) = 0 \quad \text{for } x \in \Omega, \quad \gamma_1^{\text{int}} u(x) = g(x) \quad \text{for } x \in \Gamma. \tag{1.109}$$

From the representation formula (1.95), we can obtain the solution of the above boundary value problem as

$$u(x) = \int_\Gamma u_\kappa^*(x,y)g(y)ds_y - \int_\Gamma \gamma_{1,y}^{\text{int}} u_\kappa^*(x,y)\gamma_0^{\text{int}} u(y)ds_y \quad \text{for } x \in \Omega.$$

Applying the interior trace operator γ_0^{int} to the above representation formula, this gives a first boundary integral equation to find $\bar{u} = \gamma_0^{\text{int}} u \in H^{1/2}(\Gamma)$ such that

$$\frac{1}{2}\bar{u}(x) + (K_\kappa \bar{u})(x) = (V_\kappa g)(x) \quad \text{for } x \in \Gamma. \tag{1.110}$$

When applying the conormal derivative operator γ_1^{int} to the above representation formula, this gives a second boundary integral equation,

$$(D_\kappa \bar{u})(x) = \frac{1}{2}g(x) - (K_\kappa' g)(x) \quad \text{for } x \in \Gamma. \tag{1.111}$$

Hence, $\bar{u} \in H^{1/2}(\Gamma)$ is a solution of the variational problem

$$\left\langle D_\kappa \bar{u}, v \right\rangle_\Gamma = \left\langle \left(\frac{1}{2}I - K_\kappa'\right)g, v \right\rangle_\Gamma \quad \text{for all } v \in H^{1/2}(\Gamma). \tag{1.112}$$

To investigate the unique solvability of the variational problem (1.112), and, therefore, of the boundary integral equations (1.110) and (1.111), we now consider the Neumann eigenvalue problem for the Laplace operator,

$$-\Delta u(x) = \mu u(x) \quad \text{for } x \in \Omega, \quad \gamma_1^{\text{int}} u(x) = 0 \quad \text{for } x \in \Gamma. \tag{1.113}$$

Let $\mu \in \mathbb{R}_+$ be a certain eigenvalue, and let u_μ be the corresponding eigenfunction. Since the eigenvalue problem (1.113) can be seen as the Helmholtz equation with the wave number κ satisfying $\kappa^2 = \mu$, we then obtain the boundary integral equations

$$(D_\kappa \bar{u}_\mu)(x) = \frac{1}{2}\gamma_1^{\text{int}} u_\mu(x) - (K_\kappa' \gamma_1^{\text{int}} u_\mu)(x) = 0 \quad \text{for } x \in \Gamma$$

and

$$\frac{1}{2}\bar{u}_\mu(x) + (K_\kappa \bar{u}_\mu)(x) = (V_\kappa \gamma_1^{\text{int}} u_\mu)(x) = 0 \quad \text{for } x \in \Gamma.$$

Thus, the boundary integral operators D_κ and $1/2\,I + K_\kappa$ are singular and, therefore, not invertible if $\kappa^2 = \mu$ is an eigenvalue of the Neumann eigenvalue problem (1.113). On the other hand, if κ^2 is not an eigenvalue for the Neumann eigenvalue problem (1.113), the hypersingular boundary integral operator D_κ is injective and coercive, and, therefore, invertible.

1.4.3 Exterior Dirichlet Boundary Value Problem

The exterior Dirichlet boundary value problem for the Helmholtz equation reads

$$-\Delta u(x) - \kappa^2 u(x) = 0 \quad \text{for } x \in \Omega^e, \quad \gamma_0^{\text{ext}} u(x) = g(x) \quad \text{for } x \in \Gamma, \,(1.114)$$

where, in addition, we have to require the Sommerfeld radiation condition (1.101),

$$\left| \left(\frac{x}{|x|}, \nabla u(x) \right) - \imath \kappa u(x) \right| = \mathcal{O}\left(\frac{1}{|x|^2} \right) \quad \text{as } |x| \to \infty.$$

Note that the exterior Dirichlet boundary value problem is uniquely solvable due to the radiation condition. The solution of the above problem is given by the representation formula (1.103),

$$u(x) = -\int_\Gamma u_\kappa^*(x,y) \gamma_1^{\text{ext}} u(y) ds_y + \int_\Gamma \gamma_{1,y}^{\text{ext}} u_\kappa^*(x,y) g(y) ds_y \quad \text{for } x \in \Omega^e .$$

To find the yet unknown Neumann datum $t = \gamma_1^{\text{ext}} u$, we consider the boundary integral equation which results from the representation formula when applying the exterior trace operator γ_0^{ext},

$$(V_\kappa t)(x) = -\frac{1}{2} g(x) + (K_\kappa g)(x) \quad \text{for } x \in \Gamma. \tag{1.115}$$

This boundary integral equation is equivalent to a variational problem to find $t \in H^{-1/2}(\Gamma)$ such that

$$\left\langle V_\kappa t, w \right\rangle_\Gamma = \left\langle \left(-\frac{1}{2} I + K_\kappa \right) g, w \right\rangle_\Gamma \quad \text{for all } w \in H^{-1/2}(\Gamma). \tag{1.116}$$

Since the single layer potential V_κ of the exterior Dirichlet boundary value problem coincides with the single layer potential of the interior Dirichlet boundary value problem, V_κ is not invertible when $\kappa^2 = \lambda$ is an eigenvalue of the Dirichlet eigenvalue problem (1.108). However, we have

$$\left\langle \left(-\frac{1}{2} I + K_\kappa \right) g, t_\lambda \right\rangle_\Gamma = -\left\langle g, \left(\frac{1}{2} I - K'_{-\kappa} \right) t_\lambda \right\rangle_\Gamma = 0,$$

and, therefore,

$$\left(-\frac{1}{2} I + K_\kappa \right) g \in \text{Im } V_\kappa.$$

In fact, the variational problem (1.116) of the direct approach is solvable, but the solution is not unique. As for the Neumann problem (1.21) for the Laplace equation, we can use a stabilised variational formulation to obtain a unique solution $t \in H^{-1/2}(\Gamma)$ satisfying some prescribed side condition, e.g.,

$$\langle V_0 t, t_\lambda \rangle_\Gamma = 0,$$

where $V_0 : H^{-1/2}(\Gamma) \to H^{1/2}(\Gamma)$ is the single layer potential operator of the Laplace equation. Instead of the variational problem (1.116), we then have to find the function $t \in H^{-1/2}(\Gamma)$ as the unique solution of the stabilised variational problem

$$\left\langle V_\kappa t, w \right\rangle_\Gamma + \left\langle V_0 t, t_\lambda \right\rangle_\Gamma \left\langle V_0 w, t_\lambda \right\rangle_\Gamma = \left\langle \left(-\frac{1}{2} I + K_\kappa \right) g, w \right\rangle_\Gamma$$

for all $w \in H^{-1/2}(\Gamma)$. Since this formulation requires the a priori knowledge of the eigensolution t_λ, this approach does not seem to be applicable in general.

If κ^2 is not an eigenvalue of the interior Dirichlet eigenvalue problem (1.108), then the unique solvability of the boundary integral equation (1.115) follows, since V_κ is coercive and injective.

Instead of a direct approach, we may also consider an indirect single layer potential approach,

$$u(x) = (\widetilde{V}_\kappa w)(x) = \int_\Gamma u_\kappa^*(x, y) w(y) ds_y \quad \text{for } x \in \Omega^e.$$

Then, applying the exterior trace operator, this leads to a boundary integral equation to find $w \in H^{-1/2}(\Gamma)$ such that

$$(V_\kappa w)(x) = g(x) \quad \text{for } x \in \Gamma. \tag{1.117}$$

Again, we have unique solvability of the boundary integral equation (1.117) only for those wave numbers κ^2, which are not eigenvalues of the interior Dirichlet eigenvalue problem (1.108).

When using an indirect double layer potential approach,

$$u(x) = (W_\kappa v)(x) = \int_\Gamma \gamma_{1,y}^{\text{ext}} u_\kappa^*(x, y) v(y) ds_y \quad \text{for } x \in \Omega^e,$$

this leads to a boundary integral equation to find $v \in H^{1/2}(\Gamma)$ such that

$$\frac{1}{2} v(x) + (K_\kappa v)(x) = g(x) \quad \text{for } x \in \Gamma. \tag{1.118}$$

The boundary integral operator $1/2\, I + K_\kappa$ is singular, and, therefore, not invertible when κ^2 is an eigenvalue of the interior Neumann boundary value problem (1.113). If κ^2 is not an eigenvalue of the interior Neumann eigenvalue problem (1.113), the unique solvability of the boundary integral equation (1.118) follows.

Although the exterior Dirichlet boundary value problem for the Helmholtz equation is uniquely solvable, the related boundary integral equations may not be solvable, in particular, when κ^2 either coincides with an eigenvalue of the interior Dirichlet boundary value problem or with an eigenvalue of the interior Neumann boundary value problem. However, in any case at least one of the boundary integral equations (1.117) or (1.118) is uniquely solvable since κ^2 can not be an eigenvalue of both the interior Dirichlet and the interior Neumann boundary value problem. Thus, we may combine both, the indirect single and double layer potential formulations to derive a boundary integral equation, which is uniquely solvable for arbitrary wave numbers. This leads to the well known Brakhage–Werner formulation (see [13])

$$u(x) = (W_\kappa w)(x) + \imath\,\eta(\widetilde{V}_\kappa w)(x) \quad \text{for } x \in \Omega^e,$$

which leads to the boundary integral equation

$$\frac{1}{2}w(x) + (K_\kappa w)(x) + \imath\,\eta(V_\kappa w)(x) = g(x) \quad \text{for } x \in \Gamma. \tag{1.119}$$

Here, $\eta \in \mathbb{R}$ is some real parameter. Note that this equation is usually considered in the $L_2(\Gamma)$ sense. The numerical analysis to investigate the unique solvability of the combined boundary integral equation (1.119) is based on the coercivity of the underlying boundary integral operator, and, therefore, on some compactness argument. In general, this may require more regularity assumptions for the boundary surface under consideration (cf. [23]). Instead of considering the boundary integral equation (1.119) in $L_2(\Gamma)$, one may formulate some modified boundary integral equations to be considered in the energy spaces $H^{1/2}(\Gamma)$ or $H^{-1/2}(\Gamma)$, see [17, 18].

1.4.4 Exterior Neumann Boundary Value Problem

Finally, we consider the exterior Neumann boundary value problem

$$-\Delta u(x) - \kappa^2 u(x) = 0 \quad \text{for } x \in \Omega^e, \quad \gamma_1^{\text{int}} u(x) = g(x) \quad \text{for } x \in \Gamma, \tag{1.120}$$

where we have to require the Sommerfeld radiation condition (1.101),

$$\left| \left(\frac{x}{|x|}, \nabla u(x) \right) - \imath\,\kappa u(x) \right| = \mathcal{O}\left(\frac{1}{|x|^2} \right) \quad \text{as } |x| \to \infty.$$

Due to the radiation condition, the exterior Neumann boundary value problem is uniquely solvable. The solution of the above boundary value problem is given by the representation formula (1.103),

$$u(x) = -\int_\Gamma u_\kappa^*(x,y)g(y)ds_y + \int_\Gamma \gamma_{1,y}^{\text{ext}} u_\kappa^*(x,y)\gamma_0^{\text{ext}} u(y)ds_y \quad \text{for } x \in \Omega^e.$$

To find the yet unknown Dirichlet datum $\overline{u} = \gamma_0^{\text{ext}} u$, we consider the boundary integral equation which results from the representation formula when applying the exterior conormal derivative γ_1^{ext},

$$(D_\kappa \overline{u})(x) = -\frac{1}{2}g(x) - (K_\kappa' g)(x) \quad \text{for } x \in \Gamma. \tag{1.121}$$

This hypersingular boundary integral equation is equivalent to the variational problem to find $\bar{u} \in H^{1/2}(\Gamma)$ such that

$$\left\langle D_\kappa \bar{u}, v \right\rangle_\Gamma = -\left\langle \left(\frac{1}{2}I + K_\kappa'\right)g, v \right\rangle_\Gamma \quad \text{for all } v \in H^{1/2}(\Gamma). \qquad (1.122)$$

Since the hypersingular boundary integral operator D_κ of the exterior Neumann boundary value problem coincides with the operator which is related to the interior Neumann boundary value problem, D_κ is not invertible when $\kappa^2 = \mu$ is an eigenvalue of the interior Neumann eigenvalue problem (1.113). However, we have

$$\left\langle \left(\frac{1}{2}I + K_\kappa'\right)g, u_\lambda \right\rangle_\Gamma = \left\langle g, \left(\frac{1}{2}I + K_{-\kappa}\right)u_\lambda \right\rangle_\Gamma = 0,$$

and, therefore,

$$\left(\frac{1}{2}I + K_\kappa'\right)g \in \operatorname{Im} D_\kappa.$$

In fact, the variational problem (1.122) of the direct approach is solvable, but the solution is not unique. Again, one can formulate a suitable stabilised variational problem; we skip the details.

If κ^2 is not an eigenvalue of the Neumann eigenvalue problem (1.113), then the unique solvability of the variational problem (1.122) follows, since D_κ is coercive and injective.

When applying the exterior trace operator γ_0^{ext} to the representation formula, this gives a second kind boundary integral equation to be solved,

$$-\frac{1}{2}\bar{u}(x) + (K_\kappa \bar{u})(x) = (V_\kappa g)(x) \quad \text{for } x \in \Gamma. \qquad (1.123)$$

If $\kappa^2 = \mu$ is an eigenvalue of the Dirichlet eigenvalue problem (1.108), the operator

$$\frac{1}{2}I - K_{-\kappa}' : H^{-1/2}(\Gamma) \to H^{-1/2}(\Gamma)$$

is singular and, therefore, not invertible. Then, the adjoint operator

$$\frac{1}{2}I - K_\kappa : H^{1/2}(\Gamma) \to H^{1/2}(\Gamma)$$

is also not invertible.

As for the exterior Dirichlet boundary value problem, one may formulate a combined boundary integral equation in $L_2(\Gamma)$, i.e., a linear combination of the boundary integral equations (1.121) and (1.123) gives (cf. [19])

$$-\frac{1}{2}\bar{u}(x) + (K_\kappa' \bar{u})(x) + i\eta(D_\kappa \bar{u})(x) = (V_\kappa g)(x) - i\eta\left(\frac{1}{2}g(x) + (K_\kappa' g)(x)\right)$$

for $x \in \Gamma$, which is uniquely solvable due to the coercivity of the underlying boundary integral operators when assuming sufficient smoothness of the boundary Γ.

1.5 Bibliographic Remarks

The history of using surface potentials to describe solutions of partial differential equations goes back to the middle of the 19th century. Already C. F. Gauß [33, 34] proposed to solve the Dirichlet boundary value problem for the Laplace equation in a sufficiently smoothly bounded domain by using an indirect double layer potential. To find the yet unknown density function, a second kind boundary integral equation has to be solved. C. Neumann [80] applied a series representation to construct this solution, and he showed the convergence, i.e. a contraction property, when the domain is convex. These results were then extended by several authors, see also the discussion in [108], where the solvability of second kind boundary integral equations was considered for domains with non–smooth boundaries. This proof is based on different representations of the boundary integral operators which follow from the Calderon projection property. In particular, the symmetry of the double layer potential, with respect to an inner product induced by the single layer potential, was already observed for a simple model problem by J. Plemelj [88]. A different view on the historical development of those results was given recently in [25]. For a general review on the history of boundary integral and boundary element methods, see, for example, [22].

For a long time, direct and indirect boundary integral formulations have been a standard approach to describe solutions of partial differential equations in mathematical physics, see, for example, [29, 55, 59, 61, 62, 70, 72, 74, 91].

While second kind boundary integral equation methods [6] resulting from an indirect approach have a long tradition in both the analysis and numerical treatment of boundary value problems [81, 82], direct formulations and first kind boundary integral equation methods became more popular in the last decades. This is mainly due to the rigorous mathematical analysis of boundary integral formulations and related numerical approximation schemes, which is available for first kind equations in the setting of energy spaces. First results were obtained simultaneously by J. C. Nédélec and J. Planchard [79] and by G. C. Hsiao and W. L. Wendland [57]. More general results on the mapping properties of boundary integral operators in Sobolev spaces were later given by M. Costabel and W. L. Wendland [24, 26], see also the monograph [71] by W. McLean.

While for boundary value problems with pure Dirichlet or pure Neumann boundary conditions one may use either first or second kind boundary integral equations, the situation becomes more complicated when considering boundary value problems with mixed boundary conditions. Direct formulations, which are based on the weakly singular boundary integral equation only, then lead to systems combining boundary integral operators of both the first and the second kind, see, e.g., [56]. Today, the symmetric formulation of boundary integral equations [103] seems to be more popular, see also [109, 114].

Alternative representations of boundary integral operators are important for both analytical and numerical considerations. In particular, by using integration by parts, the bilinear form of the hypersingular boundary integral operator, which is the conormal derivative of the double layer potential, can be transformed into a linear combination of weakly singular forms, see [77] for the Laplace and for the Helmholtz operator. In fact, this also remains true for the system of linear elastostatics [47]. Moreover, also the double layer potential of linear elastostatics, which is defined as a Cauchy singular integral operator, can be written as a combination of weakly singular boundary integral operators [62].

The use of boundary integral equation methods to describe solutions of boundary value problems is essentially based on the knowledge of a fundamental solution of the underlying partial differential operator. In fact, a fundamental solution is a solution of the partial differential equation with a Dirac impulse as the right hand side. While the existence of such a fundamental solution can be ensured for a quite large class of partial differential operators, in particular for partial differential operators with constant coefficients [30, 53, 73], the explicit construction can be a complicated task in general, see for example [66, 85, 86]. For more general partial differential operators, i.e. with variable coefficients, the concept of a parametrix, also known as a Levi function [73], was introduced by D. Hilbert [52]. A Levi function is a solution of the partial differential equation where the right hand side is given by a Dirac impulse and some more regular remainder. For example, such an approach was used in [89] to model shells by using a boundary–domain integral method.

2

Boundary Element Methods

The numerical approximation of boundary integral equations leads to boundary element methods in general. Since already the formulation of boundary integral equations is not unique, the choice of an appropriate discretisation scheme gives even more variety. The most common approximation methods are the Collocation scheme and the Galerkin method. In this chapter we first introduce boundary element spaces of piecewise constant piecewise linear basis functions. Then we describe some discretisation methods for different boundary integral formulations and we discuss the corresponding error estimates.

2.1 Boundary Elements

Let $\Gamma = \partial\Omega$ be the boundary of a Lipschitz domain $\Omega \subset \mathbb{R}^3$. For $N \in \mathbb{N}$, we consider a sequence of boundary element meshes

$$\Gamma_N = \bigcup_{\ell=1}^{N} \overline{\tau}_\ell. \tag{2.1}$$

In the most simple case, we assume that Γ is piecewise polyhedral and that each boundary element mesh (2.1) consists of N plane triangular boundary elements τ_ℓ with mid points x_ℓ^*. Using the reference element

$$\tau = \left\{ \xi \in \mathbb{R}^2 \, : \, 0 < \xi_1 < 1, \quad 0 < \xi_2 < 1 - \xi_1 \right\},$$

the boundary element $\tau_\ell = \chi_\ell(\tau)$ with nodes x_{ℓ_i} for $i = 1, 2, 3$ can be described via the parametrisation

$$x(\xi) = \chi_\ell(\xi) = x_{\ell_1} + \xi_1(x_{\ell_2} - x_{\ell_1}) + \xi_2(x_{\ell_3} - x_{\ell_1}) \in \tau_\ell \quad \text{for } \xi \in \tau.$$

For the area Δ_ℓ of the boundary element τ_ℓ, we then obtain

$$\Delta_\ell = \int_{\tau_\ell} ds_x = \int_{\tau} \sqrt{EG - F^2} \, d\xi = \frac{1}{2}\sqrt{EG - F^2},$$

where

$$E = \sum_{i=1}^{3} \left(\frac{\partial}{\partial \xi_1} x_i(\xi) \right)^2 = |x_{\ell_2} - x_{\ell_1}|^2,$$

$$G = \sum_{i=1}^{3} \left(\frac{\partial}{\partial \xi_2} x_i(\xi) \right)^2 = |x_{\ell_3} - x_{\ell_1}|^2,$$

$$F = \sum_{i=1}^{3} \frac{\partial}{\partial \xi_1} x_i(\xi) \frac{\partial}{\partial \xi_2} x_i(\xi) = (x_{\ell_2} - x_{\ell_1}, x_{\ell_3} - x_{\ell_1}).$$

Using Δ_ℓ, we define the local mesh size of the boundary element τ_ℓ as

$$h_\ell = \sqrt{\Delta_\ell} \quad \text{for } \ell = 1, \ldots, N$$

implying the global mesh sizes

$$h = h_{\max} = \max_{1 \leq \ell \leq N} h_\ell, \quad h_{\min} = \min_{1 \leq \ell \leq N} h_\ell. \tag{2.2}$$

The sequence of boundary element meshes (2.1) is called globally quasi uniform if the mesh ratio

$$\frac{h_{\max}}{h_{\min}} \leq c_{\mathrm{G}}$$

is uniformly bounded by a constant c_{G} which is independent of $N \in \mathbb{N}$. Finally, we introduce the element diameter

$$d_\ell = \sup_{x,y \in \tau_\ell} |x - y|.$$

We assume that all boundary elements τ_ℓ are uniformly shape regular, i.e., there exists a global constant c_{B} independent of N such that

$$d_\ell \leq c_{\mathrm{B}} h_\ell \quad \text{for all } \ell = 1, \ldots, N.$$

With

$$J_\ell = \begin{pmatrix} x_{\ell_2,1} - x_{\ell_1,1} & x_{\ell_3,1} - x_{\ell_1,1} \\ x_{\ell_2,2} - x_{\ell_1,2} & x_{\ell_3,2} - x_{\ell_1,2} \\ x_{\ell_2,3} - x_{\ell_1,3} & x_{\ell_3,3} - x_{\ell_1,3} \end{pmatrix} \in \mathbb{R}^{3 \times 2}$$

and using the parametrisation $\tau_\ell = \chi_\ell(\tau)$, a function v defined on τ_ℓ can be interpreted as a function $\widetilde{v}_\ell$ with respect to the reference element τ,

$$v(x) = v(x_{\ell_1} + J_\ell \xi) = \widetilde{v}_\ell(\xi) \quad \text{for } \xi \in \tau, \quad x = \chi_\ell(\xi).$$

Vice versa, a function $\widetilde{v}$ defined in the parameter domain τ implies a function v_ℓ on the boundary element τ_ℓ,

$$\widetilde{v}(\xi) = v(x_{\ell_1} + J_\ell \xi) = v_\ell(x) \quad \text{for } \xi \in \tau, \quad x = \chi_\ell(\xi).$$

Hence, we can define boundary element basis functions on τ_ℓ by defining associated shape functions on the reference element τ.

2.2 Basis Functions

Piecewise Constant Basis Functions

The piecewise constant shape function

$$\psi^0(\xi) = 1 \quad \text{for } \xi \in \tau$$

implies the piecewise constant basis functions on Γ

$$\psi_\ell(x) = \begin{cases} 1 & \text{for } x \in \tau_\ell, \\ 0 & \text{elsewhere} \end{cases} \tag{2.3}$$

for $\ell = 1, \ldots, N$, and, therefore, the global trial space

$$S_h^0(\Gamma) = \text{span}\big\{ \psi_\ell \big\}_{\ell=1}^N, \quad \dim S_h^0(\Gamma) = N.$$

Note that any $w_h \in S_h^0(\Gamma)$ can be written as

$$w_h = \sum_{\ell=1}^N w_\ell \psi_\ell \in S_h^0(\Gamma), \quad w_\ell \in \mathbb{R} \quad \text{for } \ell = 1, \ldots, N.$$

Moreover, a function $w_h \in S_h^0(\Gamma)$ can be identified with the vector $\underline{w} \in \mathbb{R}^N$ defined by the components w_ℓ for $\ell = 1, \ldots, N$.

In what follows, we will consider the approximation property of the trial space $S_h^0(\Gamma) \subset L_2(\Gamma)$. For this, we introduce the L_2 projection of a given function $w \in L_2(\Gamma)$,

$$Q_h w = \sum_{\ell=1}^N w_\ell \psi_\ell \in S_h^0(\Gamma),$$

which minimises the error $w - Q_h w$ in the $L_2(\Gamma)$–norm,

$$Q_h w = \arg \min_{w_h \in S_h^0(\Gamma)} \| w - w_h \|_{L_2(\Gamma)}^2 = \arg \min_{w_h \in S_h^0(\Gamma)} \int_\Gamma \Big(w(x) - w_h(x) \Big)^2 ds_x.$$

Note that $Q_h w$ is the unique solution of the variational problem

$$\int_\Gamma (Q_h w)(x) \psi_k(x) ds_x = \int_\Gamma w(x) \psi_k(x) ds_x \quad \text{for } k = 1, \ldots, N,$$

or,

$$\sum_{\ell=1}^N w_\ell \int_\Gamma \psi_\ell(x) \psi_k(x) ds_x = \int_\Gamma w(x) \psi_k(x) ds_x \quad \text{for } k = 1, \ldots, N.$$

Due to

$$\int_\Gamma \psi_\ell(x)\psi_k(x)ds_x = \begin{cases} \Delta_\ell & \text{for } k = \ell, \\ 0 & \text{for } k \neq \ell \end{cases},$$

we obtain

$$w_\ell = \frac{1}{\Delta_\ell} \int_{\tau_\ell} w(x)ds_x \quad \text{for } \ell = 1,\ldots,N.$$

From this explicit representation of w_ℓ, one can prove the error estimate, see Appendix B.2,

$$\|w - Q_h w\|_{L_2(\Gamma)}^2 \leq c \sum_{\ell=1}^N h_\ell^{2s} |w|_{H^s(\tau_\ell)}^2 \leq c h^{2s} |w|_{H_{\text{pw}}^s(\Gamma)}^2 \qquad (2.4)$$

for a sufficiently regular function $w \in H_{\text{pw}}^s(\Gamma)$ and $s \in (0,1]$. The semi-norm in (2.4) is defined as

$$|w|_{H^s(\tau_\ell)}^2 = \int_{\tau_\ell} \int_{\tau_\ell} \frac{|w(x) - w(y)|^2}{|x - y|^{2+2s}} ds_x ds_y \quad \text{for } s \in (0,1)$$

and

$$|w|_{H^1(\tau_\ell)}^2 = \int_\tau |\nabla_\xi w(\chi_\ell(\xi))|^2 d\xi \quad \text{for } s = 1.$$

From the above variational formulation, we conclude the Galerkin orthogonality

$$\int_\Gamma \left(w(x) - (Q_h w)(x) \right) v_h(x) ds_x = 0 \quad \text{for all } v_h \in S_h^0(\Gamma),$$

and, therefore, the trivial error estimate

$$\|w - Q_h w\|_{L_2(\Gamma)} \leq \|w\|_{L_2(\Gamma)}.$$

Using a duality argument, we further obtain

$$\|w - Q_h w\|_{H^\sigma(\Gamma)} \leq c h^{s-\sigma} |w|_{H_{\text{pw}}^s(\Gamma)}$$

for $\sigma \in [-1, 0]$ and $s \in [0, 1]$.

Summarising the above, we obtain the following approximation property in $S_h^0(\Gamma)$.

Theorem 2.1. *Let $w \in H_{\text{pw}}^s(\Gamma)$ for some $s \in [0,1]$. Then there holds*

$$\inf_{w_h \in S_h^0(\Gamma)} \|w - w_h\|_{H^\sigma(\Gamma)} \leq c h^{s-\sigma} |w|_{H_{\text{pw}}^s(\Gamma)} \qquad (2.5)$$

for all $\sigma \in [-1, 0]$.

Moreover, the approximation property (2.5) remains valid for all $\sigma \leq s \leq 1$ with $\sigma < 1/2$.

Piecewise Linear Discontinuous Basis Functions

With respect to the reference element τ, we may also define local polynomial shape functions of higher order. In particular, we introduce the linear shape functions

$$\psi_1^1(\xi) = 1 - \xi_1 - \xi_2, \quad \psi_2^1(\xi) = \xi_1, \quad \psi_3^1(\xi) = \xi_2 \quad \text{for } \xi \in \tau. \quad (2.6)$$

These shape functions imply globally discontinuous piecewise linear basis functions

$$\psi_{\ell,i}(x) = \begin{cases} \psi_i^1(\xi) & \text{for } x = \chi_\ell(\xi) \in \tau_\ell, \\ 0 & \text{elsewhere} \end{cases}$$

for $\ell = 1, \ldots, N$, $i = 1, 2, 3$, and, therefore, the global trial space

$$S_h^{1,-1}(\Gamma) = \text{span}\left\{ \psi_{\ell,1}(x), \psi_{\ell,2}(x), \psi_{\ell,3}(x) \right\}_{\ell=1}^N, \quad \dim S_h^{1,-1}(\Gamma) = 3N.$$

Any function $w_h \in S_h^{1,-1}(\Gamma)$ can be written as

$$w_h = \sum_{\ell=1}^N \sum_{i=1}^3 w_{\ell,i}\, \psi_{\ell,i} \in S_h^{1,-1}(\Gamma).$$

Moreover, a function $w_h \in S_h^{1,-1}(\Gamma)$ can be identified with the vector $\underline{w} \in \mathbb{R}^{3N}$ which is defined by the coefficients $w_{\ell,i}$ for $i = 1, 2, 3$ and $\ell = 1, \ldots, N$.

As for piecewise constant basis functions, we may also define the corresponding L_2 projection $Q_h w \in S_h^{1,-1}(\Gamma) \subset L_2(\Gamma)$,

$$Q_h w = \sum_{\ell=1}^N \sum_{i=1}^3 w_{\ell,i}\, \psi_{\ell,i} \in S_h^{1,-1}(\Gamma),$$

as the unique solution of the variational problem

$$\int_\Gamma (Q_h w)(x)\psi_{k,j}(x)ds_x = \int_\Gamma w(x)\psi_{k,j}(x)ds_x, \quad j = 1,2,3,\ k = 1,\ldots,N$$

satisfying the error estimate

$$\|w - Q_h w\|_{L_2(\Gamma)}^2 \le c \sum_{\ell=1}^N h_\ell^4 |w|_{H^2(\tau_\ell)}^2 \le c\, h^4 |w|_{H_{\text{pw}}^2(\Gamma)}^2$$

when assuming $w \in H_{\text{pw}}^2(\Gamma)$. Combining this with the trivial error estimate

$$\|w - Q_h w\|_{L_2(\Gamma)} \le \|w\|_{L_2(\Gamma)},$$

and using an interpolation argument, the final error estimate

$$\|w - Q_h w\|_{L_2(\Gamma)} \leq c\,h^s\,|w|_{H^s_{\text{pw}}(\Gamma)}$$

follows when assuming $w \in H^s_{\text{pw}}(\Gamma)$ for some $s \in [0, 2]$. Using again a duality argument, we finally obtain

$$\|w - Q_h w\|_{H^\sigma(\Gamma)} \leq c\,h^{s-\sigma}\,|w|_{H^s_{\text{pw}}(\Gamma)} \tag{2.7}$$

for $\sigma \in [-2, 0]$ and $s \in [0, 2]$.

Summarising the above, we obtain the approximation property in $S_h^{1,-1}(\Gamma)$.

Theorem 2.2. *Let* $w \in H^s_{pw}(\Gamma)$ *for some* $s \in [0, 2]$. *Then there holds*

$$\inf_{w_h \in S_h^{1,-1}(\Gamma)} \|w - w_h\|_{H^\sigma(\Gamma)} \leq c\,h^{s-\sigma}\,|w|_{H^s_{pw}(\Gamma)} \tag{2.8}$$

for all $\sigma \in [-2, 0]$. *Moreover, the approximation property* (2.8) *remains valid for all* $\sigma \leq s \leq 2$ *with* $\sigma < 1/2$.

Piecewise Linear Continuous Basis Functions

Up to now, we have considered only globally discontinuous basis functions which do not require any admissibility condition of the triangulation (2.1). But such a condition is needed to define globally continuous basis functions. Let $\{x_j\}_{j=1}^M$ be the set of all nodes of the triangulation (2.1). A boundary element mesh consisting of plane triangular elements is called admissible, if the intersection of two neighboured elements $\bar\tau_\ell$ and $\bar\tau_k$ is just one common edge or one common node. Then $I(j)$ is the index set of all boundary elements τ_ℓ containing the node x_j while $J(\ell)$ is the three–dimensional index set of the nodes defining the triangular element τ_ℓ.

For $j = 1, \ldots, M$, one can define globally continuous piecewise linear basis functions φ_j with

$$\varphi_j(x) = \begin{cases} 1 & \text{for } x = x_j, \\ 0 & \text{for } x = x_i \neq x_j, \\ \text{piecewise linear} & \text{elsewhere.} \end{cases}$$

Note that the restrictions of φ_j onto a boundary element τ_k for $k \in I(j)$ can be represented by the linear shape functions $\psi_{j_k}^1$,

$$\varphi_j(x) = \psi_{j_k}^1(\xi) \quad \text{for } x = \chi_k(\xi) \in \tau_k. \tag{2.9}$$

The basis functions φ_j are used to define the trial space

$$S_h^1(\Gamma) = \text{span}\left\{\varphi_j\right\}_{j=1}^M, \quad \dim S_h^1(\Gamma) = M.$$

The piecewise linear continuous L_2 projection $Q_h w \in S_h^1(\Gamma)$ is then defined as the unique solution of the variational problem

$$\int_\Gamma Q_h w(x)\varphi_j(x)ds_x = \int_\Gamma w(x)\varphi_j(x)ds_x \quad \text{for } j = 1,\dots,M.$$

Due to $S_h^1(\Gamma) \subset S_h^{1,-1}(\Gamma)$ we immediately find the error estimate

$$\|w - Q_h w\|_{H^\sigma(\Gamma)} \le c\,h^{s-\sigma}\,|w|_{H_{\mathrm{pw}}^s(\Gamma)}$$

when assuming $w \in H_{\mathrm{pw}}^s(\Gamma)$, $\sigma \in [-2,0]$, $s \in [0,2]$.

Defining $P_h u \in S_h^1(\Gamma)$ as the unique solution of the variational problem

$$\langle P_h w, v_h \rangle_{H^1(\Gamma)} = \langle w, v_h \rangle_\Gamma \quad \text{for all } v_h \in S_h^1(\Gamma)$$

we can show the error estimate

$$\|w - P_h w\|_{H^\sigma(\Gamma)} \le c\,h^{s-\sigma}\,|w|_{H_{\mathrm{pw}}^s(\Gamma)}$$

when assuming $w \in H_{\mathrm{pw}}^s(\Gamma)$ and $\sigma \in (0,1]$, $s \in [1,2]$.

Hence, we have the following result.

Theorem 2.3. *Let $v \in H_{pw}^s(\Gamma)$ for some $s \in [1,2]$. Then there holds*

$$\inf_{v_h \in S_h^1(\Gamma)} \|v - v_h\|_{H^\sigma(\Gamma)} \le c\,h^{s-\sigma}\,|v|_{H_{pw}^s(\Gamma)} \tag{2.10}$$

for all $\sigma \in [-2,1]$. Moreover, the approximation property (2.10) remains valid for all $\sigma \le s \le 2$ with $\sigma < 3/2$.

2.3 Laplace Equation

2.3.1 Interior Dirichlet Boundary Value Problem

The solution of the interior Dirichlet boundary value problem (cf. (1.14))

$$-\Delta u(x) = 0 \quad \text{for } x \in \Omega, \quad \gamma_0^{\mathrm{int}} u(x) = g(x) \quad \text{for } x \in \Gamma,$$

is given by the representation formula (cf. (1.6))

$$u(x) = \int_\Gamma u^*(x,y)t(y)ds_y - \int_\Gamma \gamma_{1,y}^{\mathrm{int}} u^*(x,y)g(y)ds_y \quad \text{for } x \in \Omega,$$

where the unknown Neumann datum $t = \gamma_1^{\mathrm{int}} u \in H^{-1/2}(\Gamma)$ is the unique solution of the boundary integral equation (cf. (1.15))

$$\frac{1}{4\pi}\int_\Gamma \frac{1}{|x-y|}t(y)ds_y = \frac{1}{2}g(x) + \frac{1}{4\pi}\int_\Gamma \frac{(x-y,n(y))}{|x-y|^3}g(y)ds_y \quad \text{for } x \in \Gamma.$$

Replacing $t \in H^{-1/2}(\Gamma)$ by a piecewise constant approximation

$$t_h = \sum_{\ell=1}^N t_\ell\,\psi_\ell \in S_h^0(\Gamma), \tag{2.11}$$

we have to find the unknown coefficient vector $\underline{t} \in \mathbb{R}^N$ from some appropriate system of linear equations.

Collocation Method

Inserting (2.11) into the boundary integral equation (1.15), and choosing the boundary element mid points x_k^* as collocation nodes, we have to solve the collocation equations

$$\frac{1}{4\pi} \int_\Gamma \frac{1}{|x_k^* - y|} t_h(y) ds_y = \frac{1}{2} g(x_k^*) + \frac{1}{4\pi} \int_\Gamma \frac{(x_k^* - y, \underline{n}(y))}{|x_k^* - y|^3} g(y) ds_y \quad (2.12)$$

for $k = 1, \ldots, N$, or using the definition (2.3) of the piecewise constant basis functions ψ_ℓ,

$$\sum_{\ell=1}^N t_\ell \frac{1}{4\pi} \int_{\tau_\ell} \frac{1}{|x_k^* - y|} ds_y = \frac{1}{2} g(x_k^*) + \frac{1}{4\pi} \int_\Gamma \frac{(x_k^* - y, \underline{n}(y))}{|x_k^* - y|^3} g(y) ds_y$$

for $k = 1, \ldots, N$. With

$$V_h[k, \ell] = \frac{1}{4\pi} \int_{\tau_\ell} \frac{1}{|x_k^* - y|} ds_y$$

for $k, \ell = 1, \ldots, N$, and

$$f_k = \frac{1}{2} g(x_k^*) + \frac{1}{4\pi} \int_\Gamma \frac{(x_k^* - y, \underline{n}(y))}{|x_k^* - y|^3} g(y) ds_y$$

for $k = 1, \ldots, N$, this results in a linear system of equations,

$$V_h \underline{t} = \underline{f}.$$

The stiffness matrix V_h of the collocation method is in general non–symmetric and the stability of the collocation scheme (2.12) and therefore the invertibility of the stiffness matrix V_h is still an open problem when Γ is the boundary of a general Lipschitz domain $\Omega \subset \mathbb{R}^3$. When assuming the stability of the collocation scheme (2.12), the quasi optimal error estimate, i.e., Cea's lemma,

$$\|t - t_h\|_{H^{-1/2}(\Gamma)} \le c \inf_{w_h \in S_h^0(\Gamma)} \|t - w_h\|_{H^{-1/2}(\Gamma)}$$

follows. Combining this with the approximation property (2.5) for $\sigma = -1/2$, we get the error estimate

$$\|t - t_h\|_{H^{-1/2}(\Gamma)} \le c h^{s+1/2} |t|_{H_{pw}^s(\Gamma)},$$

when assuming $t \in H_{pw}^s(\Gamma)$ for some $s \in [0, 1]$. Applying the Aubin–Nitsche trick (for $\sigma < -1/2$) and an inverse inequality argument (for $\sigma \in (-1/2, 0]$), we also obtain the error estimate

$$\|t - t_h\|_{H^\sigma(\Gamma)} \leq c\,h^{s-\sigma}\,|t|_{H^s_{\mathrm{pw}}(\Gamma)}, \qquad (2.13)$$

when assuming $t \in H^s_{\mathrm{pw}}(\Gamma)$ for some $s \in [0,1]$, and $\sigma \in [-1,0]$. Note that the lower bound $\sigma \geq -1$ is due to the collocation approach, independently of the degree of the used polynomial basis functions.

Inserting the computed solution t_h into the representation formula (1.6), this gives an approximate representation formula

$$\widetilde{u}(x) = \int_\Gamma u^*(x,y)t_h(y)ds_y - \int_\Gamma \gamma^{\mathrm{int}}_{1,y}u^*(x,y)g(y)ds_y$$

for $x \in \Omega$, describing an approximate solution of the Dirichlet boundary value problem (1.14). Note that $\widetilde{u}$ is harmonic, satisfying the Laplace equation, but the Dirichlet boundary conditions are satisfied only approximately. For an arbitrary $x \in \Omega$, the error is given by

$$u(x) - \widetilde{u}(x) = \int_\Gamma u^*(x,y)\Big(t(y) - t_h(y)\Big)ds_y\,.$$

Using a duality argument, the error estimate

$$|u(x) - \widetilde{u}(x)| \leq \|u^*(x,\cdot)\|_{H^{-\sigma}(\Gamma)}\|t - t_h\|_{H^\sigma(\Gamma)}$$

for some $\sigma \in \mathbb{R}$ follows. Combining this with the error estimate (2.13) for the minimal possible value $\sigma = -1$, we obtain the pointwise error estimate

$$|u(x) - \widetilde{u}(x)| \leq c\,h^{s+1}\,\|u^*(x,\cdot)\|_{H^1(\Gamma)}\,|t|_{H^s_{\mathrm{pw}}(\Gamma)},$$

when assuming $t \in H^s_{\mathrm{pw}}(\Gamma)$ for some $s \in [0,1]$. Hence, if t is sufficiently smooth, i.e. $t \in H^1_{\mathrm{pw}}(\Gamma)$, we obtain as the optimal order of convergence for $s = 1$

$$|u(x) - \widetilde{u}(x)| \leq c\,h^2\,\|u^*(x,\cdot)\|_{H^1(\Gamma)}|t|_{H^1_{\mathrm{pw}}(\Gamma)}\,. \qquad (2.14)$$

Note that the error estimate (2.14) involves the position of the observation point $x \in \Omega$. In particular, the error estimate (2.14) does not hold in the limiting case $x \in \Gamma$.

Galerkin Method

The boundary integral equation (cf. (1.15))

$$\frac{1}{4\pi}\int_\Gamma \frac{1}{|x-y|}t(y)ds_y = \frac{1}{2}g(x) + \frac{1}{4\pi}\int_\Gamma \frac{(x-y,\underline{n}(y))}{|x-y|^3}g(y)ds_y \quad \text{for } x \in \Gamma$$

is equivalent to the variational problem (1.16),

$$\langle Vt, w \rangle_{\Gamma} = \left\langle \left(\frac{1}{2}I + K\right)g, w \right\rangle_{\Gamma} \quad \text{for all } w \in H^{-1/2}(\Gamma),$$

and to the minimisation problem

$$F(t) = \min_{w \in H^{-1/2}(\Gamma)} F(w)$$

with

$$F(w) = \frac{1}{2}\langle Vw, w \rangle_{\Gamma} - \left\langle \left(\frac{1}{2}I + K\right)g, w \right\rangle_{\Gamma}.$$

Using a sequence of finite dimensional subspaces $S_h^0(\Gamma)$ spanned by piecewise constant basis functions, associated approximate solutions

$$t_h = \sum_{\ell=1}^{N} t_\ell \psi_\ell \in S_h^0(\Gamma)$$

are obtained from the minimisation problem

$$F(t_h) = \min_{w_h \in S_h^0(\Gamma)} F(w_h).$$

The solution $t_h \in S_h^0(\Gamma)$ of the above minimisation problem is defined via the Galerkin equations

$$\langle Vt_h, \psi_k \rangle_{\Gamma} = \left\langle \left(\frac{1}{2}I + K\right)g, \psi_k \right\rangle_{\Gamma} \quad \text{for } k = 1, \dots, N. \tag{2.15}$$

With (2.11) and by using the definition (2.3) of the piecewise constant basis functions ψ_ℓ, this is equivalent to

$$\sum_{\ell=1}^{N} t_\ell \frac{1}{4\pi} \int_{\tau_k} \int_{\tau_\ell} \frac{1}{|x-y|} ds_y ds_x =$$
$$\frac{1}{2} \int_{\tau_k} g(x) ds_x + \frac{1}{4\pi} \int_{\tau_k} \int_{\Gamma} \frac{(x-y, \underline{n}(y))}{|x-y|^3} g(y) ds_y ds_x$$

for $k = 1, \dots, N$. With

$$V_h[k, \ell] = \frac{1}{4\pi} \int_{\tau_k} \int_{\tau_\ell} \frac{1}{|x-y|} ds_y ds_x$$

for $k, \ell = 1, \dots, N$, and

$$f_k = \frac{1}{2} \int_{\tau_k} g(x) ds_x + \frac{1}{4\pi} \int_{\tau_k} \int_{\Gamma} \frac{(x-y, \underline{n}(y))}{|x-y|^3} g(y) ds_y ds_x$$

for $k = 1, \dots, N$, we find the coefficient vector $\underline{t} \in \mathbb{R}^N$ as the unique solution of the linear system

$$V_h \underline{t} = \underline{f}. \tag{2.16}$$

The Galerkin stiffness matrix V_h is symmetric and positive definite. Therefore, one may use a conjugate gradient scheme for an iterative solution of the linear system (2.16). Since the spectral condition number of V_h behaves like $\mathcal{O}(h^{-1})$, i.e.,

$$\kappa_2(V_h) = \|V_h\|_2 \|V_h^{-1}\|_2 = \frac{\lambda_{\max}(V_h)}{\lambda_{\min}(V_h)} \leq c\frac{1}{h},$$

an appropriate preconditioning is sometimes needed. Moreover, since the stiffness matrix V_h is dense, fast boundary element methods are required to construct more efficient algorithms, see Chapter 3.

From the $H^{-1/2}(\Gamma)$–ellipticity and the boundedness of the single layer potential

$$V : H^{-1/2}(\Gamma) \to H^{1/2}(\Gamma),$$

see Lemma 1.1, we conclude the unique solvability of the Galerkin variational problem (2.15), or, correspondingly, of the linear system (2.16), as well as the quasi optimal error estimate, i.e. Cea's lemma,

$$\|t - t_h\|_{H^{-1/2}(\Gamma)} \leq \frac{c_2^V}{c_1^V} \inf_{w_h \in S_h^0(\Gamma)} \|t - w_h\|_{H^{-1/2}(\Gamma)}.$$

Combining this with the approximation property (2.5) for $\sigma = -1/2$, we get

$$\|t - t_h\|_{H^{-1/2}(\Gamma)} \leq c\, h^{s+\frac{1}{2}} \, |t|_{H^s_{\mathrm{pw}}(\Gamma)},$$

when assuming $t \in H^s_{\mathrm{pw}}(\Gamma)$ and $s \in [0,1]$. Applying the Aubin–Nitsche trick (for $\sigma < -1/2$) and an inverse inequality argument (for $\sigma \in (-1/2, 0]$), we also obtain the error estimate

$$\|t - t_h\|_{H^\sigma(\Gamma)} \leq c\, h^{s-\sigma} \, |t|_{H^s_{\mathrm{pw}}(\Gamma)}, \tag{2.17}$$

when assuming $t \in H^s_{\mathrm{pw}}(\Gamma)$ for some $s \in [0,1]$ and $\sigma \in [-2,0]$.

Inserting the computed Galerkin solution $t_h \in S_h^0(\Gamma)$ into the representation formula (1.6), this gives an approximate representation formula

$$\widetilde{u}(x) = \int_{\Gamma} u^*(x,y)t_h(y)ds_y - \int_{\Gamma} \gamma_{1,y}^{\mathrm{int}} u^*(x,y)g(y)ds_y \quad \text{for } x \in \Omega, \tag{2.18}$$

describing an approximate solution of the Dirichlet boundary value problem (1.14). Note that $\widetilde{u}$ is harmonic satisfying the Laplace equation, but the Dirichlet boundary conditions are satisfied only approximately. For an arbitrary $x \in \Omega$, the error is given by

$$u(x) - \widetilde{u}(x) = \int_{\Gamma} u^*(x,y)\Big(t(y) - t_h(y)\Big) ds_y.$$

Using a duality argument, the error estimate

$$|u(x) - \tilde{u}(x)| \leq \|u^*(x, \cdot)\|_{H^{-\sigma}(\Gamma)} \|t - t_h\|_{H^{\sigma}(\Gamma)}$$

for some $\sigma \in \mathbb{R}$ follows. Combining this with the error estimate (2.17) for the minimal value $\sigma = -2$, we obtain the pointwise error estimate

$$|u(x) - \tilde{u}(x)| \leq c \, h^{s+2} \|u^*(x, \cdot)\|_{H^2(\Gamma)} \, |t|_{H^s_{pw}(\Gamma)},$$

when assuming $t \in H^s_{pw}(\Gamma)$ for some $s \in [0, 1]$. Hence, if $t \in H^1_{pw}(\Gamma)$ is sufficiently smooth, we obtain the optimal order of convergence for $s = 1$,

$$|u(x) - \tilde{u}(x)| \leq c \, h^3 \|u^*(x, \cdot)\|_{H^2(\Gamma)} \, |t|_{H^1_{pw}(\Gamma)}. \tag{2.19}$$

Note that the error estimate (2.19) involves the position of the observation point $x \in \Omega$ again. In particular, the error estimate (2.19) does not hold in the limiting case $x \in \Gamma$.

The computation of the right hand side $\underline{f}$ in the linear system (2.16) requires the evaluation of the integrals

$$f_k = \frac{1}{2} \int_{\tau_k} g(x) ds_x + \frac{1}{4\pi} \int_{\tau_k} \int_{\Gamma} \frac{(x - y, \underline{n}(y))}{|x - y|^3} g(y) ds_y ds_x$$

for $k = 1, \ldots, N$. An approximation of the given Dirichlet datum $g \in H^{1/2}(\Gamma)$ by a globally continuous and piecewise linear function

$$g_h = \sum_{j=1}^{M} g_j \, \varphi_j \in S^1_h(\Gamma)$$

can be obtained either by interpolation,

$$g_h = \sum_{j=1}^{M} g(x_j) \, \varphi_j, \tag{2.20}$$

or by the L_2 projection,

$$g_h = \sum_{j=1}^{M} g_j \, \varphi_j,$$

where the coefficients g_j, $j = 1, \ldots, M$ satisfy

$$\sum_{j=1}^{M} g_j \langle \varphi_j, \varphi_i \rangle_{L_2(\Gamma)} = \langle g, \varphi_i \rangle_{L_2(\Gamma)} \quad \text{for } i = 1, \ldots, M. \tag{2.21}$$

This leads to

$$\widetilde{f}_k = \frac{1}{2} \sum_{j=1}^{M} g_j \int_{\tau_k} \varphi_j(x) ds_x + \sum_{j=1}^{M} g_j \frac{1}{4\pi} \int_{\tau_k} \int_{\Gamma} \frac{(x-y, \underline{n}(y))}{|x-y|^3} \varphi_j(y) ds_y ds_x$$

$$= \sum_{j=1}^{M} g_j \left(\frac{1}{2} M_h[k,j] + K_h[k,j] \right)$$

with the matrix entries

$$M_h[k,j] = \int_{\tau_k} \varphi_j(x) ds_x, \quad K_h[k,j] = \frac{1}{4\pi} \int_{\tau_k} \int_{\Gamma} \frac{(x-y, \underline{n}(y))}{|x-y|^3} \varphi_j(y) ds_y ds_x$$

for $j = 1, \ldots, M$ and $k = 1, \ldots, N$. Instead of the linear system (2.16), we then have to solve a linear system with a perturbed right hand side $\widetilde{f}$, yielding a perturbed solution vector $\underline{\widetilde{t}}$, i.e., we have to solve the linear system

$$V_h \underline{\widetilde{t}} = \left(\frac{1}{2} M_h + K_h \right) \underline{g}. \tag{2.22}$$

For the perturbed boundary element solution $\widetilde{t}_h \in S_h^0(\Gamma)$, the error estimate

$$\|t - \widetilde{t}_h\|_{H^\sigma(\Gamma)} \le c_1 \|t - t_h\|_{H^\sigma(\Gamma)} + c_2 \|g - g_h\|_{H^{\sigma+1}(\Gamma)}$$

follows with $\sigma \in [-2, 0]$, when the L_2 projection (2.21) is used to define $g_h \in S_h^1(\Gamma)$. Note that $\sigma \in [-1, 0]$ in the case of the interpolation (2.20). Assuming $t \in H_{\text{pw}}^s(\Gamma)$ and $g \in H_{\text{pw}}^{s+1}(\Gamma)$ for some $s \in [0, 1]$, we then obtain the error estimate

$$\|t - \widetilde{t}_h\|_{H^\sigma(\Gamma)} \le h^{s-\sigma} \left(c_1 |t|_{H_{\text{pw}}^s(\Gamma)} + c_2 |g|_{H_{\text{pw}}^{s+1}(\Gamma)} \right).$$

For the approximate representation formula

$$\widetilde{u}(x) = \int_{\Gamma} u^*(x,y) t_h(y) ds_y - \int_{\Gamma} \gamma_{1,y}^{\text{int}} u^*(x,y) g_h(y) ds_y \quad \text{for } x \in \Omega,$$

we then obtain the optimal error estimate

$$|u(x) - \widetilde{u}(x)| \le c(x,t,g) h^3, \tag{2.23}$$

when using the L_2 projection (2.21) and when assuming $t \in H_{\text{pw}}^1(\Gamma)$ and $g \in H_{\text{pw}}^2(\Gamma)$. When using the interpolation (2.20) instead, the error estimate

$$|u(x) - \widetilde{u}(x)| \le c(x,t,g) h^2$$

follows.

2.3.2 Interior Neumann Boundary Value Problem

Let $\Omega \subset \mathbb{R}^3$ be a simply connected domain. The solution of the interior Neumann boundary value problem (cf. (1.21))

$$-\Delta u(x) = 0 \quad \text{for } x \in \Omega, \quad \gamma_1^{\mathrm{int}} u(x) = g(x) \quad \text{for } x \in \Gamma,$$

is given by the representation formula (cf. (1.6))

$$u(x) = \int_{\Gamma} u^*(x,y) g(y) ds_y - \int_{\Gamma} \gamma_{1,y}^{\mathrm{int}} u^*(x,y) \bar{u}(y) ds_y \quad \text{for } x \in \Omega,$$

where the unknown Dirichlet datum $\bar{u} = \gamma_0^{\mathrm{int}} u \in H^{1/2}(\Gamma)$ is a solution of the hypersingular boundary integral equation (cf. (1.27))

$$-\gamma_1^{\mathrm{int}} \int_{\Gamma} \gamma_{1,y}^{\mathrm{int}} u^*(x,y) \bar{u}(y) ds_y = \frac{1}{2} g(x) - \int_{\Gamma} \gamma_{1,x}^{\mathrm{int}} u^*(x,y) g(y) ds_y$$

for $x \in \Gamma$. Since the hypersingular boundary integral operator D has a non–trivial kernel, we consider the extended variational problem (cf. (1.29)) to find $\bar{u}_\alpha \in H^{1/2}(\Gamma)$ such that

$$\left\langle D\bar{u}_\alpha, v \right\rangle_{\Gamma} + \left\langle \bar{u}_\alpha, 1 \right\rangle_{\Gamma} \left\langle v, 1 \right\rangle_{\Gamma} = \left\langle \left(\frac{1}{2} I - K' \right) g, v \right\rangle_{\Gamma} + \alpha \left\langle v, 1 \right\rangle_{\Gamma}$$

is satisfied for all $v \in H^{1/2}(\Gamma)$. Note that from the solvability condition (1.22), we reproduce the scaling condition (1.25). Since the bilinear form of this variational problem is strictly positive, the variational problem is equivalent to the minimisation problem

$$F(\bar{u}_\alpha) = \min_{v \in H^{1/2}(\Gamma)} F(v)$$

with

$$F(v) = \frac{1}{2} \left(\left\langle Dv, v \right\rangle_{\Gamma} + \left\langle v, 1 \right\rangle_{\Gamma}^2 \right) - \left\langle \left(\frac{1}{2} I - K' \right) g, v \right\rangle_{\Gamma} - \alpha \left\langle v, 1 \right\rangle_{\Gamma}.$$

Using a sequence of finite dimensional subspaces $S_h^1(\Gamma) \subset H^{1/2}(\Gamma)$ spanned by piecewise linear and continuous basis functions, an associated approximate function

$$\bar{u}_{\alpha,h} = \sum_{j=1}^{M} \bar{u}_{\alpha,j} \, \varphi_j \in S_h^1(\Gamma) \tag{2.24}$$

is obtained from the minimisation problem

$$F(\bar{u}_{\alpha,h}) = \min_{v_h \in S_h^1(\Gamma)} F(v_h).$$

The solution $\bar{u}_{\alpha,h} \in S_h^1(\Gamma)$ of the above minimisation problem is then defined via the Galerkin equations

$$\left\langle D\bar{u}_{\alpha,h}, \varphi_i \right\rangle_\Gamma + \left\langle \bar{u}_{\alpha,h}, 1 \right\rangle_\Gamma \left\langle \varphi_i, 1 \right\rangle_\Gamma =$$
$$\left\langle \left(\frac{1}{2}I - K'\right)g, \varphi_i \right\rangle_\Gamma + \alpha \left\langle \varphi_i, 1 \right\rangle_\Gamma \qquad (2.25)$$

for $i = 1, \ldots, M$. Using (2.24), this becomes

$$\sum_{j=1}^{M} \bar{u}_{\alpha,j} \left(\left\langle D\varphi_j, \varphi_i \right\rangle_\Gamma + \left\langle \varphi_j, 1 \right\rangle_\Gamma \left\langle \varphi_i, 1 \right\rangle_\Gamma \right) =$$
$$\left\langle \left(\frac{1}{2}I - K'\right)g, \varphi_i \right\rangle_\Gamma + \alpha \left\langle \varphi_i, 1 \right\rangle_\Gamma$$

for $i = 1, \ldots, M$. With

$$D_h[i,j] = \langle D\varphi_j, \varphi_i \rangle_\Gamma = \frac{1}{4\pi} \int_\Gamma \int_\Gamma \frac{(\operatorname{curl}_\Gamma \varphi_j(y), \operatorname{curl}_\Gamma \varphi_i(x))}{|x-y|} ds_x ds_y,$$

$$a_i = \langle \varphi_i, 1 \rangle_\Gamma = \int_\Gamma \varphi_i(x) ds_x,$$

$$f_i = \left\langle \left(\frac{1}{2}I - K'\right)g, \varphi_i \right\rangle_\Gamma$$
$$= \frac{1}{2} \int_\Gamma g(x)\varphi_i(x) ds_x - \int_\Gamma \varphi_i(x) \int_\Gamma \gamma_{1,x}^{int} u^*(x,y)g(y) ds_y ds_x$$
$$= \frac{1}{2} \int_\Gamma g(x)\varphi_i(x) ds_x - \frac{1}{4\pi} \int_\Gamma \varphi_i(x) \int_\Gamma \frac{(y-x, n(x))}{|x-y|^3} g(y) ds_y ds_x$$

for $i, j = 1, \ldots, M$, we find the coefficient vector $\underline{\bar{u}}_\alpha \in \mathbb{R}^M$ as the unique solution of the linear system

$$\left(D_h + \underline{a}\,\underline{a}^\top\right)\underline{\bar{u}}_\alpha = \underline{f} + \alpha\,\underline{a}. \qquad (2.26)$$

The extended stiffness matrix $D_h + \underline{a}\,\underline{a}^\top$ is symmetric and positive definite. Therefore, one may use a conjugate gradient scheme for an iterative solution of the linear system (2.26). However, due to the estimate for the spectral condition number

$$\kappa_2(D_h + \underline{a}\,\underline{a}^\top) \leq c\frac{1}{h},$$

an appropriate preconditioning is sometimes needed.

Note, that instead of a direct evaluation of the hypersingular boundary integral operator D, we apply integration by parts to obtain the representation (1.9) in Lemma 1.4, where

$$\underline{\mathrm{curl}}_\Gamma \varphi_i(x) = \underline{n}(x) \times \nabla_x \widetilde{\varphi}_i(x) \quad \text{for } x \in \Gamma$$

is the surface curl operator, and $\widetilde{\varphi}_i$ is some locally defined extension of φ_i into the neighbourhood of Γ. Since φ_i is linear on every boundary element τ_k, and defining the extension $\widetilde{\varphi}_i$ to be constant along $\underline{n}(x)$, we obtain $\underline{\mathrm{curl}}_\Gamma \varphi_i$ to be a piecewise constant vector function. Hence, we get

$$D_h[i,j] = \tag{2.27}$$
$$\sum_{\tau_k \in \text{ supp } \varphi_i} \sum_{\tau_\ell \in \text{ supp } \varphi_j} \left(\underline{\mathrm{curl}}_\Gamma \varphi_{i|\tau_k} , \underline{\mathrm{curl}}_\Gamma \varphi_{j|\tau_\ell} \right) \frac{1}{4\pi} \int_{\tau_k} \int_{\tau_\ell} \frac{1}{|x-y|} ds_x ds_y .$$

Thus, the entries of the stiffness matrix D_h of the hypersingular boundary integral operator D are linear combinations of the entries $V_h[k,\ell]$ of the single layer potential matrix V_h. Hence, we can write

$$D_h = T^\top \begin{pmatrix} V_h & 0 & 0 \\ 0 & V_h & 0 \\ 0 & 0 & V_h \end{pmatrix} T$$

with some sparse transformation matrix $T \in \mathbb{R}^{M \times 3N}$.

From the $H^{1/2}(\Gamma)$–ellipticity of the extended bilinear form, i.e.,

$$\langle Dv, v \rangle_\Gamma + \langle v, 1 \rangle_\Gamma^2 \geq c_1^D \|v\|_{H^{1/2}(\Gamma)}^2 \quad \text{for all } v \in H^{1/2}(\Gamma),$$

we conclude the unique solvability of the variational problem (2.25), or correspondingly, of the linear system (2.26). Furthermore, the quasi optimal error estimate, i.e., Cea's lemma,

$$\|\bar{u}_\alpha - \bar{u}_{\alpha,h}\|_{H^{1/2}(\Gamma)} \leq c \inf_{v_h \in S_h^1(\Gamma)} \|\bar{u}_\alpha - v_h\|_{H^{1/2}(\Gamma)}$$

holds. Combining this with the approximation property (2.10) for $\sigma = 1/2$, we get

$$\|\bar{u}_\alpha - \bar{u}_{\alpha,h}\|_{H^{1/2}(\Gamma)} \leq c h^{s-1/2} |\bar{u}_\alpha|_{H_{\mathrm{pw}}^s(\Gamma)} , \tag{2.28}$$

when assuming $\bar{u}_\alpha \in H_{\mathrm{pw}}^s(\Gamma)$ for some $s \in [1/2, 2]$. Applying the Aubin–Nitsche trick, we also obtain the error estimate

$$\|\bar{u}_\alpha - \bar{u}_{\alpha,h}\|_{H^\sigma(\Gamma)} \leq c h^{s-\sigma} |\bar{u}_\alpha|_{H_{\mathrm{pw}}^s(\Gamma)} ,$$

when assuming $\bar{u}_\alpha \in H_{\mathrm{pw}}^s(\Gamma)$ for some $s \in [1/2, 2]$ and $\sigma \in [-1, 1/2]$.

Inserting the computed Galerkin solution $\bar{u}_{\alpha,h} \in S_h^1(\Gamma)$ into the representation formula (1.6), this gives an approximate representation formula

$$\widetilde{u}(x) = \int_\Gamma u^*(x,y)g(y)ds_y - \int_\Gamma \gamma_{1,y}^{\mathrm{int}} u^*(x,y)\bar{u}_{\alpha,h}(y)ds_y \quad \text{for } x \in \Omega$$

describing an approximate solution of the Neumann boundary value problem (1.21). For an arbitrary $x \in \Omega$, the error is given by

$$u(x) - \widetilde{u}(x) = \int_\Gamma \gamma_{1,y}^{\mathrm{int}} u^*(x,y)\Big(\bar{u}_{\alpha,h}(y) - \bar{u}_\alpha(y)\Big)ds_y .$$

Using a duality argument, the error estimate

$$|u(x) - \widetilde{u}(x)| \le \left\| \gamma_{1,y}^{\mathrm{int}} u^*(x,\cdot) \right\|_{H^{-\sigma}(\Gamma)} \|\bar{u}_\alpha - \bar{u}_{\alpha,h}\|_{H^\sigma(\Gamma)}$$

for some $\sigma \in \mathbb{R}$ follows. Combining this with the error estimate (2.28) for the minimal value $\sigma = -1$, we obtain the pointwise error estimate

$$|u(x) - \widetilde{u}(x)| \le ch^{s+1} \left\| \gamma_{1,y}^{\mathrm{int}} u^*(x,\cdot) \right\|_{H^1(\Gamma)} |\bar{u}_\alpha|_{H^s_{\mathrm{pw}}(\Gamma)} ,$$

when assuming $\bar{u}_\alpha \in H^s_{\mathrm{pw}}(\Gamma)$ for some $s \in [1/2, 2]$. Hence, if $\bar{u}_\alpha \in H^2_{\mathrm{pw}}(\Gamma)$ is sufficiently smooth, we obtain the optimal order of convergence for $s = 2$,

$$|u(x) - \widetilde{u}(x)| \le ch^3 \|\gamma_{1,y}^{\mathrm{int}} u^*(x,\cdot)\|_{H^1(\Gamma)} |\bar{u}_\alpha|_{H^2_{\mathrm{pw}}(\Gamma)} . \tag{2.29}$$

Again, the error estimate (2.29) involves the position of the observation point $x \in \Omega$, and, therefore, it is not valid in the limiting case $x \in \Gamma$.

As in the boundary element method for the Dirichlet boundary value problem, we may also approximate the given Neumann datum $g \in H^{-1/2}(\Gamma)$ first. If $g_h \in S_h^0(\Gamma)$ is defined by the L_2 projection, i.e. if it is the unique solution of the variational problem

$$\int_\Gamma g_h(x)\psi_k(x)\,ds_x = \int_\Gamma g(x)\psi_k(x)\,ds_x \quad \text{for } k = 1,\dots,N ,$$

then the error estimate

$$\|g - g_h\|_{H^\sigma(\Gamma)} \le ch^{s-\sigma} |g|_{H^s_{\mathrm{pw}}(\Gamma)}$$

holds, when assuming $g \in H^s_{\mathrm{pw}}(\Gamma)$ for some $s \in [0,1]$ and $\sigma \in [-1,0]$. Hence, if g is sufficiently smooth, i.e., $g \in H^1_{\mathrm{pw}}(\Gamma)$, we get the optimal error estimate

$$\|g - g_h\|_{H^{-1}(\Gamma)} \le ch^2 |g|_{H^1_{\mathrm{pw}}(\Gamma)}. \tag{2.30}$$

Then,

$$\widetilde{f}_i = \left\langle \left(\frac{1}{2}I - K'\right)g_h, \varphi_i \right\rangle_\Gamma$$

$$= \frac{1}{2}\sum_{\ell=1}^{N} g_\ell \int_{\tau_\ell} \varphi_i(x)ds_x - \sum_{\ell=1}^{N} g_\ell \int_\Gamma \varphi_i(x) \int_{\tau_\ell} \gamma_{1,x}^{\text{int}} u^*(x,y)ds_y ds_x$$

$$= \frac{1}{2}\sum_{\ell=1}^{N} g_\ell \int_{\tau_\ell} \varphi_i(x)ds_x - \sum_{\ell=1}^{N} g_\ell \frac{1}{4\pi} \int_\Gamma \varphi_i(x) \int_{\tau_\ell} \frac{(y-x,\underline{n}(x))}{|x-y|^3} ds_y ds_x$$

$$= \sum_{\ell=1}^{N} g_\ell \left(\frac{1}{2}M_h[\ell,i] - K_h[\ell,i]\right).$$

Instead of the linear system (2.26), we now have to solve a linear system with a perturbed right hand side $\widetilde{\underline{f}}$ yielding a perturbed solution vector $\widetilde{\underline{u}}_\alpha$, i.e., we have to solve the linear system

$$\left(D_h + \underline{a}\,\underline{a}^\top\right)\widetilde{\underline{u}}_\alpha = \left(\frac{1}{2}M_h^\top - K_h^\top\right)\underline{g} + \alpha\underline{a}. \tag{2.31}$$

For the associated boundary element solution $\widetilde{u}_{\alpha,h} \in S_h^1(\Gamma)$, the error estimate

$$\|\bar{u}_\alpha - \widetilde{u}_{\alpha,h}\|_{H^{1/2}(\Gamma)} \le \|\bar{u}_\alpha - \widetilde{u}_{\alpha,h}\|_{H^{1/2}(\Gamma)} + c\,\|g - g_h\|_{H^{-1/2}(\Gamma)}$$

$$\le c\,h^{3/2}\left(|\bar{u}_\alpha|_{H^2_{\text{pw}}(\Gamma)} + |g|_{H^1_{\text{pw}}(\Gamma)}\right),$$

holds, when assuming $\bar{u} \in H^2_{\text{pw}}(\Gamma)$ and $g \in H^1_{\text{pw}}(\Gamma)$. Applying the Aubin–Nitsche trick to obtain an error estimate in lower order Sobolev spaces, the restriction due to the error estimate (2.30) has to be considered. Hence, we obtain the error estimate

$$\|\bar{u} - \widetilde{u}_h\|_{H^\sigma(\Gamma)} \le c_1\,\|\bar{u} - \bar{u}_h\|_{H^\sigma(\Gamma)} + c_2\,\|g - g_h\|_{H^{\sigma-1}(\Gamma)}$$

$$\le c\,h^{2-\sigma}\left(|\bar{u}|_{H^2_{\text{pw}}(\Gamma)} + |g|_{H^1_{\text{pw}}(\Gamma)}\right),$$

when assuming $\bar{u}_\alpha \in H^2_{\text{pw}}(\Gamma)$, $g \in H^1_{\text{pw}}(\Gamma)$, and $\sigma \ge 0$. Therefore, the optimal error estimate reads

$$\|\bar{u}_\alpha - \widetilde{u}_{\alpha,h}\|_{L_2(\Gamma)} \le c\,h^2\left(|\bar{u}_\alpha|_{H^2_{\text{pw}}(\Gamma)} + |g|_{H^1_{\text{pw}}(\Gamma)}\right). \tag{2.32}$$

For the approximate representation formula

$$\widetilde{u}(x) = \int_\Gamma \gamma_{0,y}^{\text{int}} u^*(x,y)g_h(y)ds_y - \int_\Gamma \gamma_{1,y}^{\text{int}} u^*(x,y)\widetilde{u}_{\alpha,h}(y)ds_y \tag{2.33}$$

for $x \in \Omega$, we then obtain the best possible error estimate

$$|u(x) - \widetilde{u}(x)| \le c(x,g,\bar{u}_\alpha)\,h^2, \tag{2.34}$$

when assuming $\bar{u}_\alpha \in H^2_{\text{pw}}(\Gamma)$ and $g \in H^1_{\text{pw}}(\Gamma)$.

2.3.3 Mixed Boundary Value Problem

The solution of the mixed boundary value problem (cf. (1.34))

$$-\Delta u(x) = 0 \qquad \text{for } x \in \Omega,$$
$$\gamma_0^{int} u(x) = g(x) \qquad \text{for } x \in \Gamma_D,$$
$$\gamma_1^{int} u(x) = f(x) \qquad \text{for } x \in \Gamma_N$$

is given by the representation formula

$$u(x) = \int_{\Gamma_D} u^*(x,y)\gamma_1^{int}u(y)ds_y + \int_{\Gamma_N} u^*(x,y)f(y)ds_y$$
$$- \int_{\Gamma_D} \gamma_{1,y}^{int}u^*(x,y)g(y)ds_y - \int_{\Gamma_N} \gamma_{1,y}^{int}u^*(x,y)\gamma_0^{int}u(y)ds_y$$

for $x \in \Omega$, where we have to find the yet unknown Cauchy data $\gamma_0^{int}u$ on Γ_N and $\gamma_1^{int}u$ on Γ_D. Let $\tilde{g} \in H^{1/2}(\Gamma)$ and $\tilde{f} \in H^{-1/2}(\Gamma)$ be some arbitrary, but fixed extensions of the given boundary data $g \in H^{1/2}(\Gamma_D)$ and $f \in H^{-1/2}(\Gamma_N)$, respectively.

The new Cauchy data

$$\tilde{u} = \gamma_0^{int}u - \tilde{g} \in \tilde{H}^{1/2}(\Gamma_N)$$

and

$$\tilde{t} = \gamma_1^{int}u - \tilde{f} \in \tilde{H}^{-1/2}(\Gamma)$$

are the unique solutions of the variational problem (cf. (1.35))

$$a(\tilde{t}, \tilde{u}; w, v) = F(w, v)$$

for all $v \in \tilde{H}^{1/2}(\Gamma_N)$ and $w \in \tilde{H}^{-1/2}(\Gamma_D)$ with the bilinear form

$$a(\tilde{t}, \tilde{u}; w, v) = \frac{1}{4\pi}\int_{\Gamma_D} w(x)\int_{\Gamma_D}\frac{1}{|x-y|}\tilde{t}(y)ds_yds_x$$
$$- \frac{1}{4\pi}\int_{\Gamma_D} w(x)\int_{\Gamma_N}\frac{(x-y,\underline{n}(y))}{|x-y|^3}\tilde{u}(y)ds_yds_x$$
$$+ \frac{1}{4\pi}\int_{\Gamma_N} v(x)\int_{\Gamma_D}\frac{(y-x,\underline{n}(x))}{|x-y|^3}\tilde{t}(y)ds_yds_x$$
$$+ \frac{1}{4\pi}\int_{\Gamma}\int_{\Gamma}\frac{(\text{curl}_\Gamma v(x),\text{curl}_\Gamma\tilde{u}(y))}{|x-y|}ds_yds_x$$

and with the linear form

$$F(w,v) = \frac{1}{2}\int_{\Gamma_D} w(x)g(x)ds_x + \frac{1}{4\pi}\int_{\Gamma_D} w(x)\int_{\Gamma} \frac{(x-y,\underline{n}(y))}{|x-y|^3}\widetilde{g}(y)ds_yds_x$$

$$-\frac{1}{4\pi}\int_{\Gamma_D} w(x)\int_{\Gamma}\frac{1}{|x-y|}\widetilde{f}(y)ds_yds_x + \frac{1}{2}\int_{\Gamma_N} v(x)f(x)ds_x$$

$$-\frac{1}{4\pi}\int_{\Gamma_N} v(x)\int_{\Gamma}\frac{(y-x,\underline{n}(x))}{|x-y|}\widetilde{f}(y)ds_yds_x$$

$$-\frac{1}{4\pi}\int_{\Gamma_N}\int_{\Gamma}\frac{(\mathrm{curl}_\Gamma v(x),\ \mathrm{curl}_\Gamma\widetilde{g}(y))}{|x-y|}ds_yds_x.$$

To be able to define approximate solutions of the above variational problem, we first define suitable trial spaces,

$$S_h^0(\Gamma_D) = S_h^0(\Gamma)\cap\widetilde{H}^{-1/2}(\Gamma_D) = \mathrm{span}\left\{\psi_\ell\right\}_{\ell=1}^{N_D},$$

$$S_h^1(\Gamma_N) = S_h^1(\Gamma)\cap\widetilde{H}^{1/2}(\Gamma_N) = \mathrm{span}\left\{\varphi_j\right\}_{j=1}^{M_N}.$$

The Galerkin formulation of the variational problem (1.35) is to find

$$\widetilde{t}_h \in S_h^0(\Gamma_D)$$

and

$$\widetilde{u}_h \in S_h^1(\Gamma_N)$$

such that

$$a(\widetilde{t}_h, \widetilde{u}_h\,;\, w_h, v_h) = F(w_h, v_h) \tag{2.35}$$

is satisfied for all $w_h \in S_h^0(\Gamma_D)$ and $v_h \in S_h^1(\Gamma_N)$. This formulation is equivalent to a linear system of equations

$$\begin{pmatrix} V_h & -K_h \\ K_h^\top & D_h \end{pmatrix}\begin{pmatrix} \underline{\widetilde{t}} \\ \underline{\widetilde{u}} \end{pmatrix} = \begin{pmatrix} \underline{g} \\ \underline{f} \end{pmatrix} \tag{2.36}$$

with the following blocks:

$$V_h \in \mathbb{R}^{N_D\times N_D}, \quad K_h \in \mathbb{R}^{N_D\times M_N}, \quad D_h \in \mathbb{R}^{M_N\times M_N}.$$

The matrix entries of these blocks are defined by

$$V_h[k,\ell] = \frac{1}{4\pi}\int_{\tau_k}\int_{\tau_\ell}\frac{1}{|x-y|}ds_yds_x,$$

$$K_h[k,j] = \frac{1}{4\pi}\int_{\tau_k}\int_{\Gamma}\frac{(x-y,\underline{n}(y))}{|x-y|^3}\varphi_j(y)ds_yds_x,$$

$$D_h[i,j] = \frac{1}{4\pi}\int_{\Gamma}\int_{\Gamma}\frac{(\mathrm{curl}_\Gamma\varphi_j(y),\mathrm{curl}_\Gamma\varphi_i(x))}{|x-y|}ds_yds_x$$

for all $k, \ell = 1, \ldots, N_D$ and $i, j = 1, \ldots, M_N$. The components of the right hand side, $\underline{g} \in \mathbb{R}^{N_D}$ and $\underline{f} \in \mathbb{R}^{M_N}$, are given by

$$g_k = \frac{1}{2} \int\limits_{\tau_k} g(x) ds_x + \frac{1}{4\pi} \int\limits_{\tau_k} \int\limits_{\Gamma} \frac{(x - y, \underline{n}(y))}{|x - y|^3} \widetilde{g}(y) ds_y ds_x$$

$$- \frac{1}{4\pi} \int\limits_{\tau_k} \int\limits_{\Gamma} \frac{1}{|x - y|} \widetilde{f}(y) ds_y ds_x \,,$$

$$f_i = \frac{1}{2} \int\limits_{\Gamma_N} \varphi_i(x) f(x) ds_x - \frac{1}{4\pi} \int\limits_{\Gamma_N} \varphi_i(x) \int\limits_{\Gamma} \frac{(y - x, \underline{n}(x))}{|x - y|} \widetilde{f}(y) ds_y ds_x$$

$$- \frac{1}{4\pi} \int\limits_{\Gamma_N} \int\limits_{\Gamma} \frac{(\operatorname{curl}_\Gamma \widetilde{g}(y), \operatorname{curl}_\Gamma \varphi_i(x))}{|x - y|} ds_y ds_x$$

for all $k = 1, \ldots, N_D$ and $i = 1, \ldots, M_N$.

Since the trial spaces $S_h^0(\Gamma_D) \subset S_h^0(\Gamma)$ and $S_h^1(\Gamma_N) \subset S_h^1(\Gamma)$ are subspaces of the trial spaces already used for the Dirichlet and for the Neumann boundary value problems, the blocks of the matrix in (2.36) are submatrices of the stiffness matrices already used in (2.22) and in (2.31), respectively. In particular, the evaluation of the discrete hypersingular integral operator D_h can be reduced to the evaluation of some linear combinations of the matrix entries of the discrete single layer potential V_h.

Since the stiffness matrix in (2.36) is positive definite but block skew symmetric, we have to apply a generalised Krylov subspace method such as the Generalised Minimal Residual Method (GMRES) (see Appendix C.3) to solve (2.36) by an iterative method. Here we will describe two alternative approaches to apply the conjugate gradient scheme to solve (2.36).

Since the discrete single layer potential V_h is symmetric and positive definite and hence invertible, we can solve the first equation in (2.36) to find

$$\widetilde{\underline{t}} = V_h^{-1}\left(\underline{g} + K_h \widetilde{\underline{u}}\right).$$

Inserting this into the second equation in (2.36), this gives the Schur complement system

$$S_h \widetilde{\underline{u}} = \underline{f} - K_h^\top V_h^{-1} \underline{g} \tag{2.37}$$

with the symmetric and positive definite Schur complement matrix

$$S_h = D_h + K_h^\top V_h^{-1} K_h\,.$$

Therefore, we can apply a conjugate gradient scheme to solve (2.37), where we eventually need a suitable preconditioning matrix for S_h. Note that the matrix by vector multiplication with the Schur complement matrix S_h involves one application of the inverse single layer potential matrix V_h. This can be realised either by a direct inversion, if the dimension N_D is small, or by the application of an inner conjugate gradient scheme. Again, a suitable preconditioning matrix is eventually needed, which is spectrally equivalent to V_h.

Following [14], we can also apply a suitable transformation to (2.36) to obtain a linear system with a symmetric, positive definite matrix. In particular, the transformed matrix

$$
\begin{pmatrix} V_h C_V^{-1} - I & 0 \\ -K_h^\top C_V^{-1} & I \end{pmatrix} \begin{pmatrix} V_h & -K_h \\ K_h^\top & D_h \end{pmatrix} =
$$

$$
\begin{pmatrix} V_h C_V^{-1} V_h - V_h & (I - V_h C_V^{-1}) K_h \\ K_h^\top (I - C_V^{-1} V_h) & D_h + K_h^\top C_V^{-1} K_h \end{pmatrix}
$$

is symmetric and positive definite. Hence, instead of (2.36), we now solve the transformed linear system

$$
\begin{pmatrix} V_h C_V^{-1} V_h - V_h & (I - V_h C_V^{-1}) K_h \\ K_h^\top (I - C_V^{-1} V_h) & D_h + K_h^\top C_V^{-1} K_h \end{pmatrix} \begin{pmatrix} \widetilde{t} \\ \widetilde{u} \end{pmatrix} = \tag{2.38}
$$

$$
\begin{pmatrix} V_h C_V^{-1} - I & 0 \\ -K_h^\top C_V^{-1} & I \end{pmatrix} \begin{pmatrix} \underline{g} \\ \underline{f} \end{pmatrix}
$$

by a preconditioned conjugate gradient scheme. In the above, C_V is a suitable preconditioning matrix, which is spectrally equivalent to the discrete single layer potential V_h, i.e.,

$$
c_1 (C_V \underline{w}, \underline{w}) \leq (V_h \underline{w}, \underline{w}) \leq c_2 (C_V \underline{w}, \underline{w}) \quad \text{for all } \underline{w} \in \mathbb{R}^{N_D}.
$$

To ensure that (2.38) is equivalent to (2.36), we have to require the invertibility of

$$
V_h C_V^{-1} - I = (V_h - C_V) C_V^{-1}.
$$

Due to

$$
((V_h - C_V) \underline{w}, \underline{w}) \geq (c_1 - 1)(C_V \underline{w}, \underline{w}) \quad \text{for all } \underline{w} \in \mathbb{R}^{N_D},
$$

a sufficient condition is $c_1 > 1$, which ensures the positive definiteness of $V_h - C_V$, and, therefore, its invertibility. A suitable preconditioning matrix for (2.38) is

$$
C_M = \begin{pmatrix} V_h - C_V & 0 \\ 0 & C_S \end{pmatrix},
$$

where C_S is a preconditioning matrix for the Schur complement S_h.

From the $\widetilde{H}^{-1/2}(\Gamma_D) \times \widetilde{H}^{1/2}(\Gamma_N)$–ellipticity of the underlying bilinear form $a(\cdot, \cdot; \cdot, \cdot)$, we conclude the unique solvability of the Galerkin variational problem (2.35), and, therefore, of the linear system (2.36). In particular, we obtain the quasi optimal error estimate

$$\|\tilde{t} - \tilde{t}_h\|_{H^{-1/2}(\Gamma)}^2 + \|\tilde{u} - \tilde{u}_h\|_{H^{1/2}(\Gamma)}^2$$

$$\leq c \left(\inf_{w_h \in S_h^0(\Gamma_D)} \|\tilde{t} - w_h\|_{H^{-1/2}(\Gamma)}^2 + \inf_{v_h \in S_h^1(\Gamma_N)} \|\tilde{u} - v_h\|_{H^{1/2}(\Gamma)}^2 \right)$$

from Cea's lemma. Using the approximation property (2.5) for $\sigma = -1/2$ as well as the approximation property (2.10) for $\sigma = 1/2$, this gives

$$\|\tilde{t} - \tilde{t}_h\|_{H^{-1/2}(\Gamma)}^2 + \|\tilde{u} - \tilde{u}_h\|_{H^{1/2}(\Gamma)}^2 \leq c_1 h^{2s_1+1} |\tilde{t}|_{H_{\text{pw}}^{s_1}(\Gamma)}^2 + c_2 h^{2s_2-1} |\tilde{u}|_{H_{\text{pw}}^{s_2}(\Gamma)}^2 ,$$

when assuming $\tilde{t} \in H_{\text{pw}}^{s_1}(\Gamma)$ for some $s_1 \in [-1/2, 1]$, and $\tilde{u} \in H_{\text{pw}}^{s_2}(\Gamma)$ for some $s_2 \in [1/2, 2]$. Since, in general, those regularity estimates result from a regularity estimate for the solution $u \in H^s(\Omega)$ of the mixed boundary value problem (1.34), we obtain $\gamma_0^{\text{int}} u \in H_{\text{pw}}^{s-1/2}(\Gamma)$ and $\gamma_1^{\text{int}} u \in H_{\text{pw}}^{s-3/2}(\Gamma)$ by applying the trace theorems, and, therefore, $s_1 = s - 3/2$ and $s_2 = s - 1/2$. Thus, if $u \in H^s(\Omega)$ is the solution of the mixed boundary value problem (1.34) for some $s \in [1, 5/2]$, we then obtain the error estimate

$$\|\tilde{t} - \tilde{t}_h\|_{H^{-1/2}(\Gamma)}^2 + \|\tilde{u} - \tilde{u}_h\|_{H^{1/2}(\Gamma)}^2 \leq c h^{2(s-1)} |u|_{H^s(\Omega)}^2 .$$

As for the Dirichlet and for the Neumann boundary value problem, applying the Aubin–Nitsche trick (for $\sigma \in [-2, 1/2)$) and an inverse inequality argument (for $\sigma \in (-1/2, 0]$), we obtain the error estimate

$$\|\tilde{t} - \tilde{t}_h\|_{H^\sigma(\Gamma)}^2 + \|\tilde{u} - \tilde{u}_h\|_{H^{\sigma+1}(\Gamma)}^2 \leq c h^{2(s-\sigma)-3} |u|_{H^s(\Omega)}^2 , \qquad (2.39)$$

when assuming $u \in H^s(\Omega)$ for some $s \in [1, 5/2]$ and $\sigma \in [-2, 0]$.

Inserting the computed Galerkin solutions $\tilde{t}_h \in S_h^0(\Gamma_D)$ and $\tilde{u}_h \in S_h^1(\Gamma_N)$ into the representation formula (1.6), this gives an approximate representation formula

$$\tilde{u}(x) = \int_\Gamma u^*(x, y) \Big(\tilde{t}_h(y) + \tilde{f}(y) \Big) ds_y - \int_\Gamma \gamma_{1,y}^{\text{int}} u^*(x, y) \Big(\tilde{u}_h(y) + \tilde{g}(y) \Big) ds_y$$

for $x \in \Omega$. The above formula describes an approximate solution of the mixed boundary value problem (1.34). For an arbitrary $x \in \Omega$, the error is given by

$$u(x) - \tilde{u}(x) =$$
$$\int_{\Gamma_N} u^*(x, y) \Big(\tilde{t}(y) - \tilde{t}_h(y) \Big) ds_y - \int_{\Gamma_D} \gamma_{1,y}^{\text{int}} u^*(x, y) \Big(\tilde{u}(y) - \tilde{u}_h(y) \Big) ds_y .$$

Using a duality argument, the error estimate

$$|u(x) - \tilde{u}(x)| \leq$$
$$\|u^*(x, \cdot)\|_{H^{-\sigma_1}(\Gamma)} \|\tilde{t} - \tilde{t}_h\|_{H^{\sigma_1}(\Gamma)} + \left\| \gamma_1^{\text{int}} u^*(x, \cdot) \right\|_{H^{-\sigma_2}(\Gamma)} \|\tilde{u} - \tilde{u}_h\|_{H^{\sigma_2}(\Gamma)}$$

for some $\sigma_1, \sigma_2 \in \mathbb{R}$ follows. Combining this with the error estimate (2.39) for the minimal values $\sigma_1 = -2$ and $\sigma_2 = -1$, we obtain the pointwise error estimate

$$|u(x) - \widetilde{u}(x)| \le c\,h^{2s+1}\Big(\|u^*(x,\cdot)\|_{H^2(\Gamma)} + \|\gamma_1^{\mathrm{int}} u^*(x,\cdot)\|_{H^1(\Gamma)}\Big)|u|_{H^s(\Omega)}\,,$$

when assuming $u \in H^s(\Omega)$ for some $s \in [1, 5/2]$. In particular, for $s = 5/2$, we obtain the optimal order of convergence,

$$|u(x) - \widetilde{u}(x)| \le c(x)\,h^3\,|u|_{H^{5/2}(\Omega)}\,. \tag{2.40}$$

Note that the error estimate (2.40) is based on the exact use of the given boundary data $g \in H^{1/2}(\Gamma_D)$ and $f \in H^{-1/2}(\Gamma_N)$, and their extensions $\widetilde{g} \in H^{1/2}(\Gamma)$ and $\widetilde{f} \in H^{-1/2}(\Gamma)$.

Starting from an approximation $u_h \in S_h^1(\Gamma)$ of the complete Dirichlet datum $\gamma_0^{\mathrm{int}} u$,

$$u_h = \sum_{j=1}^{M} u_j \varphi_j = \sum_{j=1}^{M_N} u_j \varphi_j + \sum_{j=M_N+1}^{M} u_j \varphi_j = \widetilde{u}_h + g_h\,,$$

we first have to find the coefficients u_j for $j = M_N+1, \ldots, M$ of the approximate Dirichlet datum $g_h \in S_h^1(\Gamma) \cap H^{1/2}(\Gamma_N)$. This can be done, e.g., by applying the L_2 projection,

$$\sum_{j=M_N+1}^{M} u_j \int_{\Gamma_D} \varphi_j(x) \varphi_i(x) dx = \int_{\Gamma_D} g(x)\varphi_i(x) ds_x \quad \text{for } i = M_N+1, \ldots, M.$$

In a similar way, we obtain an approximation $f_h \in S_h^0(\Gamma_N)$ of the given Neumann datum $f \in H^{-1/2}(\Gamma_N)$,

$$\sum_{\ell=N_D+1}^{N} t_\ell \int_{\Gamma_N} \psi_\ell(x)\psi_k(x) dx = \int_{\Gamma_N} f(x)\psi_k(x) ds_x \quad \text{for } k = N_D+1, \ldots, N.$$

Hence, we have to find the remaining Cauchy data

$$\widetilde{t}_h \in S_h^0(\Gamma_D) \quad \text{and} \quad \widetilde{u}_h \in S_h^1(\Gamma_N)$$

from the variational problem

$$a(\widetilde{t}_h, \widetilde{u}_h\,; \psi_k, \varphi_i) = \widetilde{F}(\psi_k, \varphi_i)$$

for $k = 1, \ldots, N_D$ and $i = 1, \ldots, M_N$, where the perturbed linear form is now given by

$$\widetilde{F}(\psi_k, \varphi_i) = \frac{1}{2} \int_{\tau_k} g_h(x) ds_x + \frac{1}{4\pi} \int_{\tau_k} \int_{\Gamma} \frac{(x - y, \underline{n}(y))}{|x - y|^3} g_h(y) ds_y ds_x$$

$$- \frac{1}{4\pi} \int_{\tau_k} \int_{\Gamma_N} \frac{1}{|x - y|} f_h(y) ds_y ds_x + \frac{1}{2} \int_{\Gamma_N} f_h(x) \varphi_i(x) ds_x$$

$$- \frac{1}{4\pi} \int_{\Gamma_N} \varphi_i(x) \int_{\Gamma_N} \frac{(y - x, \underline{n}(x))}{|x - y|^3} f_h(y) ds_y ds_x$$

$$- \frac{1}{4\pi} \int_{\Gamma_N} \int_{\Gamma} \frac{(\mathrm{curl}_\Gamma \varphi_i(x), \, \mathrm{curl}_\Gamma g_h(y))}{|x - y|} ds_y ds_x.$$

The above perturbed variational problem is now equivalent to a linear system of equations

$$\begin{pmatrix} V_h & -K_h \\ K_h^\top & D_h \end{pmatrix} \begin{pmatrix} \widetilde{t} \\ \widetilde{u} \end{pmatrix} = \begin{pmatrix} -\bar{V}_h & \frac{1}{2}\bar{M}_h + \bar{K}_h \\ \frac{1}{2}\bar{M}_h^\top - \bar{K}_h^\top & -\bar{D}_h \end{pmatrix} \begin{pmatrix} f \\ g \end{pmatrix}. \tag{2.41}$$

Note that the right hand side of this system differs from the one in (2.36). The blocks on the right have the following dimensions:

$$\bar{V}_h \in \mathbb{R}^{N_D \times (N - N_D)}, \quad \bar{M}_h \in \mathbb{R}^{N_D \times (M - M_N)}, \quad \bar{K}_h \in \mathbb{R}^{N_D \times (M - M_N)}$$

and the following entries

$$\bar{V}_h[k, \ell] = \frac{1}{4\pi} \int_{\tau_k} \int_{\tau_\ell} \frac{1}{|x - y|} ds_y ds_x,$$

$$\bar{M}_h[k, j] = \int_{\tau_k} \varphi_j(x) ds_x,$$

$$\bar{K}_h[k, j] = \frac{1}{4\pi} \int_{\tau_k} \int_{\Gamma} \frac{(x - y, \underline{n}(y))}{|x - y|^3} \varphi_j(y) ds_y ds_x,$$

$$\bar{D}_h[i, j] = \frac{1}{4\pi} \int_{\Gamma} \int_{\Gamma} \frac{(\mathrm{curl}_\Gamma \varphi_j(y), \, \mathrm{curl}_\Gamma \varphi_i(x))}{|x - y|} ds_y ds_x$$

for

$$\ell = N_D + 1, \ldots, N, \; k = 1, \ldots, N_D, \; j = M_N + 1, \ldots, M, \; i = 1, \ldots, M_N.$$

Note that the matrices $\bar{V}_h$, $\bar{M}_h$, $\bar{K}_h$, and $\bar{D}_h$ are also submatrices of the stiffness matrices already used in (2.16) and (2.26) to handle the Dirichlet and Neumann boundary value problem, respectively.

The solution of the perturbed linear system (2.41) can be realised as for the linear system (2.36). The error estimates for the resulting approximations can be obtained as in the previous cases, however, the approximations of the given boundary data have to be recognised accordingly. This can be done as for the Dirichlet boundary value problem and as for the Neumann boundary value problem. In particular, the error estimate (2.39) holds for $\sigma \in [-1, 0]$, and instead of (2.40), we obtain only the pointwise error estimate

$$|u(x) - \tilde{u}(x)| \leq c(x)\, h^2\, |u|_{H^{5/2}(\Omega)} \tag{2.42}$$

for $x \in \Omega$, when assuming $u \in H^{5/2}(\Omega)$.

2.3.4 Interface Problem

We consider the interface problem (1.56)–(1.58), i.e., the system of partial differential equations (1.56),

$$-\alpha_i \Delta u_i(x) = f(x) \quad \text{for } x \in \Omega, \quad -\alpha_e \Delta u_e(x) = 0 \quad \text{for } x \in \Omega^e,$$

the transmission conditions (1.57),

$$\gamma_0^{\text{int}} u_i(x) = \gamma_0^{\text{ext}} u_e(x), \quad \alpha_i \gamma_1^{\text{int}} u_i(x) = \alpha_e \gamma_1^{\text{ext}} u_e(x) \quad \text{for } x \in \Gamma,$$

and the radiation condition (1.58) with $u_0 = 0$,

$$|u_e(x)| = \mathcal{O}\left(\frac{1}{|x|}\right) \quad \text{as } |x| \to \infty.$$

Introducing $\bar{u} = \gamma_0^{\text{int}} u_i = \gamma_0^{\text{ext}} u_e \in H^{1/2}(\Gamma)$, we have to solve the resulting variational problem (1.59),

$$\langle (\alpha_i S^{\text{int}} + \alpha_e S^{\text{ext}})\bar{u}, v \rangle_\Gamma = \langle S^{\text{int}} \gamma_0^{\text{int}} u_p - \gamma_1^{\text{int}} u_p, v \rangle_\Gamma$$

for all $v \in H^{1/2}(\Gamma)$, where u_p is a particular solution satisfying $-\Delta u_p = f$ in Ω.

Using a sequence of finite dimensional subspaces $S_h^1(\Gamma) \subset H^{1/2}(\Gamma)$ spanned by piecewise linear and continuous basis functions, an associated approximate solution

$$\bar{u}_h = \sum_{j=1}^{M} \bar{u}_j \varphi_j \in S_h^1(\Gamma)$$

can be found as the unique solution of the Galerkin equations

$$\langle (\alpha_i S^{\text{int}} + \alpha_e S^{\text{ext}})\bar{u}_h, \varphi_i \rangle_\Gamma = \langle S^{\text{int}} \gamma_0^{\text{int}} u_p - \gamma_1^{\text{int}} u_p, \varphi_i \rangle_\Gamma \tag{2.43}$$

for $i = 1, \ldots, M$. This is equivalent to a system of linear equations,

$$S_h \underline{\bar{u}} = \underline{f},$$

with $S_h \in \mathbb{R}^{M \times M}$ and $\underline{f} \in \mathbb{R}^M$ with the entries

$$S_h[i,j] = \langle (\alpha_i S^{\text{int}} + \alpha_e S^{\text{ext}}) \varphi_j, \varphi_i \rangle_\Gamma,$$

$$f_i = \langle S^{\text{int}} \gamma_0^{\text{int}} u_p - \gamma_1^{\text{int}} u_p, \varphi_i \rangle_\Gamma$$

for $i, j = 1, \dots, M$. Since the Steklov–Poincaré operators

$$(S^{\text{int}} \bar{u})(x) = V^{-1} \left(\frac{1}{2} I + K \right) \bar{u}(x)$$

$$= \left(D + \left(\frac{1}{2} I + K' \right) V^{-1} \left(\frac{1}{2} I + K \right) \right) \bar{u}(x),$$

$$(S^{\text{ext}} \bar{u})(x) = V^{-1} \left(-\frac{1}{2} I + K \right) \bar{u}(x)$$

$$= \left(D + \left(-\frac{1}{2} I + K' \right) V^{-1} \left(-\frac{1}{2} I + K \right) \right) \bar{u}(x)$$

do not allow a direct evaluation of both, the stiffness matrix and the right hand side, additional approximations are required. The application of the Steklov–Poincaré operator S^{int} related to the interior Dirichlet boundary value problem can be written as

$$(S^{\text{int}} \bar{u})(x) = \left(D + \left(\frac{1}{2} I + K' \right) V^{-1} \left(\frac{1}{2} I + K \right) \right) \bar{u}(x)$$

$$= (D\bar{u})(x) + \left(\frac{1}{2} I + K' \right) t_i(x),$$

where

$$t_i = V^{-1} \left(\frac{1}{2} I + K \right) \bar{u} \in H^{-1/2}(\Gamma)$$

is the unique solution of the variational problem

$$\langle V t_i, w \rangle_\Gamma = \left\langle \left(\frac{1}{2} I + K \right) \bar{u}, w \right\rangle_\Gamma \quad \text{for all } w \in H^{-1/2}(\Gamma).$$

Let $t_{i,h} \in S_h^0(\Gamma)$ be the unique solution of the Galerkin variational problem

$$\left\langle V t_{i,h}, w_h \right\rangle_\Gamma = \left\langle \left(\frac{1}{2} I + K \right) \bar{u}, w_h \right\rangle_\Gamma \quad \text{for all } w_h \in S_h^0(\Gamma).$$

Then,

$$(\widetilde{S}^{\text{int}} \bar{u})(x) = (D\bar{u})(x) + \left(\frac{1}{2} I + K' \right) t_{i,h}(x)$$

defines an approximate Steklov–Poincaré operator associated to the interior Dirichlet boundary value problem. In the same way, we define an approximate Steklov–Poincaré operator

$$(\widetilde{S}^{\text{ext}}\bar{u})(x) = (D\bar{u})(x) + \left(-\frac{1}{2}I + K'\right)t_{e,h}(x),$$

which is associated to the exterior Dirichlet boundary value problem, and where $t_{e,h} \in S_h^0(\Gamma)$ is the unique solution of the Galerkin equations

$$\langle Vt_{e,h}, w_h\rangle_\Gamma = \left\langle\left(-\frac{1}{2}I + K\right)\bar{u}, w_h\right\rangle_\Gamma \quad \text{for all } w_h \in S_h^0(\Gamma).$$

Now, instead of the variational problem (2.43), we consider the perturbed problem

$$\langle(\alpha_i\widetilde{S}^{\text{int}} + \alpha_e\widetilde{S}^{\text{ext}})\widetilde{u}_h, \varphi_i\rangle_\Gamma = \langle\widetilde{S}^{\text{int}}u_{p,h} - t_{p,h}, \varphi_i\rangle_\Gamma \tag{2.44}$$

for $i = 1,\ldots,M$. In (2.44), $t_{p,h} \in S_h^0(\Gamma)$ and $u_{p,h} \in S_h^1(\Gamma)$ are suitable approximations (L_2 projections) of the Cauchy data of the particular solution u_p, i.e.,

$$\langle t_{p,h}, \psi_k\rangle_{L_2(\Gamma)} = \langle\gamma_1^{\text{int}}u_p, \psi_k\rangle_{L_2(\Gamma)}$$

for $k = 1,\ldots,N$ and

$$\langle u_{p,h}, \varphi_i\rangle_{L_2(\Gamma)} = \langle\gamma_0^{\text{int}}u_p, \varphi_i\rangle_{L_2(\Gamma)}$$

for $i = 1,\ldots,M$. From (2.44), we then obtain the linear system

$$\left(\alpha_i\left(D_h + \left(\frac{1}{2}M_h^\top + K_h^\top\right)V_h^{-1}\left(\frac{1}{2}M_h + K_h\right)\right)\right) \tag{2.45}$$

$$+\alpha_e\left(D_h + \left(-\frac{1}{2}M_h^\top + K_h^\top\right)V_h^{-1}\left(-\frac{1}{2}M_h + K_h\right)\right)\right)\widetilde{\underline{u}} =$$

$$\left(D_h + \left(\frac{1}{2}M_h^\top + K_h^\top\right)V_h^{-1}\left(\frac{1}{2}M_h + K_h\right)\right)\underline{u}_p - M_h^\top\underline{t}_p,$$

where

$$V_h \in \mathbb{R}^{N\times N}, \quad M_h \in \mathbb{R}^{N\times M}, \quad K_h \in \mathbb{R}^{N\times M}, \quad D_h \in \mathbb{R}^{M\times M}$$

are the Galerkin stiffness matrices, which have already been used for the Dirichlet and for the Neumann boundary value problems. The entries of these matrices are defined as

$$V_h[k,\ell] = \frac{1}{4\pi}\int_{\tau_k}\int_{\tau_\ell}\frac{1}{|x-y|}ds_yds_x,$$

$$M_h[k,j] = \int_{\tau_k}\varphi_j(x)ds_x,$$

$$K_h[k,j] = \frac{1}{4\pi}\int_{\tau_k}\int_\Gamma\frac{(x-y,\underline{n}(y))}{|x-y|^3}\varphi_j(y)ds_yds_x,$$

$$D_h[i,j] = \frac{1}{4\pi}\int_\Gamma\int_\Gamma\frac{(\text{curl}_\Gamma\varphi_j(y),\text{curl}_\Gamma\varphi_i(x))}{|x-y|}ds_yds_x$$

for $k, \ell = 1, \ldots, N$ and $i, j = 1, \ldots, M$.

Instead of the linear system (2.45) we may also solve the equivalent coupled system

$$
\begin{pmatrix}
\alpha_i V_h & 0 & -\alpha_i(\frac{1}{2}M_h + K_h) \\
0 & \alpha_e V_h & -\alpha_e(-\frac{1}{2}M_h + K_h) \\
\alpha_i(\frac{1}{2}M_h^\top + K_h^\top) & \alpha_e(-\frac{1}{2}M_h^\top + K_h^\top) & (\alpha_i + \alpha_e)D_h
\end{pmatrix}
\begin{pmatrix}
\underline{t}_i \\
\underline{t}_e \\
\widetilde{\underline{u}}
\end{pmatrix}
$$

$$
= \begin{pmatrix}
-(\frac{1}{2}M_h + K_h)\underline{u}_p \\
\underline{0} \\
D_h\underline{u}_p - M_h^\top \underline{t}_p
\end{pmatrix}, \tag{2.46}
$$

which is of the same structure as the linear system (2.36), i.e. block skew symmetric but positive definite. Note that (2.45) is the Schur complement system of (2.46).

As for the Neumann boundary value problem, we conclude the error estimate

$$
\|\bar{u} - \widetilde{u}_h\|_{H^{1/2}(\Gamma)} \leq c_1 \inf_{v_h \in S_h^1(\Gamma)} \|\bar{u} - v_h\|_{H^{1/2}(\Gamma)}
$$

$$
+ c_2 \inf_{w_h \in S_h^0(\Gamma)} \|S^{\text{int}}\bar{u} - w_h\|_{H^{-1/2}(\Gamma)} + c_3 \inf_{w_h \in S_h^0(\Gamma)} \|S^{\text{ext}}\bar{u} - w_h\|_{H^{-1/2}(\Gamma)}.
$$

Hence, assuming $\bar{u} \in H^2_{\text{pw}}(\Gamma)$ and $S^{\text{int}/\text{ext}}\bar{u} \in H^1_{\text{pw}}, (\Gamma)$, we obtain the error estimate

$$
\|\bar{u} - \widetilde{u}_h\|_{H^{1/2}(\Gamma)} \leq c\, h^{3/2} \left(\|\bar{u}\|_{H^2_{\text{pw}}(\Gamma)} + \|S^{\text{int}}\bar{u}\|_{H^1_{\text{pw}}(\Gamma)} + \|S^{\text{ext}}\bar{u}\|_{H^1_{\text{pw}}(\Gamma)} \right),
$$

and by applying the Aubin–Nitsche trick, we get

$$
\|\bar{u} - \widetilde{u}_h\|_{L_2(\Gamma)} \leq c(\bar{u})\, h^2 .
$$

When the Dirichlet datum $\bar{u}_h$ is known, one can compute the remaining Neumann datum by solving both, the interior and exterior Dirichlet boundary value problems. Since those boundary value problems are Dirichlet boundary value problems with approximated boundary data, the corresponding error estimates are still valid.

2.4 Lame Equations

For a simply connected domain $\Omega \subset \mathbb{R}^3$, we consider the mixed boundary value problem (1.79)

$$-\sum_{j=1}^{3}\frac{\partial}{\partial x_j}\sigma_{ij}(\underline{u},x) = 0 \qquad \text{for } x \in \Omega \,,$$

$$\gamma_0^{\text{int}} u_i(x) = g_i(x) \quad \text{for } x \in \Gamma_{D,i}\,,$$

$$(\gamma_1^{\text{int}}\underline{u})_i(x) = \sum_{j=1}^{3}\sigma_{ij}(\underline{u},x)n_j(x) = f_i(x) \quad \text{for } x \in \Gamma_{N,i}\,,$$

for $i = 1, 2, 3$. Note that we assume

$$\Gamma = \overline{\Gamma}_{N,i} \cup \overline{\Gamma}_{D,i}\,, \quad \Gamma_{N,i} \cap \Gamma_{D,i} = \varnothing\,, \quad \text{meas } \Gamma_{D,i} > 0$$

for $i = 1, 2, 3$. To find the yet unknown Cauchy data $(\gamma_1^{\text{int}}\underline{u})_i$ on $\Gamma_{D,i}$ and $\gamma_0^{\text{int}} u_i$ on $\Gamma_{N,i}$, we consider the variational problem (1.80), which is related to the symmetric formulation of boundary integral equations. Hence, we have to find

$$\widetilde{t}_i = (\gamma_1^{\text{int}}\underline{u})_i - \widetilde{f}_i \in \widetilde{H}^{-1/2}(\Gamma_{D,i})$$

and

$$\widetilde{u}_i = \gamma_0^{\text{int}} u_i - \widetilde{g}_i \in \widetilde{H}^{1/2}(\Gamma_{N,i})$$

such that

$$a(\underline{\widetilde{t}}, \underline{\widetilde{u}}; \underline{w}, \underline{v}) = F(\underline{w}, \underline{v})$$

is satisfied for all $w_i \in \widetilde{H}^{-1/2}(\Gamma_{D,i})$ and $v_i \in \widetilde{H}^{1/2}(\Gamma_{N,i})$ for $i = 1, 2, 3$. Note that the bilinear form is given by

$$a(\underline{\widetilde{t}}, \underline{\widetilde{u}}; \underline{w}, \underline{v}) = \sum_{i=1}^{3}\left\langle (V^{\text{Lame}}\underline{\widetilde{t}})_i, w_i \right\rangle_{\Gamma_{D,i}} - \sum_{i=1}^{3}\left\langle (K^{\text{Lame}}\underline{\widetilde{u}})_i, w_i \right\rangle_{\Gamma_{D,i}}$$
$$+ \sum_{i=1}^{3}\left\langle \widetilde{t}_i, (K^{\text{Lame}}\underline{v})_i \right\rangle_{\Gamma_{N,i}} + \sum_{i=1}^{3}\left\langle (D^{\text{Lame}}\underline{\widetilde{u}})_i, v_i \right\rangle_{\Gamma_{N,i}}\,,$$

while the linear form is

$$F(\underline{w}, \underline{v}) =$$

$$\sum_{i=1}^{3}\left(\frac{1}{2}\left\langle g_i, w_i \right\rangle_{\Gamma_{D,i}} + \left\langle (K^{\text{Lame}}\underline{\widetilde{g}})_i, w_i \right\rangle_{\Gamma_{D,i}} - \left\langle (V^{\text{Lame}}\underline{\widetilde{f}})_i, w_i \right\rangle_{\Gamma_{D,i}}\right) +$$

$$\sum_{i=1}^{3}\left(\frac{1}{2}\left\langle f_i, v_i \right\rangle_{\Gamma_{N,i}} - \left\langle \widetilde{f}_i, (K^{\text{Lame}}\underline{v})_i \right\rangle_{\Gamma_{N,i}} - \left\langle (D^{\text{Lame}}\underline{\widetilde{g}})_i, v_i \right\rangle_{\Gamma_{N,i}}\right).$$

As for the Laplace equation, we first define suitable trial spaces,

$$S_h^0(\Gamma_{D,i}) = S_h^0(\Gamma) \cap \widetilde{H}^{-1/2}(\Gamma_{D,i}) = \text{span}\left\{\psi_\ell^i\right\}_{\ell=1}^{N_{D,i}}\,,$$

$$S_h^1(\Gamma_{N,i}) = S_h^1(\Gamma) \cap \widetilde{H}^{1/2}(\Gamma_{N,i}) = \text{span}\left\{\varphi_j^i\right\}_{j=1}^{M_{N,i}}$$

for $i = 1, 2, 3$. The Galerkin formulation of the variational problem (1.80) is to find $\widetilde{t}_{i,h} \in S_h^0(\Gamma_{D,i})$ and $\widetilde{u}_{i,h} \in S_h^1(\Gamma_{N,i})$ such that

$$a(\underline{\widetilde{t}}_h, \underline{\widetilde{u}}_h ; \underline{w}_h, \underline{v}_h) = F(\underline{w}_h, \underline{v}_h)$$

is satisfied for all $w_i \in S_h^0(\Gamma_{D,i})$ and $v_i \in S_h^1(\Gamma_{M,i})$ for $i = 1, 2, 3$. This formulation is equivalent to a linear system of equations

$$\begin{pmatrix} \bar{V}_h^{\text{Lame}} & -\bar{K}_h^{\text{Lame}} \\ \left(\bar{K}_h^{\text{Lame}}\right)^\top & \bar{D}_h^{\text{Lame}} \end{pmatrix} \begin{pmatrix} \underline{\widetilde{t}} \\ \underline{\widetilde{u}} \end{pmatrix} = \begin{pmatrix} \underline{g} \\ \underline{f} \end{pmatrix}, \qquad (2.47)$$

having the blocks

$$\bar{V}_h^{\text{Lame}} \in \mathbb{R}^{N_D \times N_D}, \quad \bar{K}_h^{\text{Lame}} \in \mathbb{R}^{N_D \times M_N}, \quad \bar{D}_h^{\text{Lame}} \in \mathbb{R}^{M_N \times M_N},$$

where

$$N_D = \sum_{i=1}^{3} N_{D,i}, \quad M_N = \sum_{i=1}^{3} M_{N,i}.$$

While the blocks in the linear system (2.47) recover only the unknown coefficients $\widetilde{t}_{i,\ell}$ and $\widetilde{u}_{i,j}$, an implementation based on the complete stiffness matrices may be advantageous. Let

$$S_h^0(\Gamma) = \text{span}\{\psi_\ell\}_{\ell=1}^{N}, \quad S_h^1(\Gamma) = \text{span}\{\varphi_j\}_{j=1}^{M}$$

be the boundary element spaces spanned by piecewise constant and piecewise linear continuous basis functions, respectively. Note that both $S_h^0(\Gamma)$ and $S_h^1(\Gamma)$ are defined with respect to a boundary element mesh of the complete surface Γ. By $P_i : \mathbb{R}^N \to \mathbb{R}^{N_{D,i}}$ and $Q_i : \mathbb{R}^M \to \mathbb{R}^{M_{N,i}}$, we denote some nodal projection operators describing the imbedding $\underline{w}^i = P_i \underline{w} \in \mathbb{R}^{N_{D,i}}$ for $\underline{w} \in \mathbb{R}^N$ with

$$w_h^i(x) = \sum_{\ell=1}^{N_{D,i}} w_\ell^i \psi_\ell^i(x) \in S_h^0(\Gamma_{D,i}), \quad w_h(x) = \sum_{\ell=1}^{N} w_\ell \psi_\ell(x) \in S_h^0(\Gamma)$$

as well as the imbedding $\underline{v}^i = Q_i \underline{v} \in \mathbb{R}^{M_{N,i}}$ for $\underline{v} \in \mathbb{R}^M$ with

$$v_h^i(x) = \sum_{j=1}^{M_{N,i}} v_j^i \varphi_j^i(x) \in S_h^1(\Gamma_{N,i}), \quad v_h(x) = \sum_{j=1}^{N} v_j \varphi_j(x) \in S_h^1(\Gamma).$$

From this we obtain the representations

$$\bar{V}_h^{\text{Lame}} = P V_h^{\text{Lame}} P^\top, \quad \bar{K}_h^{\text{Lame}} = P K_h^{\text{Lame}} Q^\top, \quad \bar{D}_h^{\text{Lame}} = Q D_h^{\text{Lame}} Q^\top,$$

where the stiffness matrices V_h^{Lame}, K_h^{Lame}, and D_h^{Lame} correspond to the Galerkin discretisation of the associated boundary integral operators V^{Lame},

K^{Lame} and D^{Lame} with respect to the boundary element spaces $[S_h^0(\Gamma)]^3$ and $[S_h^1(\Gamma)]^3$. In particular, for the discrete single layer potential $\bar{V}_h$ we have the representation

$$V_h^{\text{Lame}} = \tag{2.48}$$

$$\frac{1}{2}\frac{1}{E}\frac{1+\nu}{1-\nu}\left((3-4\nu)\begin{pmatrix} V_h & 0 & 0 \\ 0 & V_h & 0 \\ 0 & 0 & V_h \end{pmatrix} + \begin{pmatrix} V_{11,h} & V_{21,h} & V_{13,h} \\ V_{21,h} & V_{22,h} & V_{23,h} \\ V_{31,h} & V_{32,h} & V_{33,h} \end{pmatrix}\right)$$

with the matrix $V_h \in \mathbb{R}^{N \times N}$ having the entries

$$V_h[k,\ell] = \frac{1}{4\pi}\int\limits_{\tau_k}\int\limits_{\tau_\ell}\frac{1}{|x-y|}ds_y ds_x, \tag{2.49}$$

and six further matrices $V_{ij,h} \in \mathbb{R}^{N \times N}$ defined by

$$V_{ij,h}[k,\ell] = \frac{1}{4\pi}\int\limits_{\tau_k}\int\limits_{\tau_\ell}\frac{(x_i-y_i)(x_j-y_j)}{|x-y|^3}ds_y ds_x \tag{2.50}$$

$$= \frac{1}{4\pi}\int\limits_{\tau_k}\int\limits_{\tau_\ell}(x_i-y_i)\frac{\partial}{\partial y_j}\frac{1}{|x-y|}ds_y ds_x$$

for $k,\ell = 1,\ldots,N$ and $i,j = 1,2,3$. Note that V_h is just the Galerkin stiffness matrix of the single layer potential for the Laplace operator, while the matrix entries $V_{ij,h}[\ell,k]$ are similar to the Galerkin discretisation of the double layer potential for the Laplace operator.

From Lemma 1.16, we find the representation for the double layer potential K^{Lame}

$$(K^{\text{Lame}}\underline{v})(x) = (K\underline{v})(x) - \left(VM(\partial,\underline{n})\underline{v}\right)(x) + \frac{E}{1+\nu}\left(V^{\text{Lame}}M(\partial,\underline{n})\underline{v}\right)(x)$$

for $x \in \Gamma$, and, therefore, the matrix representation

$$K_h^{\text{Lame}} = \tag{2.51}$$

$$\begin{pmatrix} K_h & 0 & 0 \\ 0 & K_h & 0 \\ 0 & 0 & K_h \end{pmatrix} - \begin{pmatrix} V_h & 0 & 0 \\ 0 & V_h 0 \\ 0 & 0 & V_h \end{pmatrix}\widetilde{T} + \frac{E}{1+\nu}V_h^{\text{Lame}}\widetilde{T},$$

where V_h and K_h are the Galerkin matrices related to the single and double layer potential of the Laplace operator. Furthermore, $\widetilde{T}$ is a transformation matrix related to the matrix surface curl operator $M(\partial,\underline{n})$.

Using the representation of the bilinear form of the hypersingular boundary integral operator D^{Lame} as given in Lemma 1.18, one can derive a similar representation for the Galerkin matrix D_h^{Lame}, which is based on the transformation matrix $\widetilde{T}$ and on the Galerkin matrices related to the single layer potential of both, the Laplace operator and the system of linear elastostatics.

2.5 Helmholtz Equation

2.5.1 Interior Dirichlet Problem

The solution of the interior Dirichlet boundary value problem (cf. (1.104)),

$$-\Delta u(x) - \kappa^2 u(x) = 0 \quad \text{for } x \in \Omega, \quad \gamma_0^{\text{int}} u(x) = g(x) \quad \text{for } x \in \Gamma,$$

is given by the representation formula (cf. (1.95))

$$u(x) = \int_\Gamma u_\kappa^*(x,y)t(y)ds_y - \int_\Gamma \gamma_{1,y}^{\text{int}} u_\kappa^*(x,y)g(y)ds_y \quad \text{for } x \in \Omega,$$

where the unknown Neumann datum $t = \gamma_1^{\text{int}} u \in H^{-1/2}(\Gamma)$ is the unique solution of the boundary integral equation (cf. (1.105))

$$(V_\kappa t)(x) = \frac{1}{2}g(x) + (K_\kappa g)(x) \quad \text{for } x \in \Gamma.$$

Note that for the unique solvability, we have to assume that κ^2 is not an eigenvalue of the Dirichlet eigenvalue problem (1.108). Then, $t \in H^{-1/2}(\Gamma)$ is the unique solution of the variational problem (cf. (1.106))

$$\left\langle V_\kappa t, w \right\rangle_\Gamma = \left\langle \left(\frac{1}{2}I + K_\kappa\right)g, w \right\rangle_\Gamma \quad \text{for all } w \in H^{-1/2}(\Gamma).$$

Using a sequence of finite dimensional subspaces $S_h^0(\Gamma)$ spanned by piecewise constant basis functions, associated approximate solutions

$$t_h = \sum_{\ell=1}^N t_\ell \psi_\ell \in S_h^0(\Gamma)$$

are obtained from the Galerkin equations

$$\left\langle V_\kappa t_h, \psi_k \right\rangle_\Gamma = \left\langle \left(\frac{1}{2}I + K_\kappa\right)g, \psi_k \right\rangle_\Gamma \quad \text{for } k = 1, \ldots, N. \qquad (2.52)$$

Hence, we find the coefficient vector $\underline{t} \in \mathbb{C}^N$ as the unique solution of the linear system

$$V_{\kappa,h}\underline{t} = \underline{f}$$

with

$$V_{\kappa,h}[k,\ell] = \frac{1}{4\pi} \int_{\tau_k} \int_{\tau_\ell} \frac{e^{\iota\kappa|x-y|}}{|x-y|} ds_y ds_x, \qquad (2.53)$$

for $k, \ell = 1, \ldots, N$, and

$$f_k = \frac{1}{2} \int_{\tau_k} g(x) ds_x + \frac{1}{4\pi} \int_{\tau_k} \int_{\Gamma} (1 - \imath \kappa |x - y|) e^{\imath \kappa |x-y|} \frac{(x - y, n(y))}{|x - y|^3} g(y) ds_y ds_x$$

for $k = 1, \dots, N$.

Since the single layer potential $V_\kappa : H^{-1/2}(\Gamma) \to H^{1/2}(\Gamma)$ is coercive, i.e. V_κ satisfies (1.97), and since V_κ is injective when κ^2 is not an eigenvalue of the Dirichlet eigenvalue problem (1.108), we conclude the unique solvability of the Galerkin variational problem (2.52), as well as the quasi optimal error estimate, i.e. Cea's lemma,

$$\|t - t_h\|_{H^{-1/2}(\Gamma)} \leq c \inf_{w_h \in S_h^0(\Gamma)} \|t - w_h\|_{H^{-1/2}(\Gamma)}.$$

Combining this with the approximation property (2.5) for $\sigma = -1/2$, we get

$$\|t - t_h\|_{H^{-1/2}(\Gamma)} \leq c h^{s+\frac{1}{2}} |t|_{H^s_{pw}(\Gamma)},$$

when assuming $t \in H^s_{pw}(\Gamma)$ and $s \in [0, 1]$. Applying the Aubin–Nitsche trick (for $\sigma < -1/2$) and the inverse inequality argument (for $\sigma \in (-1/2, 0]$), we also obtain the error estimate

$$\|t - t_h\|_{H^\sigma(\Gamma)} \leq c h^{s-\sigma} |t|_{H^s_{pw}(\Gamma)}, \tag{2.54}$$

when assuming $t \in H^s_{pw}(\Gamma)$ for some $s \in [0, 1]$ and $\sigma \in [-2, 0]$.

Inserting the computed Galerkin solution $t_h \in S_h^0(\Gamma)$ into the representation formula (1.95), this gives an approximate representation formula

$$\tilde{u}(x) = \int_{\Gamma} \gamma_0^{int} u_\kappa^*(x, y) t_h(y) ds_y - \int_{\Gamma} \gamma_1^{int} u_\kappa^*(x, y) g(y) ds_y, \tag{2.55}$$

for $x \in \Omega$, describing an approximate solution of the Dirichlet boundary value problem (1.104). Note that $\tilde{u}$ satisfies the Helmholtz equation, but the Dirichlet boundary conditions are satisfied only approximately. For an arbitrary $x \in \Omega$, the error is given by

$$u(x) - \tilde{u}(x) = \int_{\Gamma} u_\kappa^*(x, y) \big(t(y) - t_h(y) \big) ds_y.$$

Using a duality argument, the error estimate

$$|u(x) - \tilde{u}(x)| \leq \|u_\kappa^*(x, \cdot)\|_{H^{-\sigma}(\Gamma)} \|t - t_h\|_{H^\sigma(\Gamma)}$$

for some $\sigma \in \mathbb{R}$ follows. Combining this with the error estimate (2.54) for the minimal value $\sigma = -2$, we obtain the pointwise error estimate

$$|u(x) - \tilde{u}(x)| \leq c h^{s+2} \|u_\kappa^*(x, \cdot)\|_{H^2(\Gamma)} |t|_{H^s_{pw}(\Gamma)}.$$

Hence, if $t \in H^1_{\mathrm{pw}}(\Gamma)$ is sufficiently smooth, we obtain the optimal order of convergence for $s = 1$,

$$|u(x) - \tilde{u}(x)| \leq c\,h^3\,\|u^*_\kappa(x,\cdot)\|_{H^2(\Gamma)}\,|t|_{H^1_{\mathrm{pw}}(\Gamma)}. \tag{2.56}$$

Again, the error estimate (2.56) involves the position of the observation point $x \in \Omega$, and, therefore, it is not valid in the limiting case $x \in \Gamma$.

As for the Dirichlet problem for the Laplace equation, the computation of f_k requires the evaluation of the integrals

$$f_k = \frac{1}{2}\int_{\tau_k} g(x)ds_x + \frac{1}{4\pi}\int_{\tau_k}\int_{\Gamma}\left(1 - \imath\,\kappa\,|x - y|\right)e^{\imath\,\kappa|x-y|}\frac{(x - y, \underline{n}(y))}{|x - y|^3}g(y)ds_y ds_x.$$

When using a piecewise linear approximation $g_h \in S^1_h(\Gamma)$ of the given Dirichlet datum $g \in H^{1/2}(\Gamma)$, we find a perturbed solution vector $\tilde{\underline{t}} \in \mathbb{C}^N$ from the linear system

$$V_{\kappa,h}\tilde{\underline{t}} = \left(\frac{1}{2}M_h + K_{\kappa,h}\right)\underline{g} \tag{2.57}$$

with additional matrices defined by the entries

$$M_h[k,j] = \int_{\tau_k}\varphi_j(x)ds_x\,,$$

and

$$K_{\kappa,h}[k,j] = \frac{1}{4\pi}\int_{\tau_k}\int_{\Gamma}\left(1 - \imath\,\kappa\,|x - y|\right)e^{\imath\,\kappa|x-y|}\frac{(x - y, \underline{n}(y))}{|x - y|^3}\varphi_j(y)ds_y ds_x \tag{2.58}$$

for $k = 1, \ldots, N$ and $j = 1, \ldots, M$. Then, the exact Galerkin solution t_h has to be replaced by the perturbed solution $\tilde{t}_h$ to obtain an approximate solution of the Dirichlet problem (1.104) for $x \in \Omega$,

$$\tilde{u}(x) = \int_{\Gamma} u^*_\kappa(x,y)\tilde{t}_h(y)ds_y - \int_{\Gamma}\gamma_1^{\mathrm{int}}u^*_\kappa(x,y)g_h(y)ds_y\,.$$

Thus, we obtain the optimal error estimate

$$|u(x) - \tilde{u}(x)| \leq c(x,t,g)\,h^3\,, \tag{2.59}$$

when using a L_2 projection to approximate the boundary conditions, and when assuming $t \in H^1_{\mathrm{pw}}(\Gamma)$ and $g \in H^2_{\mathrm{pw}}(\Gamma)$.

2.5.2 Interior Neumann Problem

Next we consider the interior Neumann boundary value problem (1.109),

$$-\Delta u(x) - \kappa^2 u(x) = 0 \quad \text{for } x \in \Omega, \quad \gamma_1^{\text{int}} u(x) = g(x) \quad \text{for } x \in \Gamma.$$

The solution is given by the representation formula for $x \in \Omega$ (cf. (1.95))

$$u(x) = \int_\Gamma u_\kappa^*(x,y) g(y) ds_y - \int_\Gamma \gamma_{1,y}^{\text{int}} u_\kappa^*(x,y) \gamma_0^{\text{int}} u(y) ds_y.$$

We assume that κ^2 is not an eigenvalue of the Neumann eigenvalue problem (1.113). In this case, the unknown Dirichlet datum $\hat{u} = \gamma_0^{\text{int}} u \in H^{1/2}(\Gamma)$ is the unique solution of the boundary integral equation (1.111),

$$(D_\kappa \hat{u})(x) = \frac{1}{2} g(x) - (K_\kappa' g)(x) \quad \text{for } x \in \Gamma,$$

or of the equivalent variational problem (1.112),

$$\left\langle D_\kappa \hat{u}, v \right\rangle_\Gamma = \left\langle \left(\frac{1}{2} I - K_\kappa'\right) g, v \right\rangle_\Gamma \quad \text{for all } v \in H^{1/2}(\Gamma).$$

Using a sequence of finite dimensional subspaces $S_h^1(\Gamma)$ spanned by piecewise linear continuous basis functions, associated approximate solutions

$$\hat{u}_h = \sum_{j=1}^M \hat{u}_j \varphi_j \in S_h^1(\Gamma)$$

are obtained from the Galerkin equations

$$\left\langle D_\kappa \hat{u}_h, \varphi_i \right\rangle_\Gamma = \left\langle \left(\frac{1}{2} I - K_\kappa'\right) g, \varphi_i \right\rangle_\Gamma \quad \text{for } i = 1, \dots, M. \tag{2.60}$$

Hence, we find the coefficient vector $\hat{\underline{u}} \in \mathbb{C}^M$ as the unique solution of the linear system

$$D_{\kappa,h} \hat{\underline{u}} = \underline{f} \tag{2.61}$$

with

$$D_{\kappa,h}[i,j] = \langle D_\kappa \varphi_j, \varphi_i \rangle_\Gamma \tag{2.62}$$

$$= \frac{1}{4\pi} \int_\Gamma \int_\Gamma \frac{e^{\imath \kappa |x-y|}}{|x-y|} (\underline{\text{curl}}_\Gamma \varphi_j(y), \underline{\text{curl}}_\Gamma \varphi_i(x)) ds_y ds_x$$

$$- \frac{\kappa^2}{4\pi} \int_\Gamma \int_\Gamma \frac{e^{\imath \kappa |x-y|}}{|x-y|} \varphi_j(y) \varphi_i(x) (\underline{n}(x), \underline{n}(y)) ds_y ds_x,$$

for $i, j = 1, \dots, M$, and

$$f_i = \frac{1}{2} \int_\Gamma g(x) \varphi_i(x) ds_x$$

$$- \frac{1}{4\pi} \int_\Gamma \varphi_i(x) \int_\Gamma (1 - \imath \kappa |x-y|) e^{\imath \kappa |x-y|} \frac{(x-y, \underline{n}(y))}{|x-y|^3} g(y) ds_y ds_x$$

for $i = 1, \ldots, M$. Note that for the computation of the matrix entries $D_{\kappa,h}[i,j]$, we can reuse the discrete single layer potential $V_{\kappa,h}$ for picewise constant basis functions, but we also need to have the Galerkin discretisation with piecewise linear continuous basis functions of the operator

$$(C_\kappa u)(x) = \int_\Gamma \frac{e^{\imath\kappa|x-y|}}{|x-y|} \, (\underline{n}(x), \underline{n}(y)) \, u(y) ds_y \,, \qquad (2.63)$$

which is similar to the single layer potential operator.

Since the hypersingular integral operator

$$D_\kappa : H^{1/2}(\Gamma) \to H^{-1/2}(\Gamma)$$

is coercive, i.e. D_κ satisfies (1.98), and since D_κ is injective when κ^2 is not an eigenvalue of the Neumann eigenvalue problem (1.113), we conclude the unique solvability of the Galerkin variational problem (2.60), as well as the quasi optimal error estimate, i.e. Cea's lemma,

$$\|\bar{u} - \bar{u}_h\|_{H^{1/2}(\Gamma)} \leq c \inf_{v_h \in S_h^1(\Gamma)} \|\bar{u} - v_h\|_{H^{1/2}(\Gamma)}.$$

Combining this with the approximation property (2.10) for $\sigma = 1/2$, we get

$$\|\bar{u} - \bar{u}_h\|_{H^{1/2}(\Gamma)} \leq c \, h^{s-\frac{1}{2}} \|\bar{u}\|_{H_{pw}^s(\Gamma)} \,,$$

when assuming $\bar{u} \in H_{pw}^s(\Gamma)$ and $s \in [1,2]$. Applying the Aubin–Nitsche trick we also obtain the error estimate

$$\|\bar{u} - \bar{u}_h\|_{H^\sigma(\Gamma)} \leq c \, h^{s-\sigma} \|\bar{u}\|_{H_{pw}^s(\Gamma)} \,, \qquad (2.64)$$

when assuming $\bar{u} \in H_{pw}^s(\Gamma)$ for some $s \in [1,2]$ and $\sigma \in [-1,1/2]$.

Inserting the computed Galerkin solution $\hat{u}_h \in S_h^1(\Gamma)$ into the representation formula (1.95), this gives an approximate representation formula for $x \in \Omega$,

$$\widetilde{u}(x) = \int_\Gamma u_\kappa^*(x,y) g(y) ds_y - \int_\Gamma \gamma_{1,y}^{int} u_\kappa^*(x,y) \hat{u}_h(y) ds_y, \qquad (2.65)$$

describing an approximate solution of the Neumann boundary value problem (1.109). Note that $\widetilde{u}$ satisfies the Helmholtz equation, but the Neumann boundary conditions are satisfied only approximately. For an arbitrary $x \in \Omega$, the error is given by

$$u(x) - \widetilde{u}(x) = \int_\Gamma \gamma_{1,y}^{int} u_\kappa^*(x,y) \Big(\hat{u}_h(y) - \hat{u}(y)\Big) ds_y.$$

Using a duality argument, the error estimate

$$|u(x) - \widetilde{u}(x)| \leq \|u_\kappa^*(x, \cdot)\|_{H^{-\sigma}(\Gamma)} \|\bar{u} - \bar{u}_h\|_{H^\sigma(\Gamma)}$$

for some $\sigma \in \mathbb{R}$ follows. Combining this with the error estimate (2.64) for the minimal value $\sigma = -1$, we obtain the pointwise error estimate

$$|u(x) - \widetilde{u}(x)| \leq c h^{s+1} \|u_\kappa^*(x, \cdot)\|_{H^1(\Gamma)} |\hat{u}|_{H^s_{\mathrm{pw}}(\Gamma)}.$$

Hence, if $\hat{u} \in H^2_{\mathrm{pw}}(\Gamma)$ is sufficiently smooth, we get the optimal order of convergence for $s = 2$,

$$|u(x) - \widetilde{u}(x)| \leq c h^3 \|u_\kappa^*(x, \cdot)\|_{H^1(\Gamma)} |\bar{u}|_{H^2_{\mathrm{pw}}(\Gamma)}. \tag{2.66}$$

Again, the error estimate (2.66) involves the position of the observation point $x \in \Omega$, and, therefore, is not valid in the limiting case $x \in \Gamma$.

When using a piecewise constant approximation $g_h \in S_h^0(\Gamma)$ of the given Neumann datum $g \in H^{-1/2}(\Gamma)$, we can compute a perturbed piecewise linear approximation $\widetilde{u}_h \in S_h^1(\Gamma)$ from the Galerkin equations

$$\left\langle D_\kappa \hat{u}_h, \varphi_i \right\rangle_\Gamma = \left\langle \left(\frac{1}{2} I - K_\kappa' \right) g_h, \varphi_i \right\rangle_\Gamma \quad \text{for } i = 1, \dots, M$$

or from the equivalent linear system

$$D_{\kappa,h} \underline{\widetilde{u}} = \left(\frac{1}{2} M_h^\top - K_{\kappa,h}' \right) \underline{g}$$

with

$$M_h^\top[i, \ell] = \int_{\tau_\ell} \varphi_i(x) ds_x = M_h[\ell, i],$$

$$K_{\kappa,h}'[i, \ell] = \frac{1}{4\pi} \int_\Gamma \varphi_i(x) \int_{\tau_\ell} (1 - \imath \kappa |x - y|) e^{\imath \kappa |x-y|} \frac{(x - y, n(y))}{|x - y|^3} ds_y ds_x .$$

An approximate solution of the interior Neumann boundary value problem is then given for $x \in \Omega$,

$$\widetilde{u}(x) = \int_\Gamma u_\kappa^*(x, y) g_h(y) ds_y - \int_\Gamma \gamma_{1,y}^{\mathrm{int}} u_\kappa^*(x, y) \widetilde{u}_h(y) ds_y.$$

As for the perturbed linear system (2.31) for the Neumann boundary value problem of the Laplace equation, we obtain the error estimate

$$|u(x) - \widetilde{u}(x)| \leq c(x, t, g) h^2, \tag{2.67}$$

when using a L_2 projection to approximate the boundary conditions, when assuming $g \in H^1_{\mathrm{pw}}(\Gamma)$ and $\bar{u} \in H^2_{\mathrm{pw}}(\Gamma)$.

2.5.3 Exterior Dirichlet Problem

The solution of the exterior Dirichlet boundary value problem (cf. (1.114))

$$-\Delta u(x) - \kappa^2 u(x) = 0 \quad \text{for } x \in \Omega^e, \quad \gamma_0^{\text{ext}} u(x) = g(x) \quad \text{for } x \in \Gamma,$$

where, in addition, we have to require the Sommerfeld radiation condition (1.101), is given by the representation formula for $x \in \Omega^e$ (cf. 1.103)

$$u(x) = -\int_\Gamma u_\kappa^*(x,y) t(y) ds_y + \int_\Gamma \gamma_{1,y}^{\text{ext}} u_\kappa^*(x,y) g(y) ds_y.$$

Again we assume that κ^2 is not an eigenvalue of the Dirichlet eigenvalue problem (1.108). The unknown Neumann datum $t = \gamma_1^{\text{ext}} \in H^{-1/2}(\Gamma)$ is then the unique solution of the boundary integral equation (cf. (1.115))

$$(V_\kappa t)(x) = -\frac{1}{2} g(x) + (K_\kappa g)(x) \quad \text{for } x \in \Gamma.$$

To compute an approximate solution of this boundary integral equation, and, therefore, of the exterior Dirichlet problem, we can proceed as in the case of the interior Dirichlet problem. In particular, when using a piecewise linear approximation $g_h \in S_h^1(\Gamma)$, we find a perturbed piecewise constant approximation $\tilde{t}_h \in S_h^0(\Gamma)$ from the Galerkin equations

$$\left\langle V_\kappa \tilde{t}_h, \psi_k \right\rangle_\Gamma = \left\langle \left(-\frac{1}{2} I + K_\kappa \right) g_h, \psi_k \right\rangle_\Gamma \quad \text{for } k = 1, \dots, N.$$

Hence, we obtain the coefficient vector $\underline{\tilde{t}} \in \mathbb{C}^N$ as the unique solution of the linear system

$$V_{\kappa,h} \underline{\tilde{t}} = \left(-\frac{1}{2} M_h + K_{\kappa,h} \right) \underline{g},$$

and an approximate solution of the exterior Dirichlet problem for $x \in \Omega$,

$$\tilde{u}(x) = -\int_\Gamma u_\kappa^*(x,y) \tilde{t}_h(y) ds_y + \int_\Gamma \gamma_{1,y}^{\text{ext}} u_\kappa^*(x,y) g_h(y) ds_y. \tag{2.68}$$

Moreover, as for the interior Dirichlet problem, there holds the optimal error estimate

$$|u(x) - \tilde{u}(x)| \le c(x,t,g) h^3, \tag{2.69}$$

when using a L_2 projection to approximate the boundary conditions, and when assuming $t \in H_{\text{pw}}^1(\Gamma)$ and $g \in H_{\text{pw}}^2(\Gamma)$.

2.5.4 Exterior Neumann Problem

The solution of the exterior Neumann boundary value problem (cf. (1.120))

$$-\Delta u(x) - \kappa^2 u(x) = 0 \quad \text{for } x \in \Omega^e, \quad \gamma_1^{\text{ext}} u(x) = g(x) \quad \text{for } x \in \Gamma,$$

where, in addition, we have to require the Sommerfeld radiation condition (1.101), is given by the representation formula for $x \in \Omega^c$ (cf. (1.95))

$$u(x) = -\int_\Gamma u_\kappa^*(x,y)g(y)ds_y + \int_\Gamma \gamma_{1,y}^{\text{ext}} u_\kappa^*(x,y)\gamma_0^{\text{ext}} u(y)ds_y.$$

Again, we assume that κ^2 is not eigenvalue of the Neumann eigenvalue problem (1.113). The unknown Dirichlet datum $\bar{u} = \gamma_0^{\text{ext}} u \in H^{1/2}(\Gamma)$ is then the unique solution of the boundary integral equation (cf. (1.121))

$$(D_\kappa \bar{u})(x) = -\frac{1}{2}g(x) - (K_\kappa' g)(x) \quad \text{for } x \in \Gamma.$$

To compute an approximate solution of this boundary integral equation, and, therefore, of the exterior Neumann problem, we can proceed as in the case of the interior Neumann problem. In particular, when using a piecewise constant approximation $g_h \in S_h^0(\Gamma)$ of the given Neumann datum g, we find a perturbed piecewise linear approximation $\tilde{u}_h \in S_h^1(\Gamma)$ from the Galerkin equations

$$\left\langle D_\kappa \tilde{u}_h, \varphi_i \right\rangle_\Gamma = \left\langle \left(-\frac{1}{2}I - K_\kappa' \right) g_h, \varphi_i \right\rangle_\Gamma \quad \text{for } i = 1, \dots, M.$$

Hence, we obtain the coefficient vector $\underline{\tilde{u}} \in \mathbb{C}^M$ as the unique solution of the linear system

$$D_{\kappa,h} \underline{\tilde{u}} = \left(-\frac{1}{2}M_h^\top - K_{\kappa,h}' \right)\underline{g},$$

and an approximate solution of the exterior Neumann problem for $x \in \Omega$,

$$\tilde{u}(x) = -\int_\Gamma u_\kappa^*(x,y)g_h(y)ds_y + \int_\Gamma \gamma_{1,y}^{\text{ext}} u_\kappa^*(x,y)\tilde{u}_h(y)ds_y.$$

Moreover, we obtain the error estimate

$$|u(x) - \tilde{u}(x)| \leq c(x,t,g)\,h^2, \tag{2.70}$$

when using the L_2 projection to approximate the boundary conditions, and when assuming $g \in H_{\text{pw}}^1(\Gamma)$ and $\hat{u} \in H_{\text{pw}}^2(\Gamma)$.

2.6 Bibliographic Remarks

The numerical analysis of boundary element methods was introduced independently by J.–C. Nédélec and J. Planchard [79] and by G. C. Hsiao and W. L. Wendland [57]. While the stability and error analysis of the Galerkin boundary element methods follow as in the case of the finite element methods, the stability of the collocation boundary element methods for general Lipschitz boundaries is still open, see [4, 5, 100, 101] for some special cases. The Aubin–Nitsche trick to obtain higher order error estimates for boundary element methods was first given in [58].

Since the implementation of boundary element methods often requires numerical integration techniques, an appropriate numerical analysis is mandatory. Galerkin collocation schemes were first discussed in [54, 68]. Further investigations on the use of numerical integration schemes were made in [45, 97, 98, 102]. In [76], the influence on an additional boundary approximation was considered.

Further references on boundary element methods are, for example, [12, 15, 21, 40, 50, 104, 117] and [99, 105].

3

Approximation of
Boundary Element Matrices

When using boundary element methods for the numerical solution of boundary value problems for three-dimensional second order partial differential equations, one has to deal with two main difficulties. First of all, almost all matrices involved are dense, i.e. all their entries do not vanish in general, leading to an asymptotically quadratic memory requirement for the whole procedure. Thus, classical boundary element realisations are applicable only for a rather moderate number N of boundary elements. Fortunately, all boundary element matrices can be decomposed into a hierarchical system of blocks which can be approximated by the use of low rank matrices. This approximation will be the main content of this chapter.

The second difficulty is the complicated form of the matrix entries to be generated. The Galerkin method, for example, requires the evaluation of double surface integrals for each matrix entry. This can not be done analytically in general. Thus, combined semi-analytical computations will be used to generate the single entries of the matrices. A more detailed description of the corresponding procedures is presented in Appendix C.

3.1 Hierarchical Matrices

The formal definition and description of hierarchical matrices as well as operations involving those matrices can be found in [41, 42]. In this section we give a more intuitive introduction to this topic.

3.1.1 Motivation

Let $K : [0, 1] \times [0, 1] \to \mathbb{R}$ be a given function of two scalar variables and let $A \in \mathbb{R}^{N \times M}$ be a given matrix having the entries

$$a_{k\ell} = K(x_k, y_\ell), \ k = 1, \ldots, N, \ \ell = 1, \ldots, M, \tag{3.1}$$

with $(x_k, y_\ell) \in [0, 1] \times [0, 1]$. It is obvious, that the asymptotic memory requirement for the dense matrix A is $\mathrm{Mem}(A) = \mathcal{O}(N\,M)$, and the asymptotic number of arithmetical operations required for a matrix-vector multiplication is $\mathrm{Op}(A\,s) = \mathcal{O}(N\,M)$ as $N, M \to \infty$. This quadratic amount is too high for modern computers, already for moderate values of N and M. However, if we agree to store just an approximation $\tilde{A}$ of the matrix A and to deal with the product $\tilde{A}\,s$ instead of the exact value $A\,s$, the situation may change. But then it is necessary to control the error, i.e. to guarantee the error bound

$$\|A - \tilde{A}\|_F \le \varepsilon \|A\|_F \,, \tag{3.2}$$

for some prescribed accuracy ε, where $\|A\|_F$ denotes the Frobenius norm of the matrix A,

$$\|A\|_F = \left(\sum_{k=1}^{N} \sum_{\ell=1}^{M} a_{k\ell}^2 \right)^{1/2} . \tag{3.3}$$

Singular value decomposition

The best possible approximation of the matrix $A \in \mathbb{R}^{N \times M}$ is given by its partial singular value decomposition

$$A \approx \tilde{A} = \tilde{A}(r) = \sum_{i=1}^{r} \sigma_i\, u_i\, v_i^\top \,, \tag{3.4}$$

where

$$\sigma_i \in \mathbb{R}_+ \,, \ u_i \in \mathbb{R}^N \,, \ v_i \in \mathbb{R}^M \,, \ i = 1, \ldots, r$$

are the biggest singular values and the corresponding singular vectors of the matrix A. The rank $r = r(\varepsilon)$ is chosen corresponding to the condition

$$\|A - \tilde{A}\|_F^2 \le \sum_{i=r+1}^{\min(N,M)} \sigma_i^2 \le \varepsilon^2 \sum_{i=1}^{\min(N,M)} \sigma_i^2 = \varepsilon^2 \|A\|_F^2 \,. \tag{3.5}$$

Unfortunately, the complete singular value decomposition of the matrix A requires $\mathcal{O}(N^3)$ arithmetical operations when assuming $N \sim M$, and, therefore, it is too expensive for practical computations. However, the singular value decomposition can be perfectly used for the illustration of the main ideas.

As an example, let us consider the following function on $[0, 1] \times [0, 1]$,

$$K(x, y) = \frac{1}{\alpha + x + y} \,, \tag{3.6}$$

where $\alpha > 0$ is some real parameter. For small values of the parameter α the function K gets an artificial "singularity" at the corner $(0, 0)$ of the square $[0, 1] \times [0, 1]$.

The domain $[0,1] \times [0,1]$ is uniformly discretised using the nodes

$$(x_k, y_\ell) = \left((k-1)h_x, (\ell-1)h_y\right), \quad h_x = \frac{1}{N-1}, \quad h_y = \frac{1}{M-1} \quad (3.7)$$

for $k = 1, \ldots, N$ and $\ell = 1, \ldots, M$. In Fig. 3.1, the logarithmic plots of the singular values of the matrix (3.1) (i.e. the quantities $\log_{10} \sigma_i$, $i = 1, \ldots, N$) for $N = M = 32$ (left plot) and $N = M = 1024$ (right plot) are presented for $\alpha = 10^{-4}$. It can easily be seen that only very few singular values are needed to represent the matrix A in its singular value decomposition (3.4). For $N = 1024$, almost all singular values are close to the computer zero, and, therefore, this matrix can be well approximated by a low rank matrix $\tilde{A}$.

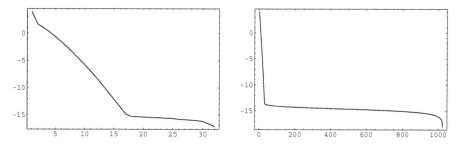

Fig. 3.1. Singular values for $N = 32$ and $N = 1024$

The number of significant singular values slowly increases with the dimension, as it can be seen in Fig. 3.2. Here the first 32 singular values (logarithmic plot) for $N = M = 32, 128, 256$ (left plot, from below) and for $N = M = 256, 512, 1024$ (right plot, from below) are shown. The accuracy of

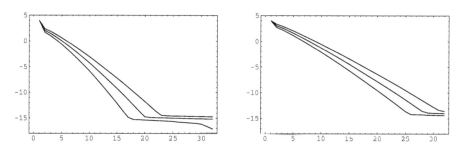

Fig. 3.2. First 32 singular values for $N = 32, 64, 128, 256, 512$, and $N = 1024$

the low rank approximation $\tilde{A}(r)$ of the matrix A is illustrated in Fig. 3.3, where the logarithmic plot of the function

$$\varepsilon(r) = \frac{\|A - \tilde{A}(r)\|_F}{\|A\|_F}$$

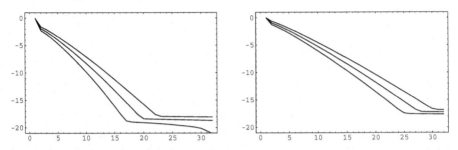

Fig. 3.3. Accuracy of the low lank approximation

for $r = 1, \ldots, 32$ is depicted. The left plot in Fig. 3.3 corresponds again to the dimensions $N = 32, 64$, and $N = 128$ (from below), while the right plot shows the results for $N = 256, 512$, and $N = 1024$ (from below). Thus, the behaviour of the singular values determines the quality of the low rank approximation (3.4). The results shown do not really depend on the parameter α. If α becomes smaller, the results are even better.

The situation changes if the "singularity" of the function K is more serious. As a further example, let us consider the following function on $[0, 1] \times [0, 1]$,

$$K(x, y) = \frac{1}{\alpha + (x - y)^2}, \tag{3.8}$$

where $\alpha > 0$ is again some real parameter. For small values of the parameter α the function K gets an artificial "singularity" along the diagonal $\{(x, x)\}$ of the square $[0, 1] \times [0, 1]$.

In Fig. 3.4 (left plot), the rank $r(\varepsilon)$ for $\varepsilon = 10^{-6}$ and $N = M = 256$ is shown as a function of the parameter α. The horizontal axis corresponds to the values $-\log_2(\alpha)$, while α changes from 2^0 till 2^{-8}. Thus, the rank of the matrix strongly depends on the parameter α. However, if we "separate" the

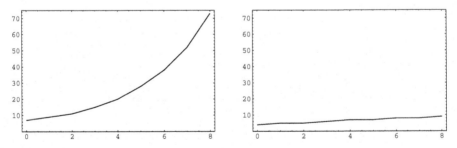

Fig. 3.4. Rang of the matrix $\tilde{A}$

variables x and y, i.e. consider only the quarter $[0, 0.5] \times [0.5, 1]$ of the square $[0, 1] \times [0, 1]$, then the situation is much better. The right plot in Fig. 3.4 shows the same curve for separated x and y, which is more or less constant now.

The logarithmic plots of the singular values of the matrix A for $\alpha = 10^{-1}$ (lower curve) and for $\alpha = 10^{-8}$ (upper curve) are shown in Fig. 3.5. The left plot in this figure corresponds to the whole square $[0,1] \times [0,1]$, while the right plot shows the results for the separated variables x and y, i.e. if we consider only the quarter $[0,0.5] \times [0.5,1]$ of the square $[0,1] \times [0,1]$. Now the main

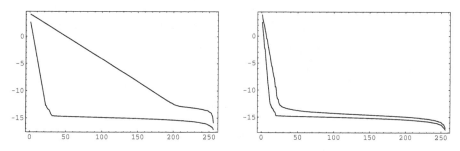

Fig. 3.5. Singular values for $N = 256$

idea of hierarchical methods is quite clear. If we decompose the whole matrix A into four blocks corresponding to the domains

$$[0,0.5] \times [0,0.5], \ [0,0.5] \times [0.5,1], \ [0.5,1] \times [0,0.5], \ [0.5,1] \times [0.5,1],$$

we will be able to approximate two of these four blocks efficiently. The two remaining, main diagonal blocks have the same structure as the original matrix, but only half of the size and their rank will be smaller. In Fig. 3.6, the left diagram corresponds to the whole matrix and its rank $r(\varepsilon) = 73$ is obtained for $\alpha = 2^{-9}$, $\varepsilon = 10^{-6}$ and $N = M = 256$. The 2×2 block matrix together with the ranks of the blocks is shown in the second diagram of Fig. 3.6. The

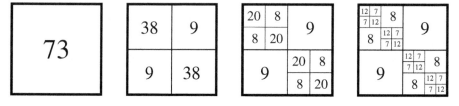

Fig. 3.6. Original matrix and its hierarchical decomposition in blocks

approximation of the separated blocks is now acceptable, and we continue to decompose only the blocks on the main diagonal. The results can be seen in the third and in the fourth diagram of Fig. 3.6. The memory requirements for these four matrices are quite different: the first matrix needs $146N$ words of memory, the second $94N$, the third $74N$, and, finally, we will need $72N$ words of memory for the last block matrix in Fig. 3.6. Thus, a hierarchical decomposition into blocks and their separate approximation using a singular value decomposition leads to a drastic reduction of memory requirements (the latter decomposition requires less than 50%) even for this rather small matrix having a "diagonal singularity". Note that the rank of the blocks on the main diagonal increases almost linear with the dimension: $12 - 20 - 38 - 73$, while the rank of the separated blocks has at most a logarithmic growth: $7 - 8 - 9$.

Thus, a hierarchical approximation of large dense matrices arising from some generating function having diagonal singularity consists of three steps:

- Construction of clusters for variables x and y,
- Finding of possible admissible blocks (i.e. blocks with separated x and y),
- Low rank approximation of admissible blocks.

In the above example, the clusters were simply the sets of points x_k which belong to smaller and smaller intervals. Fortunately, the decomposition problem is only slightly more complicated for general, three-dimensional irregular point sets. Also, the admissible blocks in the above example are very natural. They are just blocks outside of the main diagonal. In the general case, we will need some permutations of rows and columns of the matrix to construct admissible blocks. Finally, the singular value decomposition approximation we have used, is not applicable for more realistic examples. We will need more efficient algorithms, namely the Adaptive Cross Approximation (ACA), to approximate admissible blocks.

Degenerated approximation

In the above example, the approximation of the blocks for separated variables x and y is based on the smoothness of the function K for $x \neq y$. However, if the function K is degenerated, i.e. it is a finite sum of products of functions depending only on x and y,

$$K(x, y) = \sum_{i=1}^{r} p_i(x)q_i(y), \tag{3.9}$$

then the rank of the matrix A defined in (3.1) is equal to r independent of its dimension. Thus for $N, M \gg r$, the matrix A is a low rank matrix. This property is independent of the smoothness of the functions p_i, q_i in (3.9). The low rank representation of the matrix A is then

$$A = \sum_{i=1}^{r} u_i v_i^{\top},$$

with

$$(u_i)_k = p_i(x_k), \quad (v_i)_\ell = q_i(y_\ell)$$

for $k = 1, \ldots, N$ and $\ell = 1, \ldots, M$. Note that this representation is not the singular value decomposition (3.4).

Another possibility to obtain a low rank approximation of a matrix of the form (3.1) is based on the smoothness. If the function K is smooth enough, then we can use its Taylor series (cf. [44]) with respect to the variable x in some fixed point x^*,

$$K(x,y) = \sum_{i=0}^{r} \frac{1}{i!} \frac{\partial^i K(x^*, y)}{\partial x^i} (x - x^*)^i + R_r(x, y),$$

to obtain a degenerated approximation

$$A \approx \tilde{A} = \sum_{i=0}^{r} u_i v_i^\top, \tag{3.10}$$

with

$$(u_i)_k = (x_k - x^*)^i, \quad (v_i)_\ell = \frac{1}{i!} \frac{\partial^i K(x^*, y_\ell)}{\partial x^i}$$

for $k = 1, \ldots, N$ and $\ell = 1, \ldots, M$. Again, (3.10) is not the partial singular value decomposition (3.4) of the matrix A. If the remainder term R_r is uniformly bounded by the original function K satisfying

$$\left| R_r(x, y) \right| \leq \varepsilon \left| K(x, y) \right|$$

for all x and y with some $r = r(\varepsilon)$, then we can guarantee the accuracy of the low rank matrix approximation

$$\|A - \tilde{A}\|_F \leq \varepsilon \|A\|_F \tag{3.11}$$

for all dimensions N and M. The rank $r+1$ of the matrix $\tilde{A}$ is also independent of its dimension. Thus, for $N \approx M$, the matrix $\tilde{A}$ requires only $\mathrm{Mem}(\tilde{A}) = \mathcal{O}(N)$ words of computer memory. However, an efficient construction of the Taylor series for a given function in the three-dimensional case is practically impossible. Thus, it is rather an illustration for the fact that there exist low rank approximations of dense matrices, which are not based on the singular value decomposition.

A further example of a low rank approximation of a given function is the decomposition of the fundamental solution of the Laplace equation

$$u^*(x, y) = \frac{1}{4\pi} \frac{1}{|x - y|} \quad \text{for } x, y \in \mathbb{R}^3$$

(cf. (1.7)) into spherical harmonics in some point x^* with $|x - x^*| < |y - x^*|$

$$u^*(x, y) = \frac{1}{4\pi} \frac{1}{|y - x^*|} \sum_{n=0}^{\infty} \left(\frac{|x - x^*|}{|y - x^*|} \right)^n P_n \big((e_x, e_y) \big),$$

$$e_x = \frac{x - x^*}{|x - x^*|}, \quad e_y = \frac{y - x^*}{|y - x^*|},$$

where the Legendre polynomials are defined for $|u| \leq 1$ as follows,

$$P_0(u) = 1, \quad P_n(u) = \frac{1}{2^n \, n!} \frac{d^n}{du^n} (u^2 - 1)^n, \quad \text{for } n \geq 1. \tag{3.12}$$

Note that the Legendre polynomials allow the following separation of variables

$$P_n\big((e_x, e_y)\big) = \sum_{m=-n}^{n} Y_n^m(e_x) Y_n^{-m}(e_y),$$

where Y_n^m are the spherical harmonics. See [39, 93] for more details.

3.1.2 Hierarchical clustering

Large dense matrices arising from integral equations have no explicit structure in general. As a rule, because of the singularity of the kernel function on the diagonal, i.e. for $x = y$, these matrices are also not of low rank. However, it is possible to find a permutation, so that the matrix with permuted rows and columns contains rather large blocks close to some low-rank matrices with respect to the Frobenius norm (cf. (3.2)).

Cluster tree

To find a suitable permutation, a cluster tree is constructed by a recursive partitioning of some weighted, pairwise disjunct, characteristic points

$$\Big\{ (x_k, g_k), \ k = 1, \dots, N \Big\} \subset \mathbb{R}^3 \times \mathbb{R}_+ \tag{3.13}$$

and

$$\Big\{ (y_\ell, q_\ell), \ \ell = 1, \dots, M \Big\} \subset \mathbb{R}^3 \times \mathbb{R}_+ \tag{3.14}$$

in order to separate the variables x and y. A large distance between two characteristic points results in a large difference of the respective column or row numbers.

When dealing with boundary element matrices, the characteristic points can be

- the mid points x_k^* of the triangle elements τ_k with weights $g_k = \Delta_k = |\tau_k|$, when using piecewise constant basis functions ψ_k (cf. (2.3)),
- the nodes x_k of the grid with weights $g_k = |\text{supp}\,\varphi_k|$, when using piecewise linear continuous basis functions φ_k (cf. (2.9)).

A group of weighted points is called cluster if the points are "close" to each other with respect to the usual distance. A given cluster

$$Cl = \Big\{ (x_k, g_k), \ k = 1, \dots, n \Big\}$$

with $n > 1$ can be separated in two sons using the following algorithm.

Algorithm 3.1

1. Mass of the cluster

$$G = \sum_{k=1}^{n} g_k \in \mathbb{R}_+ \,,$$

2. Centre of the cluster

$$X = \frac{1}{G} \sum_{k=1}^{n} g_k \, x_k \in \mathbb{R}^3 \,,$$

3. Covariance matrix of the cluster

$$C = \sum_{k=1}^{n} g_k \, (x_k - X) \, (x_k - X)^\top \in \mathbb{R}^{3 \times 3} \,,$$

4. Eigenvalues and eigenvectors

$$C \, v_i = \lambda_i \, v_i \,, \quad i = 1, 2, 3 \,, \quad \lambda_1 \geq \lambda_2 \geq \lambda_3 \geq 0 \,,$$

5. Separation
   5.1 initialisation

$$Cl_1 := \varnothing \,, \quad Cl_2 := \varnothing \,,$$

   5.2 for $k = 1, \ldots, n$

$$\text{if } (x_k - X, v_1) \geq 0 \quad \text{then} \quad Cl_1 := Cl_1 \cup (x_k, g_k)$$
$$\text{else} \quad Cl_2 := Cl_2 \cup (x_k, g_k) \,.$$

The eigenvector v_1 of the matrix C corresponds to the largest eigenvalue of this matrix and shows in the direction of the longest expanse of the cluster. The separation plane $\{x \in \mathbb{R}^3 : (x - X, v_1) = 0\}$ goes through the centre X of the cluster and is orthogonal to the eigenvector v_1. Thus, Algorithm 3.1 divides a given arbitrary cluster of weighted points into two more or less equal sons. In Fig. 3.7, the first two separation levels of a given, rather complicated, surface are shown. The separation of a given cluster in two sons defines a permutation of the points in the cluster. The points in the first son will be numbered first and then the ones in the second son. Algorithm 3.1 will be applied recursively to the sons, until they contain less or equal than some prescribed (small and independent of N) number n_{min} of points.

Cluster pairs

Next, cluster pairs which are geometrically well separated are identified. They will be regarded as admissible cluster pairs, as e.g. the clusters in Fig. 3.8. An

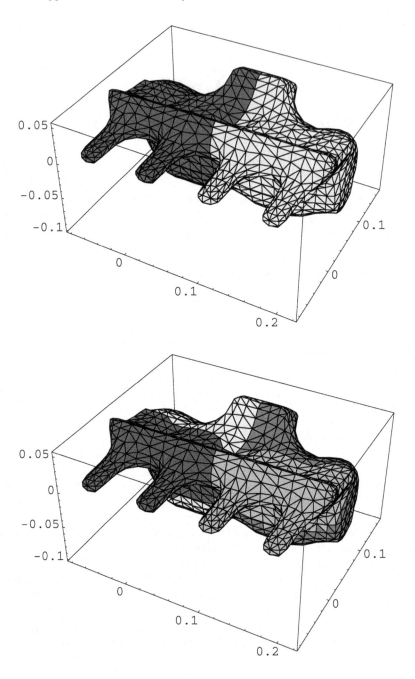

Fig. 3.7. Clusters of the first two levels

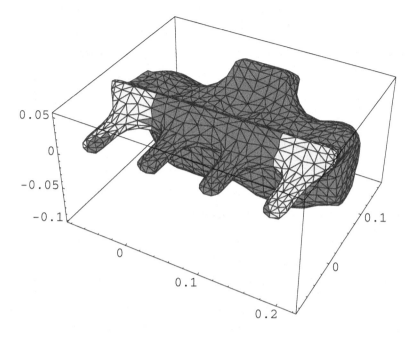

Fig. 3.8. An admissible cluster pair

appropriate admissibility criterion is the following simple geometrical condition. A pair of clusters (Cl_x, Cl_y) with $n_x > n_{min}$ and $m_y > n_{min}$ elements is admissible if

$$\min\left(\operatorname{diam}(Cl_x), \operatorname{diam}(Cl_y)\right) \leq \eta \operatorname{dist}(Cl_x, Cl_y), \tag{3.15}$$

where $0 < \eta < 1$ is a given parameter. Although the criterion (3.15) is quite simple, a rather large computational effort (quadratic with respect to the number of elements in the clusters Cl_x and Cl_y) is required for calculating the exact values

$$\operatorname{diam}(Cl_x) = \max_{k_1, k_2} |x_{k_1} - x_{k_2}|,$$
$$\operatorname{diam}(Cl_y) = \max_{\ell_1, \ell_2} |y_{\ell_1} - y_{\ell_2}|,$$
$$\operatorname{dist}(Cl_x, Cl_y) = \min_{k, \ell} |x_k - y_\ell|.$$

In practice, one can use rougher, more restrictive, but easily computable bounds

$$\operatorname{diam}(Cl_x) \leq 2 \max_k |X - x_k|,$$
$$\operatorname{diam}(Cl_y) \leq 2 \max_\ell |Y - y_\ell|,$$

$$\text{dist}(Cl_x, Cl_y) \geq |X - Y| - \frac{1}{2}\Big(\text{diam}(Cl_x) + \text{diam}(Cl_y)\Big),$$

where X and Y are the already computed centres (cf. Algorithm 3.1) of the clusters Cl_x and Cl_y, for the admissibility condition. If a cluster pair is not admissible, but $n_x > n_{min}$ and $m_y > n_{min}$ are satisfied, then there exist sons of both clusters

$$Cl_x = Cl_{x,1} \cup Cl_{x,2}, \quad Cl_y = Cl_{y,1} \cup Cl_{y,2}.$$

For simplicity, let us assume that the cluster Cl_x is bigger than Cl_y, i.e.

$$\text{diam}(Cl_x) \geq \text{diam}(Cl_y).$$

In this case, we have to check the following two new pairs

$$\Big(Cl_{x,1}, Cl_y\Big), \quad \Big(Cl_{x,2}, Cl_y\Big)$$

for admissibility, and so on. This recursive procedure stops if $n_x \leq n_{min}$ or $m_y \leq n_{min}$ is satisfied. The corresponding block of the matrix is small, and it will be computed exactly. The cluster trees for the variables x and y together with the set of the admissible cluster pairs, as well as the set of the small cluster pairs allow to split the matrix into a collection of blocks of various sizes. The hierarchical block structure of the Galerkin matrix for the single layer potential on the surface from Figs. 3.7–3.8 is shown in Fig. 3.9. The colours of the blocks indicate the "quality" of the approximation. The light grey colour corresponds to well approximated blocks, while the dark grey colour indicates a less good approximation. The small blocks are computed exactly and they are depicted in black. Thus, the remaining main problem is how to approximate the blocks which correspond to the admissible cluster pairs, without using the singular value decomposition. The corresponding procedures will be described in the following section.

3.2 Block Approximation Methods

3.2.1 Analytic Form of Adaptive Cross Approximation

Let $X, Y \subset \mathbb{R}^3$ be two non–empty domains, and let $K : X \times Y \to \mathbb{R}$ be a given function. The following abstract Adaptive Cross Approximation algorithm constructs a degenerated approximation of the function K using nodal interpolation in some points

$$\{x_1, x_2, \dots\} \subset X, \quad \{y_1, y_2, \dots\} \subset Y,$$

which will be determined during the realisation of the algorithm on an adaptive way.

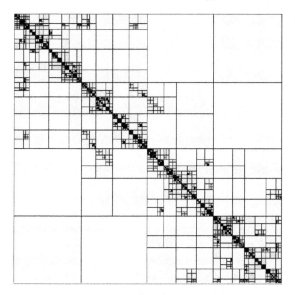

Fig. 3.9. Matrix decomposition

Algorithm 3.2

1. Initialisation

$$R_0(x,y) = K(x,y), \quad S_0 = 0.$$

2. For $i = 0, 1, 2, \ldots$ compute
   2.1. pivot element

$$(x_{i+1}, y_{i+1}) = \mathrm{ArgMax}|R_i(x,y)|,$$

   2.2. normalising constant

$$\gamma_{i+1} = (R_i(x_{i+1}, y_{i+1}))^{-1},$$

   2.3. new functions

$$u_{i+1}(x) = \gamma_{i+1} R_i(x, y_{i+1}), \quad v_{i+1}(y) = R_i(x_{i+1}, y),$$

   2.4. new residual

$$R_{i+1}(x,y) = R_i(x,y) - u_{i+1}(x)v_{i+1}(y),$$

   2.5. new approximation

$$S_{i+1}(x,y) = S_i(x,y) + u_{i+1}(x)v_{i+1}(y).$$

The stopping criterion for the above algorithm can be realised in `Step 2.1.` corresponding to the condition

$$|R_r(x,y)| \leq \varepsilon |K(x,y)| \quad \text{for } (x,y) \in X \times Y. \tag{3.16}$$

Algorithm 3.2 produces a sequence of approximations $\{S_i\}$ and an associated sequence of residuals $\{R_i\}$ possessing the approximation property

$$K(x,y) = R_i(x,y) + S_i(x,y), \quad (x,y) \in X \times Y, \ i = 0,1,\ldots,r \tag{3.17}$$

and the interpolation property

$$R_m(x_i,y) = 0, \ y \in Y, \ R_m(x,y_i) = 0, \ x \in X$$

for $i = 1,2,\ldots,m$ and for $m = 1,2,\ldots,r$. Furthermore, if the function $K(x,y)$ is harmonic for $x \neq y$ then its approximations $S_i(x,y)$ are also harmonic for all i. The residuals $\{R_i\}$ accumulate zeros, and, therefore, the sequence of the functions $\{S_i\}$ interpolates the given function $K(x,y)$ in more and more points corresponding to (3.17). If we are interested in computing an approximation $\tilde{A}$ for a matrix $A \in \mathbb{R}^{N \times M}$ having the entries

$$a_{k\ell} = K(\tilde{x}_k, \tilde{y}_\ell), \quad k = 1,\ldots,N, \ \ell = 1,\ldots,M \tag{3.18}$$

for some points $(\tilde{x}_k, \tilde{y}_\ell) \in X \times Y$, then the approximation $S_r(x,y)$ of the function $K(x,y)$ can be used to obtain

$$\tilde{a}_{k\ell} = S_r(\tilde{x}_k, \tilde{y}_\ell) \approx a_{k\ell}.$$

Due to the stopping criterion (3.16), and due to the approximation property (3.17), the following estimate obviously holds

$$\|A - \tilde{A}\|_F \leq \varepsilon \|A\|_F,$$

where $\|\cdot\|_F$ denotes the Frobenius norm of a matrix (cf. (3.3)).

Remark 3.3. `Step 2.1.` of Algorithm 3.2 should be discussed in more details. It can be very difficult, if not impossible, to solve the maximum problem formulated there. There are two possibilities to proceed. First, we can look for the maximum only in a finite set of given points $(\tilde{x}_k, \tilde{y}_\ell)$. In this case only the original entries of the matrix A will be used for its approximation. The algorithm will coincide with the algebraic fully pivoted ACA algorithm as described in Subsection 3.2.2. The other possibility is to choose some artificial points and to look for the maximum there. These points can be the zeros of the three-dimensional Chebyshev polynomials in corresponding bounding boxes for the sets X,Y. In this case the ACA approximation will be similar to the best possible polynomial interpolation.

Remark 3.4. The stopping criterion (3.16) can be applied only if the function $K(x, y)$ is smooth on (X, Y). If this is not the case, but if the function K is asymptotically smooth (cf. (3.19)), then we have to decompose the domains X and Y into two systems of clusters and to approximate the function on each admissible cluster pair (Cl_x, Cl_y) separately using Algorithm 3.2. This decomposition implies the corresponding decomposition of the matrix in a hierarchical system of blocks.

Using the theory of polynomial multidimensional interpolation, the following result was proven in [7].

Theorem 3.5. *Let the function $K(x, y)$ be asymptotically smooth with respect to y, i.e. $K(x, \cdot) \in C^\infty(\mathbb{R}^3 \backslash \{x\})$ for all $x \in \mathbb{R}^3$, satisfying*

$$|\partial_y^\alpha K(x, y)| \leq c_p |x - y|^{g-p}, \quad p = |\alpha| \tag{3.19}$$

for all multiindices $\alpha \in \mathbb{N}_0^3$ with a constant $g < 0$. Moreover, the matrix $A \in \mathbb{R}^{N \times M}$ with entries (3.18) is decomposed into blocks corresponding to the admissibility condition

$$\text{diam}(Cl_y) \leq \eta \, \text{dist}(Cl_x, Cl_y), \quad \eta < 1.$$

Then the matrix A with $M \sim N$ can be approximated up to an arbitrary given accuracy $\varepsilon > 0$ using a system of given points $(\tilde{x}_k, \tilde{y}_\ell)$,

$$\|A - \tilde{A}\|_F \leq \varepsilon \|A\|_F,$$

and

$$\text{Op}(\tilde{A}) = \text{Op}(\tilde{A}\, s) = \text{Mem}(\tilde{A}) = \mathcal{O}(N^{1+\delta} \varepsilon^{-\delta}) \quad \text{for all } \delta > 0.$$

In Theorem 3.5, $\text{Op}(\tilde{A})$ denotes the number of arithmetical operations required for the generation of the matrix $\tilde{A}$, $\text{Op}(\tilde{A}\, s)$ is the asymptotic number of arithmetical operations required for the matrix-vector multiplication with the matrix $\tilde{A}$, and $\text{Mem}(\tilde{A}) = r(M + N)$ is the asymptotic memory requirement for the matrix $\tilde{A}$ as $N \to \infty$. Thus, Theorem 3.5 states that an almost linear complexity is achieved for these important quantities.

Let us now consider matrices arising from collocation boundary element methods (cf. Chapter 2). Let Γ be a Lipschitz boundary, let

$$\left\{ \varphi_\ell : \Gamma \to \mathbb{R}, \ \ell = 1, \ldots, M \right\}$$

be a given system of basis functions, and let

$$\left\{ x_k^* \in \Gamma, \ k = 1, \ldots, N \right\}$$

be a set of collocation points. If, for example, Γ is the union of plane triangles (cf. (2.1)),

$$\Gamma = \bigcup_{\ell=1}^{N} \bar{\tau}_\ell \,,$$

then the most simple collocation method with piecewise constant basis functions

$$\psi_\ell(y) = \begin{cases} 1 \,, y \in \tau_\ell \,, \\ 0 \,, y \notin \tau_\ell \end{cases}$$

can be used. In this case, the collocation points x_k^* are the midpoints of the triangles τ_k. The corresponding collocation matrix $A \in \mathbb{R}^{N \times M}$ with

$$a_{k\ell} = \int_\Gamma K(x_k^*, y) \, \psi_\ell(y) \, ds_y \,, \quad k = 1, \ldots, N \,, \quad \ell = 1, \ldots, M \qquad (3.20)$$

for some kernel function K can be approximated using the following fully pivoted ACA algorithm.

Algorithm 3.6

1. Initialisation

$$R_0(x, y) = K(x, y), \quad S_0 = 0 \,.$$

2. For $i = 0, 1, 2, \ldots$ compute
   2.1. pivot element

$$(k_{i+1}, \ell_{i+1}) = \text{ArgMax} \left| \int_\Gamma R_i(x_k^*, y) \, \psi_\ell(y) \, ds_y \right| \,,$$

   2.2. normalising constant

$$\gamma_{i+1} = \left(\int_\Gamma R_i(x_{k_{i+1}}^*, y) \, \psi_{\ell_{i+1}}(y) \, ds_y \right)^{-1} \,,$$

   2.3. new functions

$$u_{i+1}(x) = \gamma_{i+1} \int_\Gamma R_i(x, y) \, \psi_{\ell_{i+1}}(y) \, ds_y \,, \quad v_{i+1}(y) = R_i(x_{k_{i+1}}^*, y) \,,$$

   2.4. new residual

$$R_{i+1}(x, y) = R_i(x, y) - u_{i+1}(x) v_{i+1}(y) \,,$$

   2.5. new approximation

$$S_{i+1}(x, y) = S_i(x, y) + u_{i+1}(x) v_{i+1}(y) \,.$$

Note that the approximation property (3.17) remains valid for Algorithm 3.6, while the interpolation property (3.18) remains valid only with respect to the variable x,

$$R_m(x^*_{k_i}, y) = 0, \ y \in \Gamma, \ i = 1, 2, \ldots, m, \ m = 1, 2, \ldots, r. \quad (3.21)$$

The interpolation property with respect to y changes to the orthogonality

$$\int_{\Gamma} R_m(x, y) \, \psi_{\ell_i}(y) ds_y = 0, \ x \in \Gamma, \ i = 1, \ldots, m, \ m = 1, 2, \ldots r. \quad (3.22)$$

For the analysis of Algorithm 3.6, it is useful to introduce the following functions

$$U_\ell(x) = \int_{\Gamma} K(x, y) \, \varphi_\ell(y) ds_y, \ \ell = 1, \ldots, M,$$

having the property $a_{k\ell} = U_\ell(x^*_k)$, (cf. (3.20)). Using the properties (3.21) and (3.22), we can conclude that the functions

$$\tilde{U}_\ell(x) = \int_{\Gamma} S_r(x, y) \, \psi_\ell(y) ds_y \quad (3.23)$$

coincide with U_ℓ for $\ell \in \{\ell_1, \ldots, \ell_r\}$. Moreover, all other functions U_ℓ, i.e. for $\ell \notin \{\ell_1, \ldots, \ell_r\}$, are interpolated by the functions (3.23) at points x^*_k, $k \in \{k_1, \ldots, k_r\}$. The approximation $\tilde{A}$ of the collocation matrix A is then given by the entries

$$a_{k\ell} \approx \tilde{a}_{k\ell} = \int_{\Gamma} S_r(x^*_k, y) \, \psi_\ell(y) ds_y.$$

In [9], the interpolation theory of multidimensional Chebyshev polynomials was used in order to prove Theorem 3.5 for collocation matrices.

A straightforward modification of Algorithm 3.6 leads to an algorithm for the Galerkin matrix $A \in \mathbb{R}^{N \times M}$ with elements

$$a_{k\ell} = \int_{\Gamma} \int_{\Gamma} K(x, y) \, \varphi_\ell(y) \, \psi_k(x) \, ds_y \, ds_x \quad (3.24)$$

for $k = 1, \ldots, N$ and $\ell = 1, \ldots, M$. In (3.24), a system of basis functions

$$\left\{ \varphi_\ell : \Gamma \to \mathbb{R}, \ \ell = 1, \ldots, M \right\},$$

may differ from the system of test functions

$$\left\{ \psi_k : \Gamma \to \mathbb{R}, \ k = 1, \ldots, N \right\}.$$

The Galerkin matrix A can be approximated using the following ACA algorithm.

Algorithm 3.7

1. Initialisation

$$R_0(x,y) = K(x,y), \quad S_0 = 0.$$

2. For $i = 0, 1, 2, \ldots$ compute
   2.1. pivot element

$$(k_{i+1}, \ell_{i+1}) = \text{ArgMax} \left| \int_\Gamma \int_\Gamma R_i(x,y)\, \varphi_\ell(y)\, \psi_k(x)\, ds_y\, ds_x \right|,$$

   2.2. normalising constant

$$\gamma_{i+1} = \left(\int_\Gamma \int_\Gamma R_i(x,y)\, \varphi_{\ell_{i+1}}(y)\, \psi_{k_{i+1}}(x)\, ds_y\, ds_x \right)^{-1},$$

   2.3. new functions

$$u_{i+1}(x) = \gamma_{i+1} \int_\Gamma R_i(x,y)\, \varphi_{\ell_{i+1}}(y)\, ds_y,$$

$$v_{i+1}(y) = \int_\Gamma R_i(x,y)\, \psi_{k_{i+1}}(x)\, ds_x,$$

   2.4. new residual

$$R_{i+1}(x,y) = R_i(x,y) - u_{i+1}(x)v_{i+1}(y),$$

   2.5. new approximation

$$S_{i+1}(x,y) = S_i(x,y) + u_{i+1}(x)v_{i+1}(y).$$

The approximation property (3.17) remains valid for Algorithm 3.7, which, instead of the interpolation property (3.18), possesses the following orthogonalities for $i = 1, \ldots, m$, $m = 1, 2, \ldots r$:

$$\int_\Gamma R_m(x,y)\, \varphi_{\ell_i}(y)\, ds_y = 0, \quad x \in \Gamma,$$

$$\int_\Gamma R_m(x,y)\, \psi_{k_i}(x)\, ds_x = 0, \quad y \in \Gamma.$$

It is practically impossible to compute the elements of the Galerkin matrices corresponding to (3.24) analytically in a general setting. Even in the simplest situation, e.g. for plane triangles τ_k and by using piecewise constant basis functions, some numerical integration is involved (cf. Chapter 4). If both integrals in (3.24) are computed numerically, i.e.

$$a_{k\ell} \approx \bar{a}_{k\ell} = \sum_{k_x} \sum_{k_y} \omega_{k,k_x} \omega_{\ell,k_y} K(x_{k,k_x}, y_{\ell,k_y}) \varphi_\ell(y_{\ell,k_y}) \psi_k(x_{k,k_x}), \quad (3.25)$$

where $\omega_{k,k_x}, \omega_{\ell,k_y}$ are the weights of the quadrature rule (including Jacobians) and x_{k,k_x}, y_{ℓ,k_y} are the corresponding integration points, then not the exact Galerkin matrix A, but its quadrature approximation $\bar{A}$ will be further approximated by ACA. The matrix $\bar{A}$ is, corresponding to (3.25), a finite sum of matrices as defined in (3.18), multiplied by degenerated diagonal matrices. Therefore, Theorem 3.5 remains valid for Galerkin matrices, if the relative accuracy of the numerical integration (3.25) is higher than the approximation of the matrix $\bar{A}$ by ACA.

3.2.2 Algebraic Form of Adaptive Cross Approximation

On the matrix level, all three algorithms formulated in Section 3.2.1 can be written in the fully pivoted ACA form.

Fully pivoted ACA algorithm

Let $A \in \mathbb{R}^{N \times M}$ be a given matrix.

Algorithm 3.8

1. Initialisation

$$R_0 = A, \quad S_0 = 0.$$

2. For $i = 0, 1, 2, \ldots$ compute
   2.1. pivot element

$$(k_{i+1}, \ell_{i+1}) = \text{ArgMax} \,|(R_i)_{k\ell}| \,,$$

   2.2. normalising constant

$$\gamma_{i+1} = \left((R_i)_{k_{i+1}\ell_{i+1}} \right)^{-1},$$

   2.3. new vectors

$$u_{i+1} = \gamma_{i+1} R_i e_{\ell_{i+1}}, \quad v_{i+1} = R_i^\top e_{k_{i+1}},$$

   2.4. new residual

$$R_{i+1} = R_i - u_{i+1} v_{i+1}^\top,$$

2.5. new approximation

$$S_{i+1} = S_i + u_{i+1}v_{i+1}^\top .$$

In Algorithm 3.8, e_j denotes the jth column of the identity matrix I. The whole residual matrix R_i is inspected in **Step 2.1** of Algorithm 3.8 for its maximal entry. Thus, its Frobenius norm can easily be computed in this step, and the appropriate stopping criterion for a given $\varepsilon > 0$ at step r would be

$$\|R_r\|_F \leq \varepsilon \|A\|_F .$$

Note that the crosses built from the column-row pairs with the indices k_i, ℓ_i for $i = 1, \ldots, r$ will be computed exactly

$$\left(S_m\right)_{k_i,l} = a_{k_i,l} , \quad l = 1, ..., M ,$$

$$\left(S_m\right)_{k,l_i} = a_{k,l_i} , \quad k = 1, ..., N$$

for $i = 1, \ldots, m$, $m = 1, \ldots, r$, while all other elements are approximated. The number of operations required to generate the approximation $\tilde{A} = S_r$ is $\mathcal{O}(r^2 N M)$. The memory requirement for Algorithm 3.8 is $\mathcal{O}(N M)$, since the whole matrix A is assumed to be given at the beginning. Thus, Algorithm 3.8 is much faster than a singular value decomposition, but still rather expensive for large matrices.

The efficiency of Algorithm 3.8 will now be illustrated using the following examples. First, we consider the matrix A, generated as in (3.1), for the function

$$K(x,y) = \frac{1}{\alpha + x + y} , \quad \alpha = 10^{-2}$$

(cf. (3.6)) on the uniform grid (3.7) for $N = M = 32$. In Table 3.1 the results of the application of Algorithm 3.8 are presented. The plot of the initial residual R_0, i.e. of the function K on the grid, is shown in Fig. 3.10, while the next three Figs. 3.11–3.13 show the residual R_k for $k = 3, 6$, and $k = 9$. The three-dimensional plots of the residuals $R_k(x,y)$ are presented on the left, while the corresponding matrices are depicted on the right. Note that the exactly computed crosses are shown in black, while the remaining grey scales are adapted to the actual values of the residuals, and, therefore, are different for all pictures. This example illustrates the behaviour of the fully pivoted ACA algorithm very clearly. The generation function (3.6) has only a weak "singularity" at the corner of the computational domain. This singularity is not important for the low rank approximation and is completely removed after the first two iterations. Further iterations quickly reduce the relative error of the approximation.

However, if the "singularity" is on the diagonal as for the function

$$K(x,y) = \frac{1}{\alpha + (x - y)^2} , \quad \alpha = 10^{-2} \tag{3.26}$$

Table 3.1. Fully Pivoted ACA algorithm for the function (3.6)

Step	Pivot row	Pivot column	Pivot value	Relative error
1	1	1	$1.00 \cdot 10^{+2}$	$3.43 \cdot 10^{-1}$
2	2	2	$7.91 \cdot 10^{+0}$	$1.62 \cdot 10^{-1}$
3	6	6	$1.10 \cdot 10^{+0}$	$3.66 \cdot 10^{-2}$
4	28	28	$2.25 \cdot 10^{-1}$	$2.26 \cdot 10^{-3}$
5	3	3	$6.10 \cdot 10^{-2}$	$8.40 \cdot 10^{-4}$
6	13	13	$9.87 \cdot 10^{-3}$	$2.28 \cdot 10^{-5}$
7	4	4	$3.91 \cdot 10^{-4}$	$8.85 \cdot 10^{-6}$
8	20	20	$1.02 \cdot 10^{-4}$	$2.69 \cdot 10^{-7}$
9	9	9	$6.32 \cdot 10^{-6}$	$3.30 \cdot 10^{-8}$
10	32	32	$1.97 \cdot 10^{-6}$	$1.13 \cdot 10^{-9}$

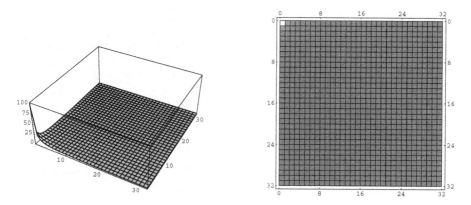

Fig. 3.10. Initial residual for the function (3.6)

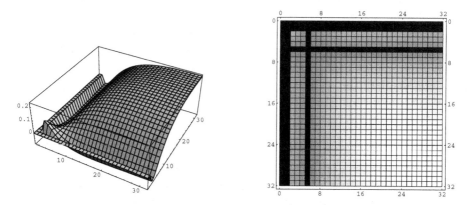

Fig. 3.11. Residual R_3 for the function (3.6)

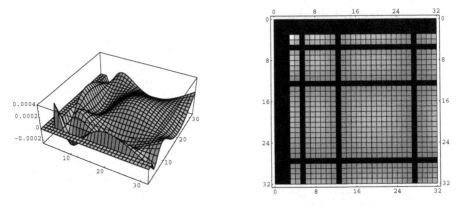

Fig. 3.12. Residual R_6 for the function (3.6)

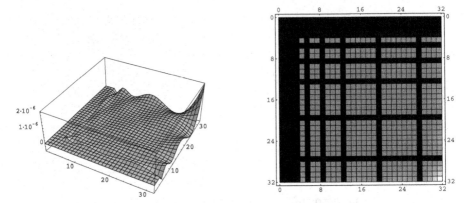

Fig. 3.13. Residual R_9 for the function (3.6)

(cf. (3.8)), then, as we have already seen, the situation changes. The results of the computations are presented in Table 3.2 and in the four Figs. 3.14–3.17.

The convergence of the ACA algorithm is now slow, the crosses chosen can not approximate the main diagonal, because they are too small there. This illustrates once again the necessity of the hierarchical clustering.

The next function we consider,

$$K(x, y) = \sin^6 \left(\pi(2x + y) \right), \tag{3.27}$$

is degenerated corresponding to the definition (3.9), having the exact low rank $r = 7$. This function is obviously infinitely smooth. But it is oscillating and the convergence of Algorithm 3.8 is slow again. The convergence is, of course, better than for the singular function (3.26), but not really sufficient. However, after exactly 7 iterations, the error is equal to computer zero. It means that Algorithm 3.8 has correctly detected the low rank of the function (3.27). The numerical results can be seen in Table 3.3 and in the four Figs. 3.18–3.21,

Table 3.2. Fully Pivoted ACA algorithm for the function (3.8)

Step	Pivot row	Pivot column	Pivot value	Relative error
1	32	32	$1.00 \cdot 10^{+2}$	$9.35 \cdot 10^{-1}$
2	1	1	$9.99 \cdot 10^{+1}$	$8.68 \cdot 10^{-1}$
3	17	17	$9.69 \cdot 10^{+1}$	$7.02 \cdot 10^{-1}$
4	9	9	$9.63 \cdot 10^{+1}$	$5.65 \cdot 10^{-1}$
5	25	25	$9.53 \cdot 10^{+1}$	$4.00 \cdot 10^{-1}$
6	5	5	$7.32 \cdot 10^{+1}$	$3.45 \cdot 10^{-1}$
7	21	21	$7.32 \cdot 10^{+1}$	$2.78 \cdot 10^{-1}$
8	13	13	$7.31 \cdot 10^{+1}$	$1.92 \cdot 10^{-1}$
9	29	29	$6.24 \cdot 10^{+1}$	$1.12 \cdot 10^{-1}$
10	3	3	$2.53 \cdot 10^{+1}$	$1.02 \cdot 10^{-1}$

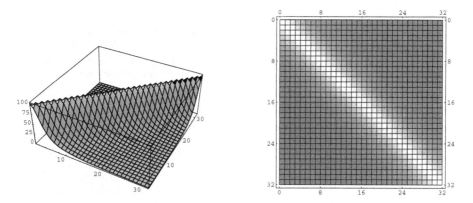

Fig. 3.14. Initial residual for the function (3.8)

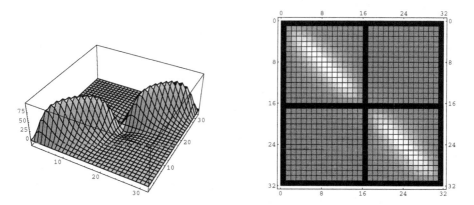

Fig. 3.15. Residual R_3 for the function (3.8)

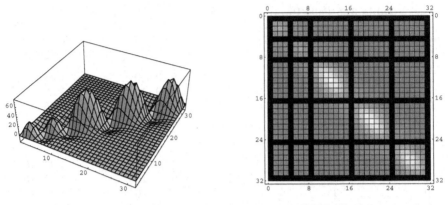

Fig. 3.16. Residual R_6 for the function (3.8)

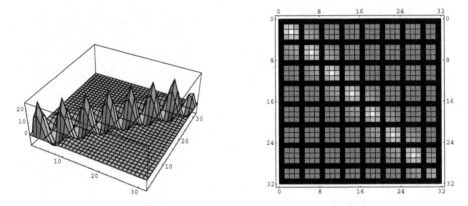

Fig. 3.17. Residual R_9 for the function (3.8)

where the initial residual and three sequential residuals (for $k = 2, 4$, and $k = 6$) are shown.

Table 3.3. Fully Pivoted ACA algorithm for the function (3.27)

Step	Pivot row	Pivot column	Pivot value	Relative error
1	32	19	$1.00 \cdot 10^{+0}$	$8.44 \cdot 10^{-1}$
2	24	3	$1.00 \cdot 10^{+0}$	$6.51 \cdot 10^{-1}$
3	28	27	$9.69 \cdot 10^{-1}$	$4.81 \cdot 10^{-1}$
4	20	11	$9.68 \cdot 10^{-1}$	$2.16 \cdot 10^{-1}$
5	30	23	$3.05 \cdot 10^{-1}$	$1.30 \cdot 10^{-1}$
6	22	7	$2.88 \cdot 10^{-1}$	$6.19 \cdot 10^{-2}$
7	26	31	$1.17 \cdot 10^{-1}$	$2.91 \cdot 10^{-15}$

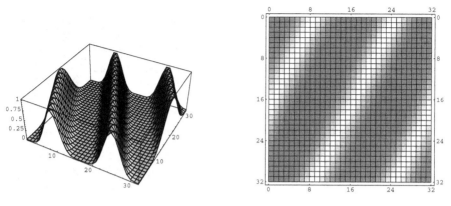

Fig. 3.18. Initial residual for the function (3.27)

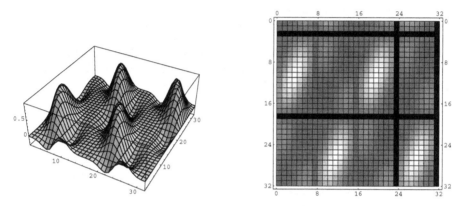

Fig. 3.19. Residual R_2 for the function (3.27)

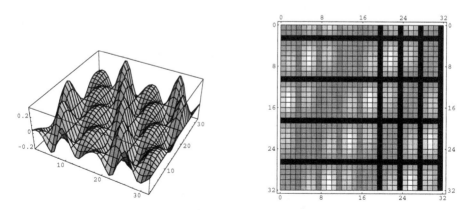

Fig. 3.20. Residual R_4 for the function (3.27)

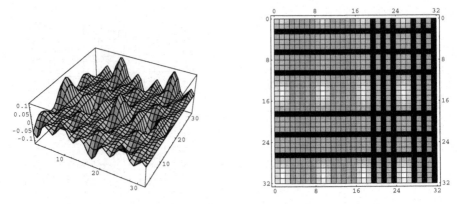

Fig. 3.21. Residual R_6 for the function (3.27)

Partially pivoted ACA algorithm

If the matrix A has not yet been generated, but if there is a possibility of generating its entries $a_{k\ell}$ individually, then the following partially pivoted ACA algorithm can be used for the approximation:

Algorithm 3.9

1. Initialisation

$$S_0 := 0, \ \mathcal{I} := \varnothing, \ \mathcal{J} := \varnothing, \ c := 0 \in \mathbb{R}^N, \ r := 0 \in \mathbb{R}^M.$$

2. Restart with the next not yet generated row
   If

$$\#\mathcal{I} = N \ \text{or} \ \#\mathcal{J} = M$$

   then STOP else

$$k_{i+1} := \min\{k : k \notin \mathcal{I}\}, \ CrossType := Row,$$

3. Generate cross
   3.1 Type of the cross
      If

$$CrossType \ == \ Row$$

      then
   3.1.1 Generate row, Update control vector

$$a := A^\top e_{k_{i+1}}, \ \mathcal{I} := \mathcal{I} \cup \{k_{i+1}\}, \ r := r + |a|,$$

   3.1.2 Test
      If $|a| = 0$ then GOTO 2.   (zero row)

3.1.3 Row of the residual and the pivot column

$$r_v := a - \sum_{m=1}^{i} (u_m)_{k_{i+1}} v_m \,,$$
$$\ell_{i+1} := \mathrm{ArgMax}\,|(r_v)_\ell| \,,$$

3.1.4 Test

If $|r_v| = 0$ then GOTO 4.   (linear depending row)

3.1.5 Normalising constant

$$\gamma_{i+1} := (r_v)_{\ell_{i+1}}^{-1} \,,$$

3.1.6 Generate column, Update control vector

$$b := A e_{\ell_{i+1}} \,, \ \ \mathcal{J} := \mathcal{J} \cup \{\ell_{i+1}\} \,, \ \ c := c + |b| \,,$$

3.1.7 Column of the residual and the pivot row

$$r_u := b - \sum_{m=1}^{i} (v_m)_{\ell_{i+1}} u_m \,,$$
$$k_{i+2} := \mathrm{ArgMax}\,|(r_u)_k| \,,$$

3.1.8 New vectors

$$u_{i+1} := r_u \,, \ \ v_{i+1} := \gamma_{i+1} r_v \,,$$

  else

3.2.1 Generate column, Update control vector

$$b := A e_{\ell_{i+1}} \,, \ \ \mathcal{J} := \mathcal{J} \cup \{\ell_{i+1}\} \,, \ \ c := c + |b| \,,$$

3.2.2 Test

If $|b| = 0$ then GOTO 2.   (zero column)

3.2.3 Column of the residual and the pivot row

$$r_u := b - \sum_{m=1}^{i} (v_m)_{\ell_{i+1}} u_m \,,$$
$$k_{i+1} := \mathrm{ArgMax}\,|(r_u)_k| \,,$$

3.2.4 Test

If $|r_u| = 0$ then GOTO 4.   (linear depending column)

3.2.5 Normalising constant

$$\gamma_{i+1} := (r_u)_{k_{i+1}}^{-1} \,,$$

3.2.6 Generate row, Update control vector

$$a := A^\top e_{k_{i+1}}, \quad \mathcal{I} := \mathcal{I} \cup \{k_{i+1}\}, \quad r := r + |a|,$$

3.2.7 Row of the residual and the pivot column

$$r_v := a - \sum_{m=1}^{i} (u_m)_{k_{i+1}} v_m,$$

$$\ell_{i+2} := \operatorname{ArgMax} |(r_v)_\ell|,$$

3.2.8 New vectors

$$u_{i+1} := \gamma_{i+1} r_u, \quad v_{i+1} := r_v,$$

3.3 New approximation

$$S_{i+1} := S_i + u_{i+1} v_{i+1}^\top,$$

3.4 Frobenius norm of the approximation

$$\|S_{i+1}\|_F^2 = \|S_i\|_F^2 + 2 \sum_{m=1}^{i} u_{i+1}^\top u_m\, v_m^\top v_{i+1} + \|u_{i+1}\|_F^2 \|v_{i+1}\|_F^2.$$

3.5 Test

 If $\|u_{i+1}\|_F \|v_{i+1}\|_F < \varepsilon \|S_{i+1}\|_F$
 then GOTO 4
 else $i := i+1$, GOTO 3

4. Check control vectors
 If $\exists i^* \notin \mathcal{I}$ and $c_{i^*} = 0$ then

$$i := i+1, \quad k_{i+1} = i^*, \quad CrossType = Row, \quad \text{GOTO 3}$$

 or
 If $\exists j^* \notin \mathcal{J}$ and $r_{j^*} = 0$ then

$$i := i+1, \quad \ell_{i+1} = j^*, \quad CrossType = Column, \quad \text{GOTO 3}$$

 else STOP

Algorithm 3.9 starts to compute an approximation for the matrix A by generating its first row. Then, the first column will be chosen automatically. If a cross is successfully computed, the next row index is prescribed, and the procedure repeats. If a zero row is generated, then it is not possible to find the column, and the algorithm restarts in **Step 2**. Since the matrix A will not be generated completely, we can use the norm of its approximant S_i to define a stopping criterion. This norm can be computed recursively as it is described in **Step 3.4**. However, since the whole matrix A will not be generated while using the partially pivoted ACA algorithm, it is necessary to check the control vectors c and r before stopping the algorithm. Note that these vectors contain the sums of absolute values of all elements generated. If, for example, there is some index $i^* \notin \mathcal{I}$ with $c_{i*} = 0$ then the row i^* has not yet contributed to the matrix. It can happen that this row contains relevant information, and, therefore, we have to restart the algorithm in **Step 3**. The same argumentation is valid for the columns. The only difference is, that the crosses after this restart will be generated on the different way: first prescribed column and then automatically chosen row. Thus, Algorithm 3.9 can be used not only for dense matrices but also for reducible, and even for sparse matrices containing only few non-zero entries.

Algorithm 3.9 requires only $\mathcal{O}(r^2(N + M))$ arithmetical operations and its memory requirement is $\mathcal{O}(r(N + M))$. Thus, this algorithm is perfect for large matrices. All approximations of boundary element matrices of the next chapter will be generated with the help of Algorithm 3.9.

3.3 Bibliographic Remarks

The history of asymptotically optimal approximations of dense matrices is now about 20 years old. It starts with the paper [93] by V. Rokhlin. The boundary value problem for a partial differential equation was transformed to a Fredholm boundary integral equation of the second kind. The Nyström method was used for the discretisation, leading to a dense large system of linear equations. This system was solved iteratively using the generalised conjugate residual algorithm. This algorithm requires matrix-vector multiplications, which were realised in a "fast" manner, leading to optimal costs of order $\mathcal{O}(N)$, or $\mathcal{O}(N \log(N))$ for the whole procedure.

Then, the method, which was called Fast Multipole Method, was developed in the papers [20, 37, 38] for large-scale particle simulations in problems of plasma physics, fluid dynamics, molecular dynamics, and celestial mechanics. The method was significantly improved in [39]. Later, the Fast Multipole Method was successfully applied to a variety of problems. In [39, 84], for example, we can find its application to the Laplace equation. In [83], the authors apply the Fast Multipole Method to the system of linear elastostatics discretised by the use of a Galerkin Boundary Element Method. Many papers on the Fast Multipole Method are devoted to the Helmholtz equation, see, for example, [2, 3], where the problem of acoustic scattering was considered, and [94].

The next method, introduced in [44], is called Panel clustering. This method was also applied to the potential problems in [43, 44, 46] and for the system of linear elastostatics in [51].

A further possibility to solve boundary integral equations on an asymptotically optimal way is based on the use of Wavelets. This research starts with the papers [1, 11], where the dense matrices arising from discretisation of integral operators were transformed into a sparse form using orthogonal or bi-orthogonal systems of compactly supported wavelets. The cost of the matrix-vector multiplication was reduced from the straightforward $\mathcal{O}(N^2)$ number of operations to $\mathcal{O}(N \log(N))$ or even $\mathcal{O}(N)$. In two papers [27, 28], the authors study the stability and the convergence of the wavelet method for pseudodifferential operator equations as well as their fast solution based on the matrix approximation. A different wavelet technique was applied in [115, 116], leading in [67] to an optimal algorithm with $\mathcal{O}(N)$ complexity. For recent results, see also [48, 49].

In [36], the authors consider an algebraic approach for the approximation of dense matrices based on the use of some their original entries.

The Adaptive Cross Approximation method was introduced in [7] for Nyström type matrices and in [9] for collocation matrices arising from boundary integral equations. This method uses a hierarchical decomposition (cf. [41, 42]) of the matrix in a system of blocks. There are several applications of this method to different problems. The ACA was applied to potential problems in [9, 87]. In [8, 10], the ACA was used for the approximation of matrices arising from the radiation heat transfer equation. The applications of the ACA to electromagnetic problems can be found in [16, 63, 64, 65]. In [16] and in [31], a comparison of the ACA method with the Fast Multipole Method was given. In [110, 111, 113, 119], the boundary integral equations arising from the Helmholtz equation were solved using the ACA method.

Finally, we refer to [69], where an algebraic multigrid preconditioners were constructed for the boundary integral formulations of potential problems approximated by using the Adaptive Cross Approximation algorithm.

4

Implementation
and Numerical Examples

4.1 Geometry Description

In this section we describe some surfaces which will be used for numerical examples in the following sections. We show the geometry of these surfaces and give the number of elements and nodes. Furthermore, the corresponding cluster structures will be shown.

4.1.1 Unit Sphere

The most simple smooth surface $\Gamma = \partial \Omega$ for $\Omega \subset \mathbb{R}^3$ is the surface of the unit sphere,

$$\Gamma = \left\{ x \in \mathbb{R}^3 \, : \, |x| = 1 \right\}. \tag{4.1}$$

As an appropriate discretisation of Γ, we consider the icosahedron that is uniformly triangulated before being projected onto the circumscribed unit sphere. On this way we obtain a sequence $\{\Gamma_N\}$ of almost uniform meshes on the unit sphere, which are shown in Figs. 4.1–4.2 for different numbers of boundary elements N. This sequence allows to study the convergence of boundary element methods for different examples. In Fig. 4.3 the clusters of the levels 1 and 2 obtained with Alg. 3.1 for $N = 1280$ are presented. In Fig. 4.4 a typical admissible cluster pair is shown.

4.1.2 TEAM Problem 10

Now we consider the TEAM problem 10 (cf. [75]). TEAM is an acronym for Testing Electromagnetic Analysis Methods, which is a community that creates benchmark problems to test finite element analysis software. An exciting coil is set between two steel channels and a thin steel plate is inserted between the channels. Thus, the domian consists of four disconnected parts. The coarsest

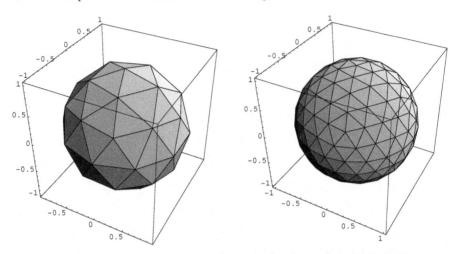

Fig. 4.1. Discretisation of the unit sphere for $N = 80$ and $N = 320$

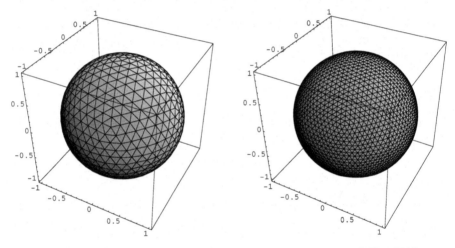

Fig. 4.2. Discretisation of the unit sphere for $N = 1280$ and $N = 5120$

mesh of this model contains $N = 4928$ elements. We perform two uniform mesh refinements in order to get meshes with $N = 19712$ and $N = 78848$ elements, respectively. The initial mesh for $N = 4928$ is shown in Fig. 4.5. The speciality of this model is an extremely thin chink (less then 0.2% of the model size) between the steel plate and the channels. Another speciality of it is the very fine discretisation close to the edges of the channels. In Fig. 4.6 the clusters of the levels 1 and 2 for $N = 4928$ are shown.

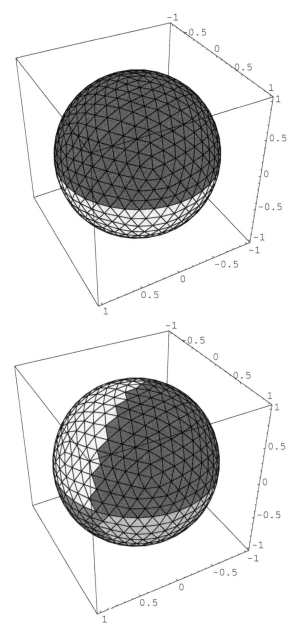

Fig. 4.3. Clusters of the level 1 and 2 for $N = 1280$

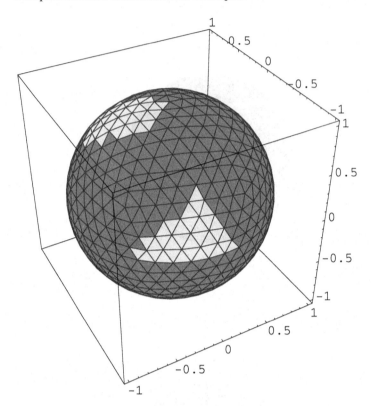

Fig. 4.4. An admissible cluster pair for $N = 1280$

4.1.3 TEAM Problem 24

The test rig consists of a rotor and a stator (cf. [92]). The stator poles are fitted with coils as shown in Fig. 4.7. The rotor is locked at 22° with respect to the stator, providing only a small overlap between the poles. The coarsest mesh of this model is shown in Fig. 4.7 and contains $N = 4640$ elements. We perform two uniform mesh refinements in order to get meshes with $N = 18560$ and $N = 74240$ elements, respectively. This model consists of four independent parts. In Fig. 4.8 the clusters of the levels 1 and 2 for $N = 4928$ are shown.

4.1.4 Relay

The simply connected domain shown in Fig. 4.9 will be considered as model for a relay. The speciality of this domain is the small air gap between the kernel and the armature. Its surface contains $N = 4944$ elements. We perform two uniform mesh refinements in order to get meshes with $N = 19776$ and $N = 79104$ elements, respectively. In Fig. 4.10 the corresponding clusters can be seen.

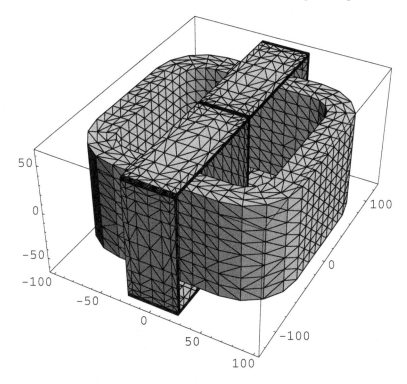

Fig. 4.5. TEAM problem 10 for $N = 4928$

4.1.5 Exhaust manifold

A simplified model of an exhaust manifold is shown in Fig. 4.11. Its surface contains $N = 2264$ elements. We perform two uniform mesh refinements in order to get meshes with $N = 9056$ and $N = 36224$ elements, respectively. The clusters of the level 1 and 2 are presented in Fig. 4.12.

4.2 Laplace Equation

In this section we consider some numerical examples for the Laplace equation

$$-\Delta u(x) = 0 \qquad (4.2)$$

where u is an analytically given harmonic function.

4.2.1 Analytical solutions

Particular solutions of the Laplace equation (4.2) are, for example,

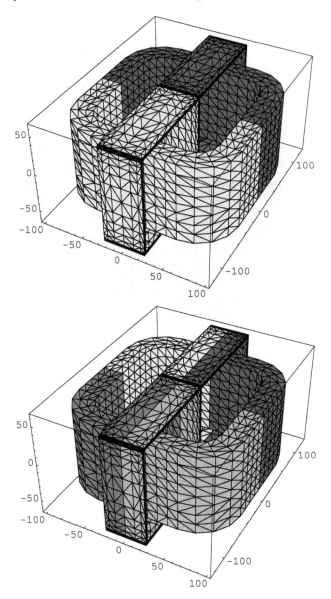

Fig. 4.6. Clusters of the level 1 and 2, TEAM problem 10 for $N = 4928$

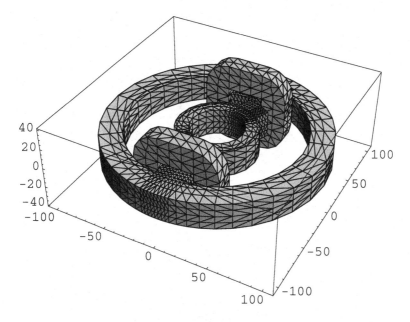

Fig. 4.7. TEAM problem 24 for $N = 4640$

$$\Phi_{k_1,k_2,k_3}(x) = \exp\left(k_1 x_1 + k_2 x_2 + k_3 x_3\right), \ k_1^2 + k_2^2 + k_3^2 = 0\,,$$

$$\Phi_{0,k_2,k_3}(x) = (a + b\,x_1)\exp\left(k_2 x_2 + k_3 x_3\right), \ k_2^2 + k_3^2 = 0\,, \qquad (4.3)$$

$$\Phi_{0,0,0}(x) = (a_1 + b_1 x_1)(a_2 + b_2 x_2)(a_3 + b_3 x_3)\,.$$

Here, k_1, k_2, and k_3 are arbitrary complex numbers satisfying the corresponding conditions. Thus, different products of real valued linear, exponential, trigonometric and hyperbolic functions can be chosen for numerical tests, if we consider interior boundary value problems in a three-dimensional, open, and bounded domain $\Omega \subset \mathbb{R}^3$. Furthermore, the fundamental solution of the Laplace equation (cf. (1.7))

$$u^*(x,\tilde{y}) = \frac{1}{4\pi}\frac{1}{|x - \tilde{y}|}\,, \qquad (4.4)$$

can be considered as a particular solution of the Laplace equation for both, interior ($x \in \Omega$, $\tilde{y} \in \Omega^e = \mathbb{R}^3 \setminus \overline{\Omega}$), and exterior ($x \in \Omega^e$, $\tilde{y} \in \Omega$) boundary value problems.

4.2.2 Discretisation, Approximation and Iterative Solution

We solve the interior Dirichlet, Neumann and mixed boundary value problems, as well as an interface problem using a Galerkin boundary element method

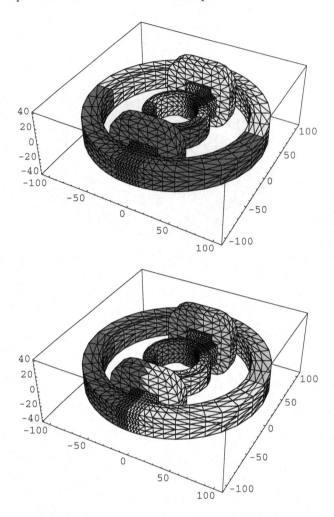

Fig. 4.8. Clusters of the level 1 and 2, TEAM problem 24 for $N = 4640$

(cf. Section 2). Piecewise linear basis functions φ_ℓ will be used for the approximation of the Dirichlet datum $\gamma_0^{\text{int}} u$ and piecewise constant basis functions ψ_k for the approximation of the Neumann datum $\gamma_1^{\text{int}} u$. We will use the L_2 projection for the approximation of the given part of the Cauchy data. The boundary element matrices V_h, K_h and D_h are generated in approximative form using the partially pivoted ACA algorithm with a variable relative accuracy ε_1. The resulting systems of linear equations are solved using some variants of the Conjugate Gradient method (CGM) with or without preconditioning up to a relative accuracy $\varepsilon_2 = 10^{-8}$.

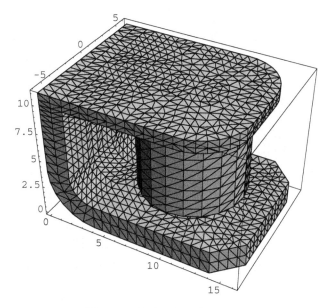

Fig. 4.9. Relay for $N = 4944$

4.2.3 Generation of Matrices

The most important matrices to be generated while using the Galerkin boundary element method are the single layer potential matrix V_h and the double layer potential matrix K_h, having the entries

$$V_h[k, \ell] = \frac{1}{4\pi} \int_{\tau_k} \int_{\tau_\ell} \frac{1}{|x - y|} \, ds_y ds_x \quad \text{for } k, \ell = 1, \ldots, N, \qquad (4.5)$$

and

$$K_h[k, j] = \frac{1}{4\pi} \int_{\tau_k} \int_{\Gamma} \frac{(x - y, \underline{n}(y))}{|x - y|^3} \, \varphi_j(y) \, ds_y ds_x \qquad (4.6)$$

for $k = 1, \ldots, N$ and $j = 1, \ldots, M$ (cf. Section 2.3). The analytical evaluation of these integrals seems to be impossible in general. Thus, some numerical quadrature rules have to be involved. These quadrature formulae produce some additional numerical errors in the whole procedure. However, it is possible to compute the inner integrals of the entries (4.5)–(4.6), namely the integrals

$$S(\tau, x) = \frac{1}{4\pi} \int_{\tau} \frac{1}{|x - y|} \, ds_y \qquad (4.7)$$

and

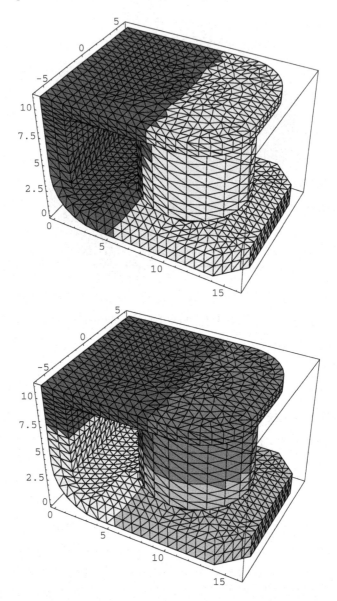

Fig. 4.10. Clusters of the level 1 and 2, Relay for $N = 4944$

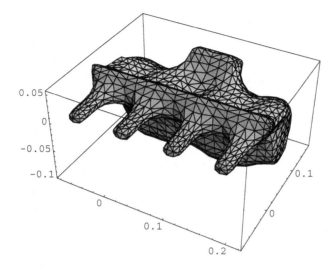

Fig. 4.11. Exhaust manifold for $N = 2264$

$$D_i(\tau, x) = \frac{1}{4\pi} \int_\tau \frac{(x - y, \underline{n}(y))}{|x - y|^3} \, \psi_{\tau,i}(y) \, ds_y \quad \text{for } i = 1, 2, 3 \,. \tag{4.8}$$

Here, $\tau \subset \mathbb{R}^3$ is a plane triangle having the nodes x_1, x_2, x_3, and $\psi_{\tau,i}$ is the piecewise linear function (2.6) which corresponds to the node x_i, i.e.

$$\psi_{\tau,i}(x_j) = \delta_{ij} \,, \quad j = 1, 2, 3 \,.$$

The explicit form of these functions can be seen in Appendix C.2. By the use of the functions (4.7)–(4.8), the matrix entries (4.5)–(4.6) can be rewritten as follows:

$$V_h[k, \ell] = \frac{1}{2} \left(\int_{\tau_k} S(\tau_\ell, x) \, ds_x + \int_{\tau_\ell} S(\tau_k, x) \, ds_x \right) \tag{4.9}$$

for $k, \ell = 1, \dots, N$ and

$$K_h[k, j] = \sum_{\tau \in I(j)} \int_{\tau_k} D_{i \,:\, x_i(\tau) = \tau_j}(\tau, x) \, ds_x \tag{4.10}$$

for $k = 1, \dots, N$, $j = 1, \dots, M$. Note that we have used the symmetrisation for the entries of the single layer potential in (4.9). In (4.10), the summation takes place over all triangles τ containing the node x_j. The index $i \in \{1, 2, 3\}$ of the function D_i to be integrated over τ_k, is chosen in such a way, that the node i of the triangle τ is x_j. We are not going to compute the latter integrals in (4.9)–(4.10) in a closed form. Thus, some numerical integration

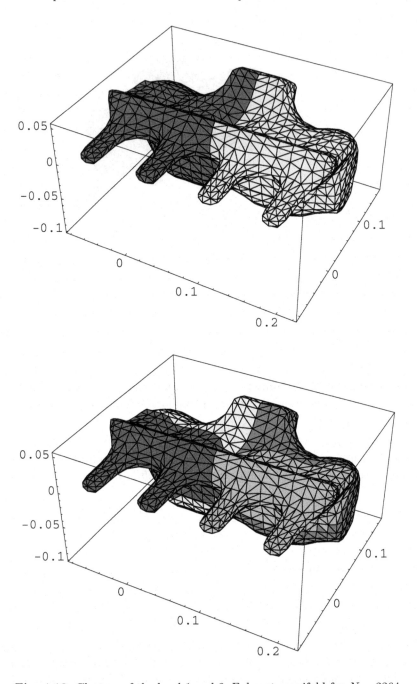

Fig. 4.12. Clusters of the level 1 and 2, Exhaust manifold for $N = 2264$

has to be involved. If we denote by N_G the number of integration points, by ω_m the weights of the quadrature, and by $x_{\tau,m}$ the integration points within the triangle τ, then the exact entries (4.9)–(4.10) can be approximated by

$$V_h[k, \ell] \approx \frac{1}{2} \left(\sum_{m=1}^{N_g} \omega_m \Big(S(\tau_\ell, x_{\tau_k, m}) + S(\tau_k, x_{\tau_\ell, m}) \Big) \right) \qquad (4.11)$$

for $k, \ell = 1, \ldots, N$ and

$$K_h[k, j] \approx \sum_{\tau \in I(j)} \sum_{m=1}^{N_g} \omega_m D_{i \,:\, x_i(\tau) = x_j}(\tau, x_{\tau_k, m}) \qquad (4.12)$$

for $k = 1, \ldots, N$ and $j = 1, \ldots, M$. For our numerical tests, we have used a 7-point quadrature rule, see Appendix C.1 for more details.

4.2.4 Interior Dirichlet Problem

Here we solve the Laplace equation (4.2) together with the boundary condition $\gamma_0^{\mathrm{int}} u(x) = g(x)$ for $x \in \Gamma$, where Γ is a given surface. The variational problem (1.16)

$$\Big\langle Vt, w \Big\rangle_\Gamma = \Big\langle \Big(\frac{1}{2}I + K\Big)g, w \Big\rangle_\Gamma \qquad \text{for all } w \in H^{-1/2}(\Gamma)$$

is discretised, which leads to a system of linear equations (2.16)

$$V_h \tilde{\underline{t}} = \Big(\frac{1}{2}M_h + K_h\Big)\underline{g}.$$

Since the matrix V_h is symmetric and positive definite, the classical Conjugate Gradient method (CGM) is used as solver.

Unit sphere

The analytical solution is taken in the form (4.3). For $x = (x_1, x_2, x_3)^\top \in \Omega$ we consider the harmonic function

$$u(x) = \mathrm{Re}\, \Phi_{0, 2\pi, \imath 2\pi} = (1 + x_1) \exp(2\pi\, x_2) \cos(2\pi\, x_3) \qquad (4.13)$$

as a test solution of the Laplace equation (4.2). The results of the computations are shown in Tables 4.1 and 4.2. The number of boundary elements is listed in the first column of these tables. The second column contains the number of nodes, while in the third column of Table 4.1, the prescribed accuracy for the ACA algorithm for the approximation of both matrices $K_h \in \mathbb{R}^{N \times M}$ and $V_h \in \mathbb{R}^{N \times N}$ is given. The fourth column of this table shows the memory requirements in MByte for the approximate double layer potential matrix K_h. The quality of this approximation in percentage of the original matrix is listed

Table 4.1. ACA approximation of the Galerkin matrices K_h and V_h

N	M	ε_1	MByte(K_h)	%	MByte(V_h)	%
80	42	$1.0 \cdot 10^{-2}$	0.03	97.8	0.02	48.7
320	162	$1.0 \cdot 10^{-3}$	0.26	65.6	0.21	27.2
1280	642	$1.0 \cdot 10^{-4}$	2.45	39.1	1.94	15.5
5120	2562	$1.0 \cdot 10^{-5}$	20.05	20.0	15.72	7.9
20480	10242	$1.0 \cdot 10^{-6}$	149.19	9.3	115.83	3.6
81920	40962	$1.0 \cdot 10^{-7}$	1085.0	4.2	837.50	1.6

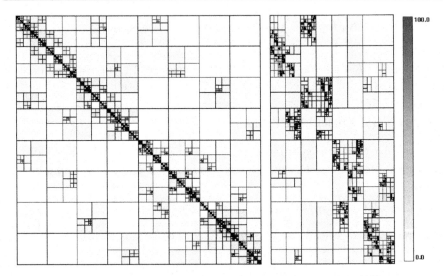

Fig. 4.13. Partitioning of the BEM matrices for $N = 5120$ and $M = 2562$

in the next column. The corresponding values for the single layer potential matrix V_h can be seen in the columns six and seven. The partitioning of the matrix for $N = 5120$ as well as the quality of the approximation of single blocks is shown in Fig. 4.13. The left diagram in Fig. 4.13 shows the symmetric single layer potential matrix V_h, while the rectangular double layer potential matrix K_h is depicted in the right diagram. The legend indicates the percentage of memory needed for the ACA approximation of the blocks compared to the full memory. Further numerical results are shown in Table 4.2. The third column in Table 4.2 shows the number of Conjugate Gradient iterations needed to reach the prescribed accuracy ε_2. The relative L_2 error for the Neumann datum

$$Error_1 = \frac{\|\gamma_1^{\text{int}} u - \tilde{t}_h\|_{L_2(\Gamma)}}{\|\gamma_1^{\text{int}} u\|_{L_2(\Gamma)}} \tag{4.14}$$

is given in the fourth column. The next column represents the rate of convergence for the Neumann datum, i.e. the quotient between the errors in two

Table 4.2. Accuracy of the Galerkin method, Dirichlet problem

N	M	$Iter$	$Error_1$	CF_1	$Error_2$	CF_2
80	42	22	$9.34 \cdot 10^{-1}$	–	$7.29 \cdot 10^{-0}$	–
320	162	32	$5.06 \cdot 10^{-1}$	1.85	$3.29 \cdot 10^{-1}$	22.16
1280	642	45	$2.23 \cdot 10^{-1}$	2.27	$3.53 \cdot 10^{-2}$	9.32
5120	2562	56	$1.04 \cdot 10^{-1}$	2.14	$3.54 \cdot 10^{-3}$	9.97
20480	10242	72	$5.11 \cdot 10^{-2}$	2.03	$4.11 \cdot 10^{-4}$	8.61
81920	40962	94	$2.53 \cdot 10^{-2}$	2.02	$4.30 \cdot 10^{-5}$	9.56

consecutive lines of column four. Finally, the last two columns show the absolute error (cf. (2.19)) in a prescribed inner point $x^* \in \Omega$,

$$Error_2 = |u(x^*) - \tilde{u}(x^*)|, \quad x^* = (0.250685, 0.417808, 0.584932)^\top, \quad (4.15)$$

for the value $\tilde{u}(x^*)$ obtained using an approximate representation formula (2.18). Table 4.2 obviously shows a linear convergence $\mathcal{O}(N^{-1/2}) = \mathcal{O}(h)$ of the Galerkin boundary element method for the Neumann datum in the L_2 norm. It should be noted that this theoretically guaranteed convergence order can already be observed when approximating the matrices K_h and V_h with much less accuracy as it was used to obtain the results in Table 4.1. However, this high accuracy is necessary in order to be able to observe the third order (or even better) pointwise convergence rate within the domain Ω presented in the last two columns of Table 4.2. Especially for $N = 81920$, a very high accuracy of $\varepsilon_1 = 1.0 \cdot 10^{-7}$ of the ACA approximation is necessary.

In Figs. 4.14-4.15, the given Dirichlet datum and computed Neumann datum for $N = 5120$ boundary elements and $M = 2562$ nodes are presented. The numerical curve obtained when using an approximate representation formula in comparison with the curve of the exact values (4.13) along the line

$$x(t) = \begin{pmatrix} -0.3 \\ -0.5 \\ -0.7 \end{pmatrix} + t \begin{pmatrix} 0.6 \\ 1.0 \\ 1.4 \end{pmatrix}, \quad 0 \le t \le 1 \qquad (4.16)$$

inside of the domain Ω is shown in Fig. 4.16 for $N = 80$ (left plot) and for $N = 320$ (right plot). The values of the numerical solution $\tilde{u}$ and of the analytical solution u have been computed in 512 points uniformly placed on the line (4.16). The thick dashed line represents in these figures the course of the analytical solution (4.13), while the thin solid line shows the course of the numerical solution $\tilde{u}$. The values of the variable x_1 along the line (4.16) are used for the axis of abscissas. The next Fig. 4.17 shows these curves for $N = 1280$ (left plot) and on the zoomed interval $[0.2, 0.3]$ (right plot) with respect to the variable x_1 in order to see the difference between them. It is almost impossible to see any optical difference between the numerical and analytical curves for higher values of N. Note that the point x^* in (4.15) is

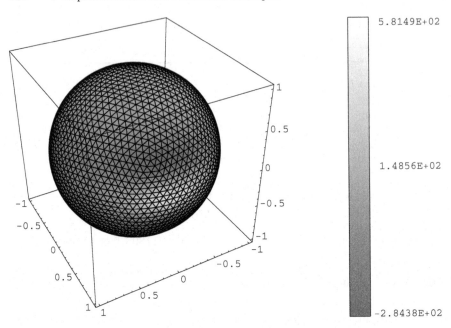

Fig. 4.14. Given Dirichlet datum for the unit sphere, $N = 5120$

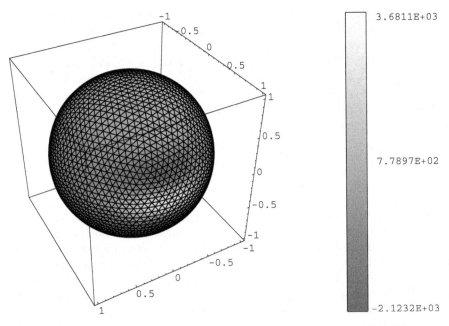

Fig. 4.15. Computed Neumann datum for the unit sphere, $N = 5120$

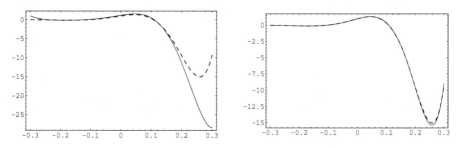

Fig. 4.16. Numerical and analytical curves for $N = 80$ and $N = 320$, Dirichlet problem

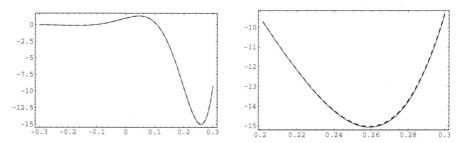

Fig. 4.17. Numerical and analytical curves for $N = 1280$, Dirichlet problem

chosen close to the minimum of the function u along the line, where the error seems to reach its maximum.

TEAM Problem 10

The analytical solution is now taken in the form (4.4) with $\tilde{y} = (0, 60, 50)^\top$. The results of the computations are shown in Tables 4.3 and 4.4. The third

Table 4.3. ACA approximation of the Galerkin matrices K_h and V_h, Dirichlet problem

N	M	ε_1	MByte(K_h)	%	MByte(V_h)	%
4928	2470	$1.0 \cdot 10^{-2}$	14.03	15.11	7.46	4.0
19712	9862	$1.0 \cdot 10^{-3}$	131.85	8.9	65.22	2.2
78848	39430	$1.0 \cdot 10^{-4}$	1190.00	5.0	604.56	1.3

column shows the number of iterations required by the Conjugate Gradient method with diagonal preconditioning

$$D = \text{diag}\left\{ |\tau_\ell|, \ \ell = 1, \ldots, N \right\}. \tag{4.17}$$

Table 4.4. Accuracy of the Galerkin method, Dirichlet problem

N	M	$Iter$	$Error_1$	CF_1	$Error_2$	CF_2
4928	2470	91	$6.02 \cdot 10^{-1}$	–	$2.55 \cdot 10^{-5}$	–
19712	9862	183	$1.99 \cdot 10^{-1}$	3.02	$4.69 \cdot 10^{-6}$	5.44
78848	39430	248	$1.13 \cdot 10^{-1}$	1.76	$5.22 \cdot 10^{-7}$	8.98

The last two columns of Table 4.4 show the absolute error (cf. (2.19)) in a prescribed inner point $x^* \in \Omega$,

$$Error_2 = |u(x^*) - \tilde{u}(x^*)|, \ x^* = (0.0, 90.0, 49.7943)^\top , \qquad (4.18)$$

for the value $\tilde{u}(x^*)$ obtained using the approximate representation formula (2.18). All other entries in these tables have the same meaning as those displayed in Tables 4.1-4.2. In Figs. 4.18–4.19 the given Dirichlet datum and the computed Neumann datum for $N = 4928$ boundary elements and $M = 2470$ nodes are presented. The numerical curve obtained when using the approxi-

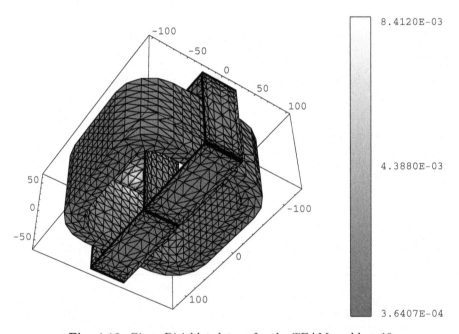

Fig. 4.18. Given Dirichlet datum for the TEAM problem 10

mate representation formula in comparison with the curve of the exact values (4.4) along the line

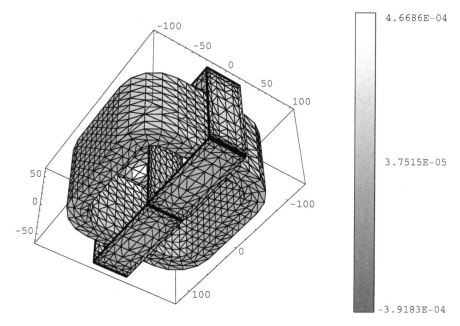

Fig. 4.19. Computed Neumann datum for the TEAM problem 10

$$x(t) = \begin{pmatrix} 0.0 \\ 90.0 \\ -49.99 \end{pmatrix} + t \begin{pmatrix} 0.0 \\ 0.0 \\ 99.98 \end{pmatrix} , \ 0 \le t \le 1 \tag{4.19}$$

inside of the domain Ω is shown in Fig. 4.20 for $N = 4928$. The values of the numerical solution $\tilde{u}$ and of the analytical solution u have been computed in 512 points uniformly placed on the line (4.19). The thick dashed line represents the course of the analytical solution (4.4), while the thin solid line shows the course of the numerical solution $\tilde{u}$. The values of the variable x_3 along the line (4.19) are used for the axis of abscissas. The right plot in this figure shows a zoomed picture on the interval $[40, 49.99]$ with respect to the variable x_3. Note that the end of the line (4.19) is very close to the boundary of the domain, which lies at $x_3 = 50$. Thus, the loss of accuracy of the numerical representation formula close to the boundary can be clearly seen. The courses of the numerical solutions obtained for $N = 19712$ (left plot) and $N = 78848$ (right plot) are shown in Fig. 4.21. They do not distinguish optically from the course of the analytical solution on the whole interval. Thus, we show only the zoomed pictures.

4.2.5 Interior Neumann Problem

We consider the interior Neumann boundary value problem with the boundary condition $\gamma_1^{\mathrm{int}} u(x) = g(x)$ for $x \in \Gamma$. The variational problem (1.29)

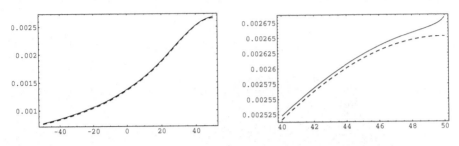

Fig. 4.20. Numerical and analytical curves for $N = 4928$, Dirichlet problem

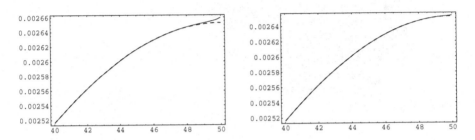

Fig. 4.21. Numerical and analytical curves for $N = 19712$ and $N = 78848$

$$\left\langle D\bar{u}, v \right\rangle_\Gamma + \left\langle \bar{u}, 1 \right\rangle_\Gamma \left\langle v, 1 \right\rangle_\Gamma = \left\langle \left(\frac{1}{2}I - K'\right)g, v \right\rangle_\Gamma + \alpha \left\langle v, 1 \right\rangle_\Gamma$$

for all $v \in H_{**}^{1/2}(\Gamma)$, is discretised and leads to a system of linear equations (cf. (2.31))

$$\left(D_h + \underline{a}\,\underline{a}^\top\right)\underline{\widetilde{u}} = \left(\frac{1}{2}M_h^\top - K_h^\top\right)\underline{g} + \alpha\,\underline{a},$$

where the vector $\underline{a} \in \mathbb{R}^M$ contains the integrals of the piecewise linear basis functions φ_ℓ over the surface Γ,

$$a_\ell = \frac{1}{3}\left|\operatorname{supp}\varphi_\ell\right|, \quad \ell = 1, \ldots, M.$$

The symmetric and positive definite system is then solved using a Conjugate Gradient method up to the relative accuracy $\varepsilon_2 = 10^{-8}$.

Unit Sphere

We consider again the harmonic function (4.13) as the exact solution. The results for the ACA approximation of the matrix $D_h \in \mathbb{R}^{M \times M}$ are presented in Table 4.5. The corresponding results for the matrix $K_h^\top$ are identical to those already presented in Table 4.1. Note that in this example, the Galerkin matrix with piecewise linear basis functions for the hypersingular operator is generated according to (1.9). The partitioning of the matrices $K_h^\top$ and D_h for

Table 4.5. ACA approximation of the Galerkin matrix D_h, Neumann problem

N	M	ε_1	MByte(D_h)	%
80	42	$1.0 \cdot 10^{-2}$	0.01	51.2
320	162	$1.0 \cdot 10^{-3}$	0.10	48.1
1280	642	$1.0 \cdot 10^{-4}$	1.02	32.3
5120	2562	$1.0 \cdot 10^{-5}$	8.67	17.3
20480	10242	$1.0 \cdot 10^{-6}$	64.75	8.09
81920	40962	$1.0 \cdot 10^{-7}$	446.13	3.49

$N = 5120$ and $M = 2562$ as well as the quality of the approximation of the single blocks are shown in Fig. 4.22. The left diagram in Fig. 4.22 shows the

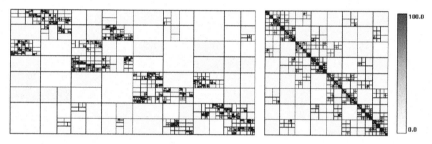

Fig. 4.22. Partitioning of the BEM matrices for $N = 5120$ and $M = 2562$.

rectangular double layer potential matrix $K_h^T \in \mathbb{R}^{M \times N}$, while the symmetric hypersingular matrix D_h is depicted in the right diagram. The legend indicates the percentage of memory needed for the ACA approximation of the blocks compared to the full memory. The accuracy obtained for the whole numerical

Table 4.6. Accuracy of the Galerkin method, Neumann problem

N	M	$Iter$	$Error_1$	CF_1	$Error_2$	CF_2
80	42	10	$7.63 \cdot 10^{-1}$	–	$9.37 \cdot 10^{-1}$	–
320	162	14	$2.72 \cdot 10^{-1}$	2.81	$1.95 \cdot 10^{-0}$	–
1280	642	17	$6.02 \cdot 10^{-2}$	4.52	$4.27 \cdot 10^{-1}$	4.56
5120	2562	25	$1.37 \cdot 10^{-2}$	4.39	$1.01 \cdot 10^{-1}$	4.22
20480	10242	35	$3.28 \cdot 10^{-3}$	4.18	$2.49 \cdot 10^{-2}$	4.08
81920	40962	51	$8.02 \cdot 10^{-4}$	4.09	$6.13 \cdot 10^{-3}$	4.06

procedure is presented in Table 4.6. The numbers in this table have the same meaning as in Table 4.2. The third column shows the number of iterations required by the Conjugate Gradient method without preconditioning. Note that the convergence of the Galerkin method for the unknown Dirichlet datum in the L_2 norm

$$Error_1 = \frac{\|\gamma_0^{int} u - \widetilde{u}_h\|_{L_2(\Gamma)}}{\|\gamma_0^{int} u\|_{L_2(\Gamma)}} \tag{4.20}$$

is now quadratic corresponding to the error estimate (2.32). Also in the inner point x^* (cf. (4.15)) we now observe the quadratic convergence (7th column) as it was predicted in (2.34) instead of the cubic order obtained for the Dirichlet problem (cf. Table 4.2). This fact is clearly illustrated in Figs. 4.23–4.24, where the convergence of the boundary element method can be seen optically. The results obtained for $N = 80$ are plotted in Fig. 4.23 (left plot). The numerical

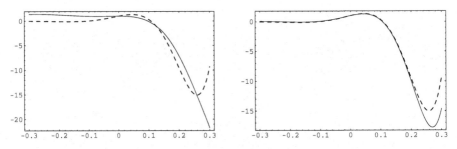

Fig. 4.23. Numerical and analytical curves for $N = 80$ and $N = 320$, Neumann problem

curve in Fig. 4.23 (right plot) is notedly better than the previous one. However, its quality is not as high as the one of the corresponding curve obtained solving the Dirichlet problem (cf. Fig. 4.16). The next Fig. 4.24 shows the same curves for $N = 1280$ and for $N = 5120$. Here, we do not need to zoom the pictures in order to see the difference between the numerical and the analytical curves.

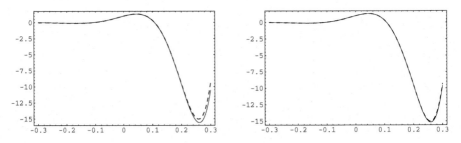

Fig. 4.24. Numerical and analytical curves for $N = 1280$ and $N = 5120$, Neumann problem

Exhaust Manifold

The analytical solution is taken in the form (4.4) with $\tilde{y} = (0, 0, 0.06)^\top$. The results of the computations are reported in Tables 4.7 and 4.8. The third

Table 4.7. ACA approximation of the Galerkin matrices K_h and V_h, Neumann problem

N	M	ε_1	MByte(K_h)	%	MByte(V_h)	%
2264	1134	$1.0 \cdot 10^{-3}$	6.99	35.7	3.94	10.1
9056	4530	$1.0 \cdot 10^{-4}$	62.27	19.9	34.99	5.6
36224	18114	$1.0 \cdot 10^{-5}$	500.66	10.0	282.44	2.8

Table 4.8. Accuracy of the Galerkin method, Neumann problem

N	M	$Iter$	$Error_1$	CF_1	$Error_2$	CF_2
2264	1134	72	$2.12 \cdot 10^{-2}$	–	$2.69 \cdot 10^{-3}$	–
9056	4530	110	$4.95 \cdot 10^{-3}$	4.3	$5.36 \cdot 10^{-4}$	5.0
36224	18114	163	$1.13 \cdot 10^{-3}$	4.4	$1.07 \cdot 10^{-4}$	5.0

column shows the number of iterations required by the Conjugate Gradient method with diagonal preconditioning,

$$D = \text{diag}\left\{ |\text{supp}\, \varphi_\ell| \,, \ \ell = 1, \ldots, M \right\}. \tag{4.21}$$

The last two columns of Table 4.8 show the absolute error (cf. (2.34)) in a prescribed inner point $x^* \in \Omega$,

$$Error_2 = |u(x^*) - \tilde{u}(x^*)| \,, \ x^* = (-0.0112524, 0.1, -0.05)^\top, \tag{4.22}$$

for the value $\tilde{u}(x^*)$ obtained using an approximate representation formula (2.33). All other entries in these tables have the usual meaning. The quadratic convergence of the Dirichlet datum in the L_2 norm as well as the quadratic convergence (or even slightly better) in the inner point x^* can be observed again.

In Figs. 4.25–4.26 the given Dirichlet datum and computed Neumann datum for $N = 2264$ boundary elements and $M = 1134$ nodes are presented.

The numerical curve obtained when using an approximate representation formula in comparison with the curve of the exact values (4.4) along the line

$$x(t) = \begin{pmatrix} -0.05 \\ 0.1 \\ -0.05 \end{pmatrix} + t \begin{pmatrix} 0.2 \\ 0.0 \\ 0.0 \end{pmatrix}, \ 0 \leq t \leq 1 \tag{4.23}$$

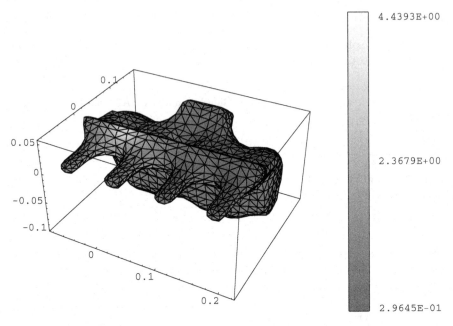

Fig. 4.25. Computed Dirichlet datum, exhaust manifold

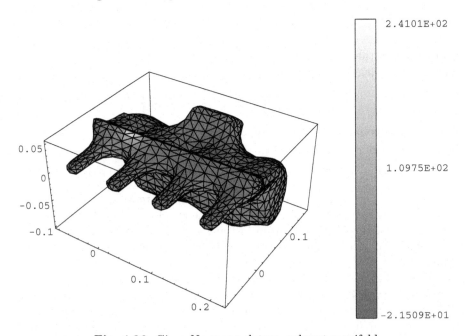

Fig. 4.26. Given Neumann datum, exhaust manifold

inside of the domain Ω is shown in Fig. 4.27 for $N = 2254$ and $N = 9056$, while Fig. 4.28 shows the results obtained for $N = 36224$ (left plot). The right plot in this figure presents the same curve on a zoomed interval $[-0.05, 0.05]$. The values of the numerical solution $\tilde{u}$ and of the analytical solution u have been computed in 512 points uniformly placed on the line (4.23). The thick dashed line represents the course of the analytical solution (4.4), while the thin solid line shows the course of the numerical solution $\tilde{u}$. The values of the variable x_1 along the line (4.23) are used for the axis of abscissas. Again, a quite high accuracy of the Galerkin BEM can be observed.

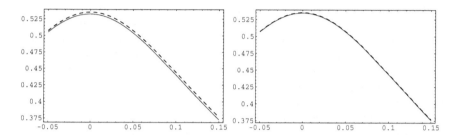

Fig. 4.27. Numerical and analytical curves for $N = 4944$ and $N = 19776$

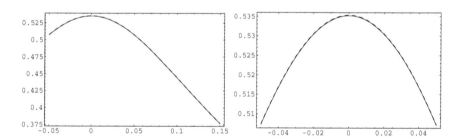

Fig. 4.28. Numerical and analytical curves for $N = 36224$

4.2.6 Interior Mixed Problem

Here, we consider the interior mixed boundary value problem (cf. (1.34))

$$
\begin{aligned}
-\Delta u(x) &= 0 && \text{for } x \in \Omega\,, \\
\gamma_0^{\text{int}} u(x) &= g(x) && \text{for } x \in \Gamma_D\,, \\
\gamma_1^{\text{int}} u(x) &= f(x) && \text{for } x \in \Gamma_N\,,
\end{aligned}
\qquad (4.24)
$$

where the function g is the interior trace of the exact solution on the boundary Γ, while the function f denotes its interior conormal derivative on Γ. After discretisation, the variational problem (1.35) leads to the skew symmetric system of linear equations (cf. (2.36))

$$
\begin{pmatrix} V_h & -K_h \\ K_h^\top & D_h \end{pmatrix} \begin{pmatrix} \tilde{t} \\ \tilde{u} \end{pmatrix} = \begin{pmatrix} -\bar{V}_h & \frac{1}{2}\bar{M}_h + \bar{K}_h \\ \frac{1}{2}\bar{M}_h^\top - \bar{K}_h^\top & -\bar{D}_h \end{pmatrix} \begin{pmatrix} f \\ g \end{pmatrix}
$$

The matrix of the single layer potential V_h and the matrix of the double layer potential K_h are generated in an approximative form using the partially pivoted ACA algorithm 3.9 with increasing accuracy. The system of linear equations (2.41) is then solved using the Conjugate Gradient method for the Schur complement system as in (2.37) up to the relative accuracy $\varepsilon_2 = 10^{-8}$. Note that this realisation requires an additional solution of a linear system with the single layer potential matrix in each iteration step. This system is solved again using a Conjugate Gradient method up to the relative accuracy $\varepsilon_2 = 10^{-8}$. The matrix of the hypersingular operator is not generated explicitly. Its multiplication with a vector is realised using the matrix of the single layer potential as it is described in (2.27).

Unit Sphere

In the first example, we prescribe the Dirichlet datum on the upper part of the unit sphere $\Gamma_D = \{x \in \Gamma : x_3 \geq 0\}$ and the Neumann datum on the lower part $\Gamma_N = \{x \in \Gamma : x_3 < 0\}$. The analytical solution is taken in the form (4.13) and the numerical results are presented in Tables 4.9–4.10. Of course,

Table 4.9. Accuracy of the Galerkin method on the boundary, mixed problem

N	M	$Iter_1$	$Iter_2$	$D - Error_1$	CF_D	$N - Error_1$	CF_N
80	42	8	18-19	$5.97 \cdot 10^{-1}$	–	$8.78 \cdot 10^{-1}$	–
320	162	11	25-27	$2.33 \cdot 10^{-1}$	2.56	$4.54 \cdot 10^{-1}$	1.93
1280	642	16	36-38	$5.00 \cdot 10^{-2}$	4.66	$2.11 \cdot 10^{-1}$	2.15
5120	2562	24	49-53	$1.14 \cdot 10^{-2}$	4.39	$1.02 \cdot 10^{-1}$	2.07
20480	10242	36	64-75	$2.73 \cdot 10^{-3}$	4.18	$5.07 \cdot 10^{-2}$	2.01

the results for the ACA approximation of the matrices V_h and K_h are the same as for the Dirichlet problem (cf. Table 4.1). The accuracy obtained for the mixed boundary value problem is presented in Table 4.9. The numbers in this table have the following meaning: The third column in Table 4.9 shows the number of Conjugate Gradient iterations without preconditioning needed to reach the prescribed accuracy ε_2 for the Schur complement, while the fourth column indicates the numbers of Conjugate Gradient iterations without preconditioning needed in each iteration step for the system with the single layer potential matrix. Thus, the total number of iterations when solving a mixed boundary value problem is much higher than for solving a pure Dirichlet or Neumann boundary value problem. The error for the Dirichlet datum and the convergence factor are shown in columns 5 and 6, while the corresponding error for the Neumann datum can be seen in columns 7 and 8. Note that the convergence of the Galerkin method for the unknown Dirichlet datum in the L_2 norm (4.20) is quadratic (6th column), while the convergence of the Neumann datum (4.14) is linear (8th column). Corresponding to Table 4.1, the matrices K_h and V_h together with some additional memory will require more than 2 Gbyte of memory for $N = 81902$. Thus, we are not able to store both these matrices on a regular workstation. The error in the inner point x^*

Table 4.10. Accuracy of the Galerkin method in the inner point x^*, mixed problem

N	M	$Error_2$	CF_2
80	42	$5.82 \cdot 10^{-0}$	–
320	162	$4.94 \cdot 10^{-1}$	11.77
1280	642	$1.43 \cdot 10^{-1}$	3.46
5120	2562	$3.68 \cdot 10^{-2}$	3.88
20480	10242	$9.31 \cdot 10^{-3}$	3.95

(cf. (4.15)) can be observed in Table 4.10. Here, quadratic convergence can be observed, at least asymptotically.

TEAM Problem 10

Here, we consider the mixed boundary value problem for the Laplace equation in the domain presented in Fig. 4.5. The Dirichlet part of the boundary Γ is defined by $\Gamma_D = \{x \in \Gamma : x_3 = 50\}$. Thus, Dirichlet boundary conditions are given only on the "top" of the coil. The analytical solution is taken in the form (4.4) with $\tilde{y} = (0.0, 60.0, 50.0)^\top \notin \overline{\Omega}$ and the numerical results are presented in Table 4.11. The meaning of the values presented in this table is

Table 4.11. Accuracy of the Galerkin method on the boundary, TEAM problem 10

N	M	$Iter_1$	$Iter_2$	$D - Error_1$	CF_D	$N - Error_1$	CF_N
4928	2470	1061	90-95	$3.04 \cdot 10^{-2}$	–	$6.04 \cdot 10^{-1}$	–
19712	9862	1732	118-123	$7.60 \cdot 10^{-3}$	4.00	$2.03 \cdot 10^{-1}$	2.98

the same as in Table 4.9. The third column in Table 4.11 shows the number of Conjugate Gradient iterations with diagonal preconditioning (4.21) needed to reach the prescribed accuracy ε_2 for the Schur complement, while the fourth column indicates the number of Conjugate Gradient iterations with diagonal preconditioning (4.17) needed in each iteration step for the system with the single layer potential matrix. Note that the number of iterations is rather high for this geometrically very complicated example. Thus, a more effective preconditioning is required. The courses of the numerical solution inside of the domain Ω along the line (4.19) is very similar to those presented in Figs. 4.20–4.21. The Cauchy data can be seen in Figs. 4.18–4.19.

TEAM Problem 24

Here, we consider the mixed boundary value problem for the Laplace equation in the domain presented in Fig. 4.7. The Dirichlet part of the boundary Γ is defined by $\Gamma_D = \{x \in \Gamma : x_3 \geq 0\}$. Thus, the Dirichlet boundary condition is given on the upper part of the symmetric surface Γ. The analytical solution is taken in the form (4.4) with $\tilde{y} = (0.0, -80.0, 20.0)^\top \notin \overline{\Omega}$ and the numerical results are presented in Table 4.12 The meaning of the values presented in this

Table 4.12. Accuracy of the Galerkin method on the boundary, TEAM problem 24

N	M	$Iter_1$	$Iter_2$	$D - Error_1$	CF_D	$N - Error_1$	CF_N
4640	2320	48	72-78	$2.56 \cdot 10^{-2}$	–	$3.08 \cdot 10^{-1}$	–
18560	9280	71	93-102	$4.67 \cdot 10^{-3}$	5.48	$1.48 \cdot 10^{-1}$	2.08

table is the same as in Table 4.11. We have used the same preconditioning as in the previous example. The number of iterations reported in the third and in the fourth columns of the table is now much less. In Figs. 4.29–4.30 the Dirichlet datum and the Neumann datum for $N = 4944$ boundary elements and $M = 2474$ nodes are presented. The numerical curve obtained when using an approximate representation formula in comparison with the curve of the exact values (4.4) along the line

$$x(t) = \begin{pmatrix} 0.0 \\ -100.0 \\ 0 \end{pmatrix} + t \begin{pmatrix} 0.0 \\ 40.0 \\ 0 \end{pmatrix}, \ 0 \leq t \leq 1 \qquad (4.25)$$

inside of the domain Ω is shown in Fig. 4.31 for $N = 4640$. The values of the numerical solution $\tilde{u}$ and of the analytical solution u have been computed in 512 points uniformly placed on the line (4.25). The thick dashed line represents the course of the analytical solution (4.4), while the thin solid line shows the course of the numerical solution $\tilde{u}$. The values of the variable x_2 along the the line (4.25) are used for the axis of abscissas. The course of the numerical approximation does not optically distinguish from the exact solution on the left plot of Fig. 4.31. Thus, we show the zoom of these curves on the interval $[-84, -76]$ (right plot).

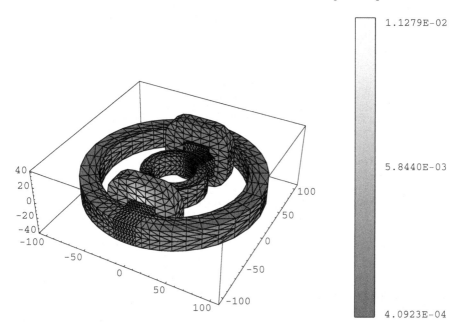

Fig. 4.29. Computed Dirichlet datum for the TEAM problem 24

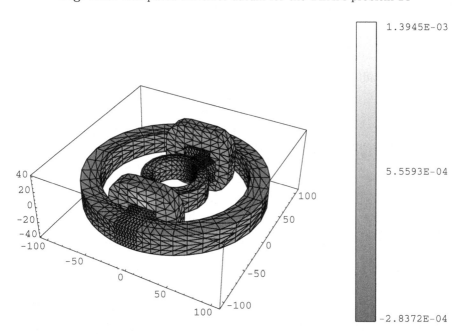

Fig. 4.30. Computed Neumann datum for the TEAM problem 24

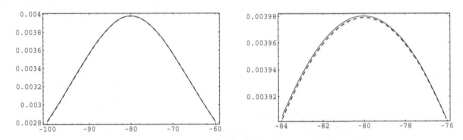

Fig. 4.31. Numerical and analytical curves for $N = 4640$, Mixed problem

4.2.7 Inhomogeneous Interface Problem

The purpose of this subsection is twofold. The first goal is to illustrate the numerical solution of the Poisson equation by the use of a particular solution (cf. Subsection 1.1.7). The second goal is to illustrate the numerical solution for an interface problem by Boundary Element Methods (cf. Subsection 1.1.8).

We consider the following interface problem

$$-\alpha_i \Delta u_i(x) = f_i(x) \quad \text{for } x \in \Omega, \quad -\alpha_e \Delta u_e(x) = 0 \quad \text{for } x \in \Omega^e, \qquad (4.26)$$

with transmission conditions describing the continuity of the potential and of the flux, respectively,

$$\gamma_0^{\text{int}} u_i(x) = \gamma_0^{\text{ext}} u_e(x), \; \alpha_i \gamma_1^{\text{int}} u_i(x) = \alpha_e \gamma_1^{\text{ext}} u_e(x) \quad \text{for } x \in \Gamma, \qquad (4.27)$$

and with the radiation condition

$$|u_e(x)| = \mathcal{O}\left(\frac{1}{|x|}\right) \quad \text{as } |x| \to \infty.$$

If a particular solution u_i^p of the interior Poisson equation is known,

$$-\alpha_i \Delta u_i^p(x) = f_i(x) \quad \text{for } x \in \Omega,$$

then the above interface problem can be reformulated as follows (cf. Subsection 1.1.8). Introduce a new unknown function u_i^* by

$$u_i = u_i^* + u_i^p$$

and rewrite the interface problem in terms of the functions u_i^* and u_e

$$-\alpha_i \Delta u_i^*(x) = 0 \quad \text{for } x \in \Omega, \quad -\alpha_e \Delta u_e(x) = 0 \quad \text{for } x \in \Omega^e,$$

with new transmission conditions

$$\gamma_0^{\text{int}} u_i^*(x) = \gamma_0^{\text{ext}} u_e(x) - \gamma_0^{\text{int}} u_i^p(x),$$

$$\alpha_i \gamma_1^{\text{int}} u_i^*(x) = \alpha_e \gamma_1^{\text{ext}} u_e(x) - \alpha_i \gamma_1^{\text{int}} u_i^p(x) \quad \text{for } x \in \Gamma.$$

Then, by the use of the interior and exterior Steklov-Poincaré operators S^{int} and S^{ext} (cf. (1.13), (1.46))

$$\gamma_1^{\text{int}} u_i^* = S^{\text{int}} \gamma_0^{\text{int}} u_i^* \,, \quad \gamma_1^{\text{ext}} u_e = -S^{\text{ext}} \gamma_0^{\text{ext}} u_e \,,$$

we rewrite the interface problem as (cf. (1.59))

$$\left(\alpha_i S^{\text{int}} + \alpha_e S^{\text{ext}} \right) \gamma_0^{\text{int}} u_i^* = -\alpha_i \gamma_1^{\text{int}} u_i^p - \alpha_e S^{\text{ext}} \gamma_0^{\text{int}} u_i^p \,. \tag{4.28}$$

Once the Dirichlet datum $\gamma_0^{\text{int}} u_i^*$ is found, we solve the interior Dirichlet boundary value problem for the Neumann datum $\gamma_1^{\text{int}} u_i^*$. The Cauchy data for the unknown functions u_i and u_e are then obtained via

$$\gamma_0^{\text{int}} u_i = \gamma_0^{\text{ext}} u_e = \gamma_0^{\text{int}} u_i^* + \gamma_0^{\text{int}} u_i^p \,, \quad \gamma_1^{\text{int}} u_i = \frac{\alpha_e}{\alpha_i} \gamma_1^{\text{ext}} u_e = \gamma_1^{\text{int}} u_i^* + \gamma_1^{\text{int}} u_i^p \,.$$

Unit Sphere

Let Γ be the surface of the unit sphere (4.1). The constants α_i, α_e and the right hand side f_i in (4.26) are

$$\alpha_i = \alpha_e = 1 \,, \quad f_i(x) = 1 \,, \quad \text{for } x \in \Omega \,.$$

The exact solution of this simple model problem is

$$u_i(x) = \frac{3 - |x|^2}{6} \,, \quad x \in \Omega \,, \quad u_e(x) = \frac{1}{3\,|x|} \,, \quad x \in \Omega^e \,. \tag{4.29}$$

Consider the function

$$u_i^p(x) = -\frac{1}{2} x_1^2 \,, \quad \text{for } x \in \Omega$$

as a particular solution of the Poisson equation.

The Galerkin method with piecewise linear basis functions φ_ℓ for the Dirichlet data $\gamma_0^{\text{int}} u_i = \gamma_0^{\text{ext}} u_e$ and $\gamma_0^{\text{int}} u_i^*$ and with piecewise constant basis functions ψ_k for the Neumann data $\gamma_1^{\text{int}} u_i = \gamma_1^{\text{ext}} u_e$ and $\gamma_1^{\text{int}} u_i^*$ will be used. The matrix of the single layer potential V_h and the matrix of the double layer potential K_h are generated in an approximative form using the partially pivoted ACA algorithm 3.9 with increasing accuracy ε_1. The resulting system of linear equations (cf. (2.45)) is then solved using the Conjugate Gradient method without preconditioning up to the relative accuracy $\varepsilon_2 = 10^{-8}$. The accuracy obtained for the analytical solution (4.29) is presented in Table 4.13. The numbers in this table have the following meaning. The third column in Table 4.13 shows the number of Conjugate Gradient iterations needed to reach the prescribed accuracy ε_2 for the linear system (2.45). The error for the Dirichlet datum and the convergence factor are shown in columns 4 and 5, while the corresponding error for the Neumann datum can be seen in columns 6 and 7. Note that the convergence of the Galerkin method for the unknown Dirichlet datum in the L_2 norm (4.20) is quadratic (5th column), while the convergence of the Neumann datum (4.14) is linear (7th column).

Table 4.13. Accuracy of the Galerkin method on the boundary, interface problem

N	M	$Iter$	$D - Error_1$	CF_D	$N - Error_1$	CF_N
80	42	8	$8.47 \cdot 10^{-2}$	–	$1.65 \cdot 10^{-1}$	–
320	162	12	$2.22 \cdot 10^{-2}$	3.82	$7.99 \cdot 10^{-2}$	2.07
1280	642	17	$5.59 \cdot 10^{-3}$	3.97	$3.88 \cdot 10^{-2}$	2.06
5120	2562	22	$1.40 \cdot 10^{-3}$	3.99	$1.92 \cdot 10^{-2}$	2.02
20480	10242	33	$3.50 \cdot 10^{-4}$	4.00	$9.55 \cdot 10^{-3}$	2.01

4.3 Linear Elastostatics

In this section we consider two numerical examples for the mixed boundary value problem of linear elastostatics (cf. (1.79))

$$
\begin{aligned}
-\sum_{j=1}^{3} \frac{\partial}{\partial x_j} \sigma_{ij}(\underline{u}, x) &= 0 && \text{for } x \in \Omega,\ i = 1, 2, 3, \\
\gamma_0^{\text{int}} \underline{u}(x) &= \underline{g}(x) && \text{for } x \in \Gamma_D, \\
\gamma_1^{\text{int}} \underline{u}(x) &= \underline{f}(x) && \text{for } x \in \Gamma_N,
\end{aligned}
\tag{4.30}
$$

where $\underline{u}$ is the displacement field of an elastic body initially occupying some bounded open domain $\Omega \in \mathbb{R}^3$ with boundary $\Gamma = \overline{\Gamma}_D \cup \overline{\Gamma}_N$.

4.3.1 Generation of Matrices

The most important matrices to be generated while using the Galerkin boundary element method for the mixed boundary value problem (4.30) are the single layer potential matrix V_h^{Lame} and the double layer potential matrix K_h^{Lame} (cf. 2.4), having the representation (2.48) and (2.51), respectively. Thus, in addition to the single and double layer potential matrices (V_h and K_h) for the Laplace operator, six additional dense matrices $V_{ij,h} \in \mathbb{R}^{N \times N}$ for $1 \leq i \leq j \leq 3$ have to be generated corresponding to (cf. (2.50))

$$
V_{ij,h}[k, \ell] = \frac{1}{4\pi} \int\limits_{\tau_k} \int\limits_{\tau_\ell} \frac{(x_i - y_i)(x_j - y_j)}{|x - y|^3} ds_y ds_x \, .
$$

Using the abbreviation (cf. Appendix C.2.3)

$$
S_{ij}(\tau, x) = \frac{1}{4\pi} \int\limits_{\tau} \frac{(x_i - y_i)(x_j - y_j)}{|x - y|^3} ds_y \, ,
$$

the above entries can be written in a symmetrised form

$$V_{ij,h}[k, \ell] = \frac{1}{2} \left(\int\limits_{\tau_k} S_{ij}(\tau_\ell, x) \, ds_x + \int\limits_{\tau_\ell} S_{ij}(\tau_k, x) \, ds_x \right).$$

The explicit form of the functions S_{ij} can be seen in Appendix C.2.3. The remaining integrals in the above symmetric form of the matrix entries $V_{ij,h}$ can be computed numerically using a 7-point quadrature rule, see Appendix C.1.

4.3.2 Relay

The geometry of the domain is shown in Fig. 4.32. The bottom of the relay is chosen to be the Dirichlet part Γ_D of the boundary Γ and the boundary condition is homogeneous, i.e.

$$\gamma_0^{int} \underline{u}(x) = 0, \quad \text{for } x \in \Gamma_D = \{ x \in \Gamma : x_3 = 0 \}.$$

The remaining part of the boundary is then considered as the Neumann boundary, where only on the top of the domain inhomogeneous boundary conditions are formulated,

$$\gamma_1^{int} \underline{u}(x) = \begin{cases} 0, & x \in \Gamma : x_3 < 10, \\ 1, & x \in \Gamma : x_3 = 10. \end{cases}$$

We choose the Young modulus $E = 114\,000$ and the Poisson ratio $\nu = 0.24$ that correspond to the values of steel. The original domain is shown in Fig. 4.32 for $N = 4944$ The matrix of the single layer potential V_h for the Laplace operator (cf. (2.49)), six matrices of the single layer potential V_h for the Lame operator (cf. (2.50)), and the matrix of the double layer potential K_h (cf. (2.51)) for the Laplace operator are generated in an approximative form using the partially pivoted ACA algorithm 3.9 with increasing accuracy. The system of linear equations is then solved using the Conjugate Gradient method for the Schur complement of the system (cf. (2.47)) up to the relative accuracy $\varepsilon_2 = 10^{-8}$. Note that this realisation requests an additional solution of a linear system with the single layer potential matrix in each iteration step. This system is solved again using Conjugate Gradient method up to the relative accuracy $\varepsilon_2 = 10^{-8}$. The matrix of the hypersingular operator is not generated explicitly. Its multiplication with a vector is realised using the matrix of the single layer potential as it is described in Section 2.4.

The results of the approximation are presented in Tables 4.14–4.16. The number of boundary elements is listed in the first column of these tables. The second column contains the number of nodes, while the prescribed accuracy for the ACA algorithm for the approximation of all matrices $K_h \in \mathbb{R}^{N \times M}$ and $V_h, V_{k\ell,h} \in \mathbb{R}^{N \times N}$, $k, \ell = 1, 2, 3$ is given in the third column. The pairs of further columns of these tables show the memory requirements in MByte and the percentage of memory compared to the original matrix. The deformed

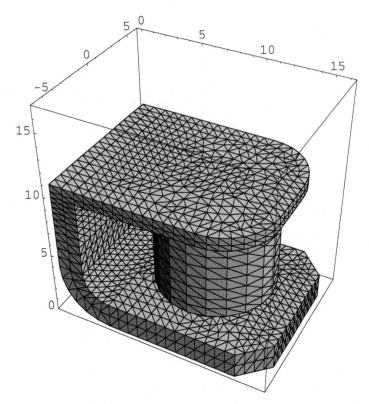

Fig. 4.32. Relay for $N = 4944$

Table 4.14. ACA approximation of the Galerkin matrices V_h and K_h

N	M	ε_1	V_h	%	K_h	%
4944	2474	$1.0 \cdot 10^{-4}$	37.24	20.0	52.96	56.8
19776	9890	$1.0 \cdot 10^{-5}$	258.65	8.7	326.45	10.9

Table 4.15. ACA approximation of the Galerkin matrices $V_{11,h}, V_{12,h}$ and $V_{13,h}$

N	M	ε_1	$V_{11,h}$	%	$V_{12,h}$	%	$V_{13,h}$	%
4944	2474	$1.0 \cdot 10^{-4}$	46.03	24.7	46.75	25.1	45.74	24.5
19776	9890	$1.0 \cdot 10^{-5}$	435.61	14.6	433.91	14.5	402.00	13.5

Table 4.16. ACA approximation of the Galerkin matrices $V_{22,h}$, $V_{23,h}$ and $V_{33,h}$

N	M	ε_1	$V_{22,h}$	%	$V_{23,h}$	%	$V_{33,h}$	%
4944	2474	$1.0 \cdot 10^{-4}$	47.21	25.3	46.78	25.1	45.75	24.5
19776	9890	$1.0 \cdot 10^{-5}$	463.43	15.5	415.98	13.9	421.66	14.1

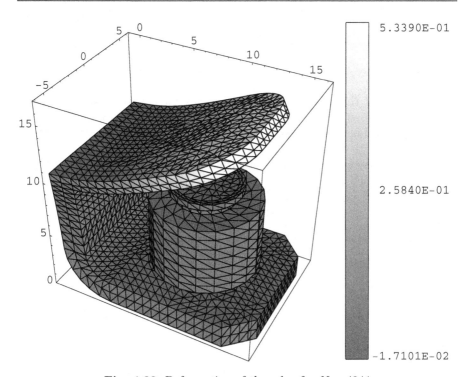

Fig. 4.33. Deformation of the relay for $N = 4944$

Table 4.17. Number of iterations, Relay problem

N	M	$Iter_1$	$Iter_2$
4944	2474	286	26-28
19776	9890	368	25-29

domain can be seen from the same point of view in Fig. 4.33. In this figure, the real deformation is amplified by a factor 10. In Table 4.17, the number of iterations required by the Conjugate Gradient method is shown. The third column of this table shows the number of iterations for the Schur complement equation (2.47), while the fourth column shows the number of iterations required for the iterative solution of the linear system for the single layer potential in each iteration step. The required accuracy was $\varepsilon_2 = 10^{-8}$ for both systems.

4.3.3 Foam

The geometry of the domain, which is a model for a metal foam, is shown in Fig. 4.34. The speciality of this domain is its multiple connectivity and rather small volume compared to its surface. There is only one discretisation of the domain with $N = 28952$ surface elements. The bottom and the top of the foam are chosen to be the Dirichlet part Γ_D of the boundary Γ, and the boundary condition is homogeneous on the bottom, i.e.

$$\gamma_0^{\text{int}} \underline{u}(x) = 0\,, \quad \text{for } x \in \Gamma : x_3 = 0\,,$$

while a prescribed constant displacement is posed on the top, i.e.

$$\gamma_0^{\text{int}} \underline{u}(x) = (0,0,0.1)^\top\,, \quad \text{for } x \in \Gamma : x_3 = 15\,.$$

The remaining part of the boundary is then considered as the Neumann boundary, where homogeneous boundary conditions are formulated:

$$\gamma_1^{\text{int}} \underline{u}(x) = 0\,, \quad \text{for } x \in \Gamma : 0 < x_3 < 15\,.$$

We choose the Young modulus $E = 114\,000$ and the Poisson ratio $\nu = 0.24$ that correspond to the values of steel. The original domain is shown in Fig. 4.34 for $N = 28952$. The matrix of the single layer potential V_h for the Laplace operator (cf. (2.49)), six matrices of the single layer potential V_h for the Lame operator (cf. (2.50)), and the matrix of the double layer potential K_h (cf. (2.51)) for the Laplace operator are generated in an approximative form using the partially pivoted ACA algorithm 3.9. The system of linear equations is then solved using a Conjugate Gradient method for the Schur complement system (cf. (2.47)) up to the relative accuracy $\varepsilon_2 = 10^{-8}$. Note that this realisation requires an additional solution of a linear system with the single layer potential matrix in each iteration step. This system is solved again using a Conjugate Gradient method up to the relative accuracy $\varepsilon_2 = 10^{-8}$. The matrix of the hypersingular operator is not generated explicitly. Its multiplication with a vector is realised using the matrix of the single layer potential as it is described in Section 2.4.

The results of the approximation are presented in Tables 4.18–4.20. The number of boundary elements is listed in the first column of these tables.

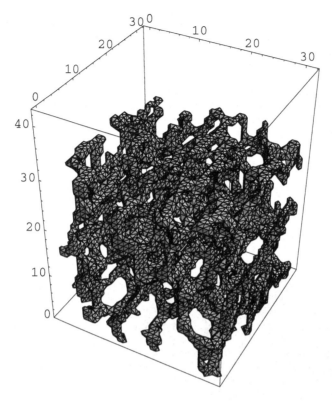

Fig. 4.34. Foam for $N = 28952$

The second column contains the number of nodes while in the third column
the prescribed accuracy for the ACA algorithm for the approximation of all
matrices $K_h \in \mathbb{R}^{N \times M}$ and $V_h, V_{k\ell,h} \in \mathbb{R}^{N \times N}$, $k, \ell = 1, 2, 3$ is given. The
pairs of further columns of these tables show the memory requirements in
MByte and the percentage of memory compared to the original matrix. The

Table 4.18. ACA approximation of the Galerkin matrices V_h and K_h

N	M	ε_1	V_h	%	K_h	%
28952	14152	$1.0 \cdot 10^{-4}$	260.66	4.1	496.58	15.9

deformed domain can be seen from the same point of view in Fig. 4.35. In
this figure, the real deformation is amplified by a factor 100. The number of
iterations required by the Conjugate Gradient method is shown in Table 4.21.
In this table, the third column shows the number of iterations for the Schur
complement equation (2.47), while the number of iterations required for the

Table 4.19. ACA approximation of the Galerkin matrices $V_{11,h}, V_{12,h}$ and $V_{13,h}$

N	M	ε_1	$V_{11,h}$	%	$V_{12,h}$	%	$V_{13,h}$	%
28952	14152	$1.0 \cdot 10^{-4}$	398.36	6.2	417.94	6.5	418.55	6.5

Table 4.20. ACA approximation of the Galerkin matrices $V_{22,h}, V_{23,h}$ and $V_{33,h}$

N	M	ε_1	$V_{22,h}$	%	$V_{23,h}$	%	$V_{33,h}$	%
28952	14152	$1.0 \cdot 10^{-4}$	402.36	6.3	415.22	6.5	398.93	6.2

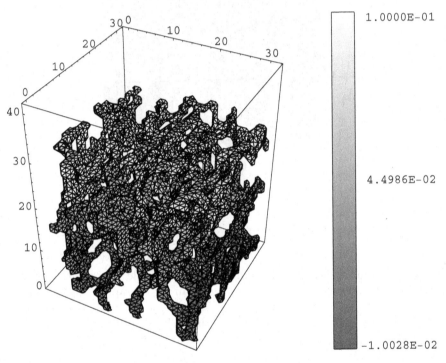

Fig. 4.35. Deformation of the foam for $N = 28952$

iterative solution of the linear system for the single layer potential in each iteration step can be seen in the fourth column. The required accuracy was $\varepsilon_2 = 10^{-8}$ for both systems.

4.4 Helmholtz Equation

In this section we consider some numerical examples for the Helmholtz equation

Table 4.21. The number of iterations, Foam problem

N	M	$Iter_1$	$Iter_2$
28952	14152	253	19-21

$$-\Delta u(x) - \kappa^2 u(x) = 0, \tag{4.31}$$

where u is an analytically given function.

4.4.1 Analytical Solutions

Particular solutions of the Helmholtz equation (4.31) are, for example,

$$\Phi_{k_1,k_2,k_3}(x) = \exp\left(\imath\left(k_1 x_1 + k_2 x_2 + k_3 x_3\right)\right), \quad k_1^2 + k_2^2 + k_3^2 = \kappa^2,$$

$$\Phi_{0,k_2,k_3}(x) = (a + b\,x_1)\exp\left(\imath\left(k_2 x_2 + k_3 x_3\right)\right), \quad k_2^2 + k_3^2 = \kappa^2, \tag{4.32}$$

$$\Phi_{0,0,0}(x) = (a_1 + b_1 x_1)(a_2 + b_2 x_2)\exp\left(\imath\,\kappa\,x_3\right).$$

Here k_1, k_2 and k_3 are arbitrary complex numbers satisfying the corresponding conditions. Thus, different products of linear, exponential, trigonometric, and hyperbolic functions can be chosen for numerical tests, if we consider interior boundary value problems in a three-dimensional open bounded domain $\Omega \subset \mathbb{R}^3$. Furthermore, the fundamental solution

$$u^*(x, \tilde{y}) = \frac{1}{4\pi} \frac{e^{\imath\,\kappa\,|x-\tilde{y}|}}{|x - \tilde{y}|} \tag{4.33}$$

can be considered as a particular solution of the Helmholtz equation (4.31) for both, interior ($x \in \Omega$, $\tilde{y} \in \Omega^e = \mathbb{R}^3 \setminus \overline{\Omega}$), and exterior ($x \in \Omega^e$, $\tilde{y} \in \Omega$) boundary value problems.

4.4.2 Discretisation, Approximation and Iterative Solution

We solve the interior and exterior Dirichlet and Neumann boundary value problems using a Galerkin boundary element method (cf. Section 2). Piecewise linear basis functions φ_ℓ will be used for the approximation of the Dirichlet datum $\gamma_0^{\mathrm{int}} u$ and piecewise constant basis functions ψ_k for the approximation of the Neumann datum $\gamma_1^{\mathrm{int}} u$. We will use the L_2 projection for the approximation of the given part of the Cauchy data. The boundary element matrices V_h, K_h, and C_h are generated in an approximative form using the complex valued version of the partially pivoted ACA algorithm 3.9 with a variable relative accuracy ε_1. The resulting systems of linear equations are solved using the GMRES method with or without preconditioning up to a relative accuracy $\varepsilon_2 = 10^{-8}$.

4.4.3 Generation of Matrices

The most important matrices to be generated while using the Galerkin boundary element method for boundary value problems for the Helmholtz equation (4.31) are the single layer potential matrix $V_{\kappa,h}$ (cf. (2.53)),

$$V_{\kappa,h}[k,\ell] = \frac{1}{4\pi} \int\limits_{\tau_k} \int\limits_{\tau_\ell} \frac{e^{i\kappa|x-y|}}{|x-y|} ds_y ds_x \,,$$

and the double layer potential matrix $K_{\kappa,h}$ (cf. 2.58),

$$K_{\kappa,h}[k,j] = \frac{1}{4\pi} \int\limits_{\tau_k} \int\limits_{\Gamma} (1 - i\kappa|x-y|) e^{i\kappa|x-y|} \frac{(x-y, \underline{n}(y))}{|x-y|^3} \varphi_j(y) ds_y ds_x \,.$$

Furthermore, when solving the Neumann boundary value problem for the Helmholtz equation, the matrix of the hypersingular operator (cf. (2.62))

$$D_{\kappa,h}[i,j] =$$

$$\frac{1}{4\pi} \int\limits_{\Gamma} \int\limits_{\Gamma} \frac{e^{i\kappa|x-y|}}{|x-y|} (\underline{\mathrm{curl}}_\Gamma \varphi_j(y), \underline{\mathrm{curl}}_\Gamma \varphi_i(x)) ds_y ds_x -$$

$$\frac{\kappa^2}{4\pi} \int\limits_{\Gamma} \int\limits_{\Gamma} \frac{e^{i\kappa|x-y|}}{|x-y|} \varphi_j(y) \varphi_i(x) (\underline{n}(x), \underline{n}(y)) ds_y ds_x$$

has to be involved. The first part of this formula corresponds for $\kappa = 0$ to the hypersingular operator for the Laplace equation, and, therefore, can be handled in the same way, i.e. these entries are some linear combinations of the entries of the matrix of the single layer potential $V_{\kappa,h}$. It remains to generate an additional matrix $C_{\kappa,h}$, having the entries

$$C_{\kappa,h}[i,j] = \int\limits_{\Gamma} \int\limits_{\Gamma} \frac{e^{i\kappa|x-y|}}{|x-y|} \varphi_j(y) \varphi_i(x) (\underline{n}(x), \underline{n}(y)) ds_y ds_x \,. \qquad (4.34)$$

To generate the entries of the single layer potential matrix $V_{\kappa,h}$ numerically, we rewrite (2.53) as follows:

$$V_{\kappa,h}[k,\ell] = V_{0,h}[k,\ell] + \frac{1}{4\pi} \int\limits_{\tau_k} \int\limits_{\tau_\ell} \frac{e^{i\kappa|x-y|} - 1}{|x-y|} ds_y ds_x \,.$$

In the above, the entries $V_{0,h}[k,\ell]$ are the entries of the single layer potential matrix of the Laplace operator, and, therefore, can be computed as it was discussed in Section 4.2.3. The remaining double integral has no singularity for $x \to y$, and can be computed numerically, using the 7-point quadrature rule (cf. Appendix C.1) for each triangle.

For the double layer potential matrix, the same idea leads to the following decomposition:

$$K_{\kappa,h}[k,j] =$$
$$K_{0,h}[k,j] + \frac{1}{4\pi} \int_{\tau_k} \int_{\Gamma} \left((1 - \imath\kappa|x-y|)e^{\imath\kappa|x-y|} - 1 \right) \frac{(x-y,\underline{n}(y))}{|x-y|^3} \varphi_j(y) ds_y ds_x \,.$$

Again, the first part of this decomposition belongs to the double layer potential matrix of the Laplace operator, while the second part has no singularity for $x \to y$. Thus, the 7-point quadrature rule can be applied again. However, the numerical integration with respect to the variable y has to be done over all triangles in the support of the basis function φ_j for each integration point with respect to the variable x. Therefore, the generation of the matrix entries for the double layer potential matrix for the Helmholtz equation is by far more complicated than for the Laplace equation.

However, the most complicated numerical procedure is required when generating the entries of the matrix $C_{\kappa,h}$ corresponding to (4.34). Using the same decomposition as for the previous matrices, we get

$$C_{\kappa,h}[i,j] = C_{0,h}[i,j] + \int_{\Gamma} \int_{\Gamma} \frac{e^{\imath\kappa|x-y|} - 1}{|x-y|} \varphi_j(y)\varphi_i(x)(\underline{n}(x),\underline{n}(y)) ds_y ds_x \,.$$

In the above, the second summand has no singularity for $x \to y$, and the 7-point quadrature rule can be applied again. Note that in this case, the quadrature rule has to be applied to every triangle in the support of the function φ_i, and, for each of its integration points, to every triangle in the support of the function φ_j. Furthermore, a symmetrisation is necessary in order to keep the symmetry of the matrix $C_{\kappa,h}$. Fortunately, the first summand $C_{0,h}[i,j]$ does not require some additional analytical work. Since the normal vectors $\underline{n}(x)$ and $\underline{n}(y)$ are constant within the single triangles in the supports of the functions φ_i and φ_j, these integrals can be computed using the symmetrised combination of the analytical integration and of the 7-point quadrature rule.

4.4.4 Interior Dirichlet Problem

Here we solve the Helmholtz equation (4.31) together with the Dirichlet boundary condition $\gamma_0^{int}u(x) = g(x)$ for $x \in \Gamma$, where Γ is a given surface. The variational problem (1.106)

$$\langle V_\kappa t, w \rangle_\Gamma = \left\langle \left(\frac{1}{2}I + K_\kappa\right)g, w \right\rangle_\Gamma \quad \text{for all} \quad w \in H^{-1/2}(\Gamma)$$

is discretised and leads to a system of linear equations

$$V_{\kappa,h}\underline{\tilde{t}} = \left(\frac{1}{2}M_h + K_{\kappa,h}\right)\underline{g} \,. \tag{4.35}$$

The matrix $V_{\kappa,h}$ of this system is symmetric. This property can be used in order to save computer memory while generating the matrix. However, the matrix is not selfadjoint, and, therefore, the Conjugate Gradient method can not be used. Thus, for an iterative solution of the system (4.35), the complex GMRES method will be used instead.

Unit Sphere

The analytical solution is taken in the form (4.32). For $x = (x_1, x_2, x_3)^\top \in \Omega$ we consider the function

$$u(x) = \Phi_{0,-i\,2\sqrt{3},4} = 4\,x_1 \exp(2\sqrt{3}\,x_2) \exp(i\,4\,x_3)\,, \qquad (4.36)$$

which satisfies the Helmholtz equation (4.31) for $\kappa = 2$. The results of the computations for this rather moderate wave number are shown in Tables 4.22 and 4.23. The number of boundary elements is listed in the first column of

Table 4.22. ACA approximation of the Galerkin matrices $K_{\kappa,h}$ and $V_{\kappa,h}$

N	M	ε_1	MByte(K_h)	%	MByte(V_h)	%
80	42	$1.0 \cdot 10^{-2}$	0.05	99.9	0.05	50.4
320	162	$1.0 \cdot 10^{-3}$	0.66	84.0	0.55	35.0
1280	642	$1.0 \cdot 10^{-4}$	6.08	48.5	5.02	20.1
5120	2562	$1.0 \cdot 10^{-5}$	49.05	24.5	39.46	9.86
20480	10242	$1.0 \cdot 10^{-6}$	357.90	11.2	280.60	4.47

these tables. The second column contains the number of nodes, while in the third column of Table 4.22 the prescribed accuracy for the ACA algorithm for the approximation of both matrices $K_{\kappa,h} \in \mathbb{C}^{N \times M}$ and $V_{\kappa,h} \in \mathbb{C}^{N \times N}$ is given. In this table, the fourth column shows the memory requirements in MByte for the approximate double layer potential matrix $K_{\kappa,h}$. The quality of this approximation in percentage of the original matrix is listed in the next column, whereas the corresponding values for the single layer potential matrix $V_{\kappa,h}$ can be seen in the columns six and seven. The third column in Table 4.23 shows the number of GMRES iterations needed to reach the prescribed accuracy ε_2, while the relative L_2 error for the Neumann datum,

$$Error_1 = \frac{\|\gamma_1^{int} u - \tilde{t}_h\|_{L_2(\Gamma)}}{\|\gamma_1^{int} u\|_{L_2(\Gamma)}}\,,$$

is given in the fourth column. The next column represents the rate of convergence for the Neumann datum, i.e. the quotient of the errors in two consecutive lines of column four. Finally, the last two columns show the absolute error (cf. (2.56)) in a prescribed inner point $x^* \in \Omega$,

Table 4.23. Accuracy of the Galerkin method, Dirichlet problem

N	M	Iter	$Error_1$	CF_1	$Error_2$	CF_2
80	42	21	$6.98 \cdot 10^{-1}$	–	$3.12 \cdot 10^{-0}$	–
320	162	29	$3.13 \cdot 10^{-1}$	2.23	$8.68 \cdot 10^{-2}$	36.00
1280	642	37	$1.45 \cdot 10^{-1}$	2.16	$7.42 \cdot 10^{-3}$	11.70
5120	2562	46	$7.03 \cdot 10^{-2}$	2.06	$7.37 \cdot 10^{-4}$	10.06
20480	10242	57	$3.48 \cdot 10^{-2}$	2.02	$7.74 \cdot 10^{-5}$	9.52

$$Error_2 = |u(x^*) - \tilde{u}(x^*)|, \ x^* = (0.28591, 0.476517, 0.667123)^\top, \quad (4.37)$$

for the value $\tilde{u}(x^*)$ obtained using the approximate representation formula (2.55). Table 4.23 obviously shows a linear convergence $\mathcal{O}(N^{-1/2}) = \mathcal{O}(h)$ of the Galerkin boundary element method for the Neumann datum in the L_2 norm. It should be noted that this theoretically guaranteed convergence order can already be observed when approximating the matrices $K_{\kappa,h}$ and $V_{\kappa,h}$ with much less accuracy as it was used to obtain the results in Table 4.22. However, this high accuracy is necessary in order to be able to observe the third order (or even better) pointwise convergence rate within the domain Ω presented in the last two columns of Table 4.23.

In Figs. 4.36–4.37, the given Dirichlet datum (real and imaginary parts) for $N = 1280$ boundary elements is presented. The computed Neumann datum is presented in Figs. 4.38 (real part) and 4.39 (imaginary part). The numerical curves obtained when using an approximate representation formula in comparison with the curve of the exact values (4.36) along the line

$$x(t) = \begin{pmatrix} -0.3 \\ -0.5 \\ -0.7 \end{pmatrix} + t \begin{pmatrix} 0.6 \\ 1.0 \\ 1.4 \end{pmatrix}, \ 0 \le t \le 1 \quad (4.38)$$

inside of the domain Ω are shown in Fig. 4.40 for $N = 80$ and in Fig. 4.41 for $N = 320$. The values of the numerical solution $\tilde{u}$ and of the analytical solution u have been computed in 512 points uniformly placed on the line (4.38). In these figures, the thick dashed line represents the course of the analytical solution (4.36), while the thin solid line shows the course of the numerical solution $\tilde{u}$. The values of the variable x_2 along the the line (4.38) are used for the axis of abscissas. The left plots in these figures correspond to the real parts of the solutions while the imaginary parts are shown on the right. The next Fig. 4.42 shows these curves for $N = 1280$, but on the zoomed interval $[0.3, 0.5]$ with respect to the variable x_2 in order to see the very small difference between them. It is almost impossible to see any optical difference between the numerical and analytical curves for higher values of N. Note that the point x^* in (4.37) is chosen close to the maximum of the function $\operatorname{Im} u$ along the line where the error seems to reach its maximum.

Thus, for the moderate value of the wave number $\kappa = 2$, the quality of numerical results on the unit sphere is almost the same as for the Laplace

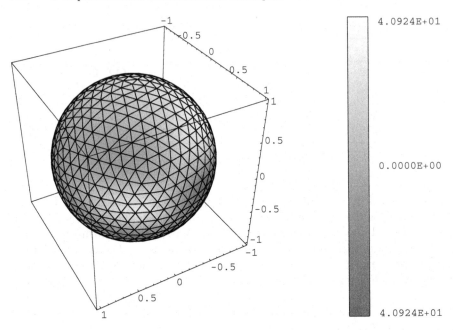

Fig. 4.36. Given Dirichlet datum (real part) for the unit sphere, $N = 1280$

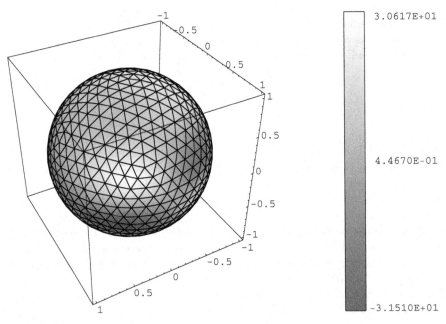

Fig. 4.37. Given Dirichlet datum (imaginary part) for the unit sphere, $N = 1280$

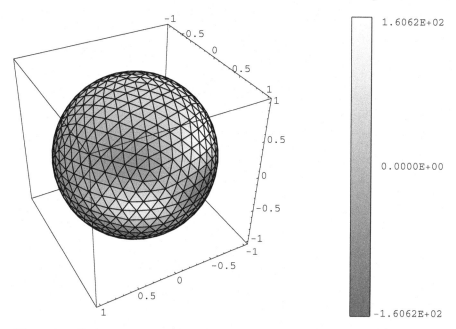

Fig. 4.38. Computed Neumann datum (real part) for the unit sphere, $N = 1280$

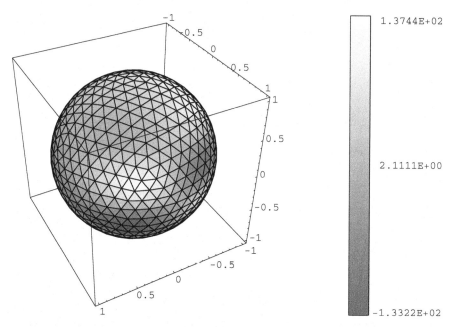

Fig. 4.39. Computed Neumann datum (imaginary part) for the unit sphere, $N = 1280$

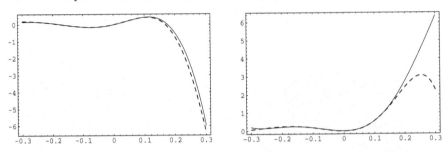

Fig. 4.40. Numerical and analytical curves for $N = 80$, Dirichlet problem

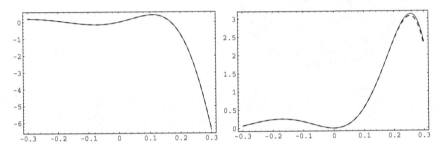

Fig. 4.41. Numerical and analytical curves for $N = 320$, Dirichlet problem

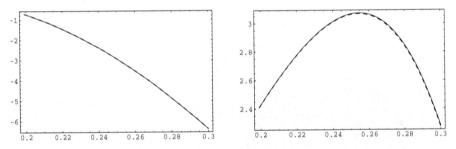

Fig. 4.42. Numerical and analytical curves for $N = 1280$, Dirichlet problem

equation. The ACA approximation is good, the number of GMRES iterations is low without any preconditioning, and it grows corresponding to the theory and, finally, the theoretical linear convergence order of the Neumann datum on the surface Γ as well as the cubic convergence order in the inner points of the domain Ω are perfectly illustrated.

Unit Sphere. Multifrequency Analysis

Since the Helmholtz equation provides an additional parameter, the wave number κ, it is especially interesting and important to study the behaviour of our numerical methods with respect to this parameter. The quality of the matrix approximation, the number of iterations needed to solve the corresponding linear systems, and, of course, the accuracy of the whole procedure are of special interest. We will now solve the Helmholtz equation for a fixed discretisation of the surface Γ but for a sequence of wave numbers $\kappa \in [\kappa_{min}, \kappa_{max}]$.

If $\operatorname{Im} \kappa \neq 0$, then the inner Dirichlet boundary value problem is uniquely solvable. The situation is different for $\operatorname{Im} \kappa = 0$. In this case the uniqueness holds only if κ^2 is not an eigenvalue of the Laplace operator subjected to homogeneous Dirichlet boundary conditions,

$$-\Delta u(x) = \lambda\, u(x) \quad \text{for } x \in \Omega, \quad \gamma_0^{\text{int}} u(x) = 0 \quad \text{for } x \in \Gamma,$$

(cf. Section 1.4). For general Γ, the eigenvalues of the Laplace operator are not known and it can happen that one or even several of them belong to the interval $[\kappa_{min}, \kappa_{max}]$. In this case some difficulties will occur when solving the discrete problem. Now we are going to illustrate the situation. The exact eigenfunctions and eigenvalues on the unit ball Ω are analytically known and can be represented in spherical coordinates

$$x = \varrho \begin{pmatrix} \cos\varphi \sin\theta \\ \sin\varphi \sin\theta \\ \cos\theta \end{pmatrix}, \quad 0 \le \varrho < 1,\ 0 \le \varphi < 2\pi,\ 0 \le \theta \le \pi$$

as follows

$$u_{k,n,m}(\varrho, \varphi, \theta) = \frac{J_{n+1/2}(\mu_{n,m}\varrho)}{\sqrt{\varrho}}\, P_{n,|k|}(\cos\theta)\, e^{\imath k\varphi}, \qquad (4.39)$$

with

$$m \in \mathbb{N}, \quad n \in \mathbb{N}_0, \quad |k| \le n.$$

In (4.39), $\mu_{n,m}$ are the zeros of the Bessel functions $J_{n+1/2}$. $P_{n,k}$ are the associated Legendre polynomials

$$P_{n,k}(u) = (-1)^k \left(1 - u^2\right)^{k/2} \frac{d^k}{dx^k} P_n(u)$$

defined for

$$|u| \le 1, \quad k = 0, \dots, n, \quad n \in \mathbb{N}_0.$$

The Legendre polynomials P_n are given in (3.12). The corresponding eigenvalues are

$$\lambda_{n,m} = \mu_{n,m}^2.$$

For $n = k = 0$ and $m \in \mathbb{N}$, the eigenvalues and the eigenfunctions are of an especially simple form. In this case we use

$$J_{1/2}(z) = \sqrt{\frac{2}{\pi}}\, \frac{\sin z}{\sqrt{z}}$$

and obtain

$$u_{0,0,m}(\varrho, \varphi, \theta) = c_m \frac{\sin(\mu_{0,m}\varrho)}{\varrho}, \quad m \in \mathbb{N}. \tag{4.40}$$

Thus, the corresponding critical values of κ are

$$\kappa = \mu_{0,m} = \pi m, \quad m \in \mathbb{N}.$$

In particular, $\kappa = \pi$ is a critical value.

We solve the boundary value problem for the Dirichlet boundary value problem for the Helmholtz equation (4.31) having the analytical solution (4.33) with $\tilde{y} = (1.1, 0.0, 0.0)^\top \notin \overline{\Omega}$. We will use 17 uniformly distributed values of κ on the interval $[3.1, 3.2]$. The discretisation of the boundary will be a polyhedron having $N = 320$ boundary elements (cf. Fig. 4.1). The following figures illustrate the results: In Fig. 4.43, the L_2 error of the Neumann datum is shown as a function of κ in the left plot. The right plot shows the number of GMRES iterations needed to reach the relative accuracy $\varepsilon_2 = 10^{-8}$ of the numerical solution of the linear system (2.57). Thus, a significant jump of the

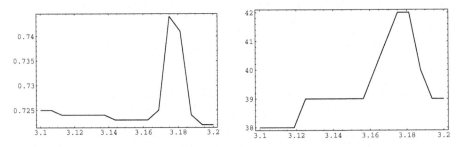

Fig. 4.43. Multifrequency computation for $N = 320$, Dirichlet problem

accuracy is displayed for the value $\kappa = 3.1750$. Also, a significant increase of the number of iterations can be seen close to the critical value $\kappa = 3.1750$. The quality of the ACA approximation of the matrices $K_{\kappa,h}$ and $V_{\kappa,h}$ is more or less the same for all values of the parameter κ on this rather small interval. Thus we can deduce that the value of the parameter $\kappa^2 = 3.1750^2$ is close to the eigenvalue of the Dirichlet boundary value problem for the Laplace equation in the polyhedron Ω_h with $N = 320$ elements. This value is remarkably close to the first eigenvalue π^2 of the continuous problem. However, the discrete values of the parameter κ will never meet the correct eigenvalue exactly and, probably, the closeness to the eigenvalue will not be detected. In this situation quite wrong results can be obtained. We illustrate this fact in the next three figures where the real parts (left plots) and the imaginary parts (right plots) of the analytical solution (thick dashed lines) and of the numerical solution (thin solid lines) are presented for three values of the parameter $\kappa = 3.15, 3.175, 3.2$. We can see that the numerical solution differs significantly from the analytical one for $\kappa = 3.175$, while the numerical results for $\kappa = 3.15$ and $\kappa = 3.2$ are quite good for this rather rough discretisation.

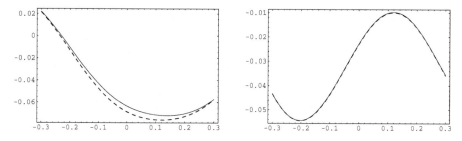

Fig. 4.44. Numerical and analytical curves for $\kappa = 3.15$ and $N = 320$, Dirichlet problem

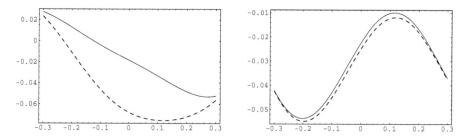

Fig. 4.45. Numerical and analytical curves for $\kappa = 3.175$ and $N = 320$, Dirichlet problem

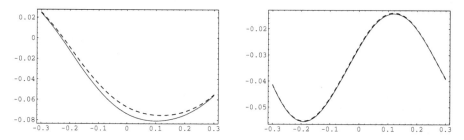

Fig. 4.46. Numerical and analytical curves for $\kappa = 3.2$ and $N = 320$, Dirichlet problem

In the next example, we solve the Dirichlet boundary value problem on the polyhedron Ω_h with $N = 1280$ elements for 65 values of the wave number κ uniformly distributed on the interval $[0, 16]$. Thus, the first value corresponds to the Laplace operator.

In Figs. 4.47–4.48 we show how the ACA approximation quality of the matrices $K_{\kappa,h}$ and $V_{\kappa,h}$ depends on the wave number. The left plots in these figures present the memory requirements in MByte, while the right plots show the same result in percentage compared to the full memory for $\varepsilon_1 = 10^{-4}$. The linear dependence of the memory requirement of the wave number is clearly indicated by these numerical tests. Also this example shows the loss

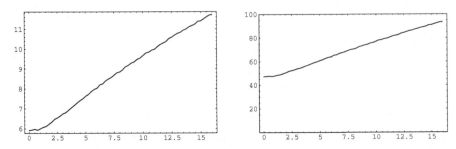

Fig. 4.47. Approximation of the double layer potential matrix $K_{\kappa,h}$ for $N = 1280$

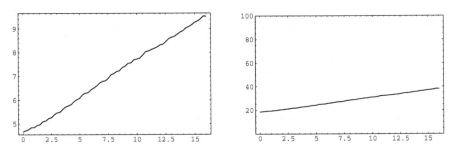

Fig. 4.48. Approximation of the single layer potential matrix $V_{\kappa,h}$ for $N = 1280$

of the accuracy close to the critical values of the parameter κ. In Fig. 4.49, we present again the L_2 norm of the error for the Neumann datum (left plot) and the number of GMRES iterations (right plot) as functions of the wave number κ. The left plot in Fig. 4.49 clearly shows a total loss of accuracy close to the

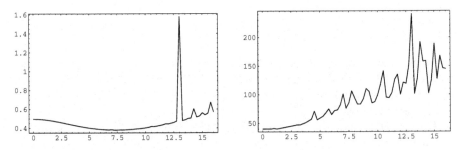

Fig. 4.49. Multifrequency computation for $N = 1280$, Dirichlet problem

wave number $\kappa = 13.0$. If we plot the analytical and the numerical values of the solution of the boundary value problem for three subsequent points 12.75, 13.0 and 13.25 for the parameter κ we can see this loss of accuracy optically. The results are presented in Figs. 4.50–4.52. It is remarkable that

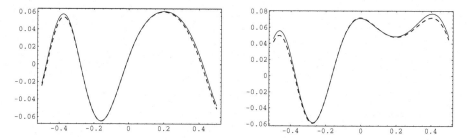

Fig. 4.50. Numerical and analytical curves for $\kappa = 12.75$ and $N = 1280$, Dirichlet problem

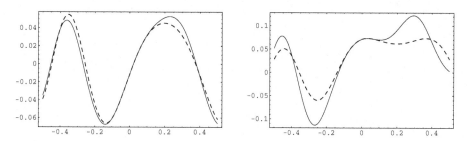

Fig. 4.51. Numerical and analytical curves for $\kappa = 13.0$ and $N = 1280$, Dirichlet problem

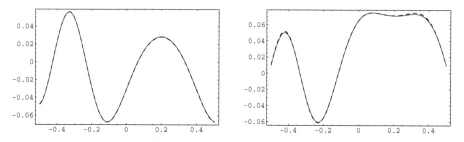

Fig. 4.52. Numerical and analytical curves for $\kappa = 13.25$ and $N = 1280$, Dirichlet problem

only one critical value (close to 4π) of the wave number κ was detected on the interval $[0, 16]$. This fact is due to the rather big step of 0.25 with respect to κ, which was used in the above example.

Exhaust Manifold

The analytical solution is taken in the form (4.33) with $\tilde{y} = (0, 0, 0.06)^\top$ and $\kappa = 80$, which is moderate compared to the rather small dimension of the domain (cf. Fig. 4.11). The results of the computations are shown in Tables 4.24 and 4.25. The third column shows the number of iterations required

Table 4.24. ACA approximation of the Galerkin matrices $K_{\kappa,h}$ and $V_{\kappa,h}$

N	M	ε_1	MByte($K_{\kappa,h}$)	%	MByte($V_{\kappa,h}$)	%
2264	1134	$1.0 \cdot 10^{-3}$	16.62	42.4	11.82	15.1
9056	4530	$1.0 \cdot 10^{-4}$	137.86	22.0	97.08	7.8
36224	18114	$1.0 \cdot 10^{-5}$	1046.80	10.5	696.65	3.5

Table 4.25. Accuracy of the Galerkin method, Dirichlet problem

N	M	$Iter$	$Error_1$	CF_1	$Error_2$	CF_2
2264	1134	177	$3.10 \cdot 10^{-1}$	-	$8.88 \cdot 10^{-3}$	-
9056	4530	208	$1.40 \cdot 10^{-1}$	2.2	$1.08 \cdot 10^{-3}$	8.2
36224	18114	244	$5.83 \cdot 10^{-2}$	2.4	$9.27 \cdot 10^{-5}$	11.7

by the GMRES method without preconditioning. The fourth column displays the L_2 error of the Neumann datum, while the next column shows its linear convergence. The last pair of columns of Table 4.25 shows the absolute error (cf. (2.56)) in a prescribed inner point $x^* \in \Omega$,

$$Error_2 = |u(x^*) - \tilde{u}(x^*)|, \quad x^* = (0.145303, 0.1, -0.05)^\top \qquad (4.41)$$

for the value $\tilde{u}(x^*)$ obtained using an approximate representation formula (2.55). Finally, the last column of this table indicates the cubic (or even better) convergence of this quantity.

In Figs. 4.53– 4.54, the real part and the imaginary part of the given Dirichlet datum are presented. The computed Neumann datum is presented in Figs. 4.55–4.56, where again the left plot corresponds to the real part of the Neumann datum, while the imaginary part is shown on the right. The numerical curve, obtained when using an approximate representation formula in comparison with the curve of the exact values (4.33) along the line

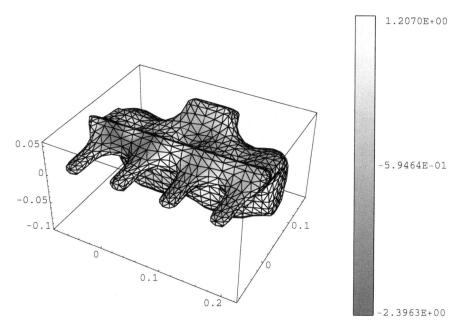

1.2070E+00

-5.9464E-01

-2.3963E+00

Fig. 4.53. Given Dirichlet datum (real part) for the exhaust manifold

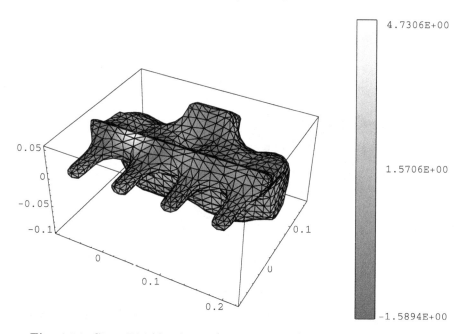

4.7306E+00

1.5706E+00

-1.5894E+00

Fig. 4.54. Given Dirichlet datum (imaginary part) for the exhaust manifold

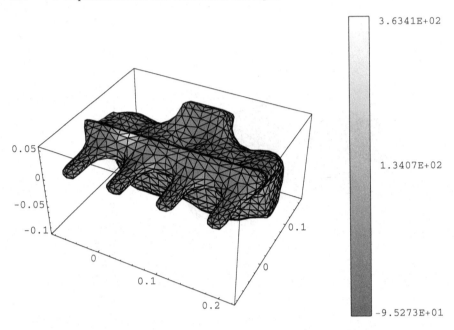

Fig. 4.55. Computed Neumann datum (real part) for the exhaust manifold

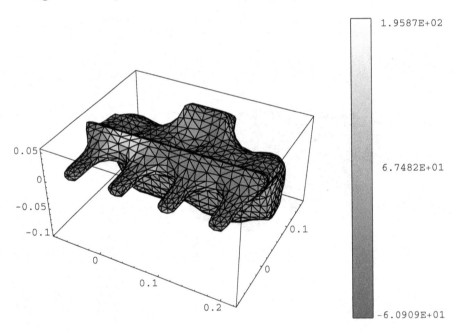

Fig. 4.56. Computed Neumann datum (imaginary part) for the exhaust manifold

$$x(t) = \begin{pmatrix} -0.05 \\ 0.1 \\ -0.05 \end{pmatrix} + t \begin{pmatrix} 0.2 \\ 0.0 \\ 0.0 \end{pmatrix} , \; 0 \le t \le 1 \qquad (4.42)$$

inside of the domain Ω is shown in Figs. 4.57-4.58 for $N = 2264$ and correspondingly for $N = 9056$. The values of the numerical solution $\tilde{u}$ and of the analytical solution u have been computed in 512 points uniformly placed on the line (4.42). The thick dashed line represents the course of the analytical solution (4.33) while the thin solid line shows the course of the numerical solution $\tilde{u}$. The values of the variable x_1 along the line (4.42) are used for the axis of abscissas. Note that the numerical solution for $N = 9056$ perfectly

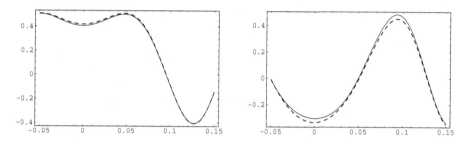

Fig. 4.57. Numerical and analytical curves for the exhaust manifold for $N = 2264$

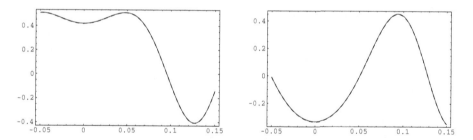

Fig. 4.58. Numerical and analytical curves for the exhaust manifold for $N = 9056$

coincides with the analytical curves.

4.4.5 Interior Neumann Problem

We consider the interior Neumann boundary value problem for the Helmholtz equation with the boundary condition $\gamma_1^{\mathrm{int}} u(x) = g(x)$ for $x \in \Gamma$. The variational problem (1.109)

$$\left\langle D_\kappa \bar{u}, v \right\rangle_\Gamma = \left\langle \left(\frac{1}{2} I - K'_\kappa \right) g, v \right\rangle_\Gamma \quad \text{for all } v \in H^{1/2}(\Gamma)$$

is discretised and leads to a system of linear equations (cf. (2.61))

$$D_{\kappa,h}\widetilde{\underline{u}} = \left(\frac{1}{2}M_h^\top - K_{\kappa,h}^\top\right)\underline{g}\,.$$

The symmetric but complex valued system is then solved using the GMRES method up to the relative accuracy $\varepsilon_2 = 10^{-8}$.

Unit Sphere

We consider again the harmonic function (4.36) as the exact solution. The results for the ACA approximation of the matrix $C_{\kappa,h} \in \mathbb{C}^{M \times M}$ are presented in Table 4.26. The corresponding results for the matrices $K_{\kappa,h}^\top$ for the computation of the right hand side of the above system and $V_{\kappa,h}$, which will be used for the multiplication with the matrix $D_{\kappa,h}$ are identical to those already presented in Table 4.22. Note that in this example the complex valued Galerkin

Table 4.26. ACA approximation of the Galerkin matrix $C_{\kappa,h}$, Neumann problem

N	M	ε_1	MByte($C_{\kappa,h}$)	%
80	42	$1.0 \cdot 10^{-2}$	0.01	51.2
320	162	$1.0 \cdot 10^{-3}$	0.20	49.1
1280	642	$1.0 \cdot 10^{-4}$	2.15	34.2
5120	2562	$1.0 \cdot 10^{-5}$	17.53	17.5
20480	10242	$1.0 \cdot 10^{-6}$	126.29	7.89

matrix $C_{\kappa,h}$ with piecewise linear basis functions is generated. This is a rather time consuming procedure. Thus, a quite good approximation of this matrix is especially important when using the ACA algorithm. The accuracy obtained

Table 4.27. Accuracy of the Galerkin method, Neumann problem

N	M	$Iter$	$Error_1$	CF_1	$Error_2$	CF_2
80	42	10	$4.94 \cdot 10^{-1}$	–	$2.15 \cdot 10^{-0}$	–
320	162	17	$1.25 \cdot 10^{-1}$	3.95	$4.72 \cdot 10^{-1}$	4.67
1280	642	25	$2.75 \cdot 10^{-2}$	4.55	$1.09 \cdot 10^{-1}$	4.32
5120	2562	37	$6.41 \cdot 10^{-3}$	4.29	$2.60 \cdot 10^{-2}$	4.19
20480	10242	46	$1.55 \cdot 10^{-3}$	4.14	$6.41 \cdot 10^{-3}$	4.05

for the whole numerical procedure is presented in Table 4.27. The numbers in this table have the usual meaning. The third column shows the number of iterations required by the GMRES method without preconditioning. Note that the convergence of the Galerkin method for the unknown Dirichlet datum in L_2 norm,

$$Error_1 = \frac{\|\gamma_0^{int}u - \widetilde{u}_h\|_{L_2(\Gamma)}}{\|\gamma_0^{int}u\|_{L_2(\Gamma)}},$$

is now quadratic corresponding to the estimate (2.67). In the inner point x^*, we now observe quadratic convergence (7th column), as it was predicted in (2.67), instead of the cubic order obtained for the Dirichlet problem (cf. Table 4.23). This fact is clearly illustrated in Figs. 4.59–4.61, where the convergence of the boundary element method can be seen optically. The results obtained for $N = 80$ are plotted in Fig. 4.59, where the left plot shows the course of the real part of the solution, while the imaginary part is presented on the right. The numerical curves in Fig. 4.60 are notedly better than the previous ones. However, their quality is not as high as of the corresponding curves obtained while solving the Dirichlet problem (cf. Fig. 4.41).

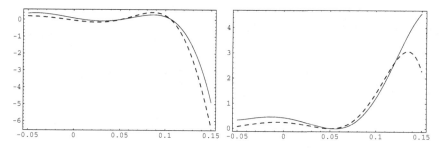

Fig. 4.59. Numerical and analytical curves for $N = 80$, Neumann problem

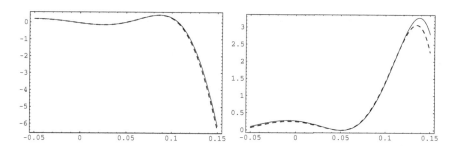

Fig. 4.60. Numerical and analytical curves for $N = 320$, Neumann problem

Exhaust Manifold

The analytical solution is taken in the form (4.33) with $\widetilde{y} = (0,0,0.06)^\top$ and $\kappa = 10$, which is rather small compared with the dimension of the domain (cf.

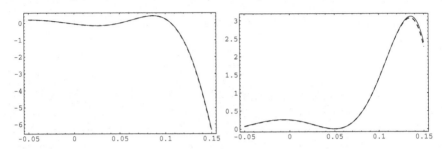

Fig. 4.61. Numerical and analytical curves for $N = 1280$, Neumann problem

Fig. 4.11). We will consider bigger values of the wave number, when studying the multifrequency behaviour of the problem. This small value of κ is well situated to demonstrate convergence properties of the Galerkin BEM for a regular value of the wave number. The results of the computations are shown in Tables 4.28 and 4.29. The quality of the ACA approximation of the ma-

Table 4.28. ACA approximation of the Galerkin matrices $K_{\kappa,h}$ and $V_{\kappa,h}$

N	M	ε_1	$MB(K_{\kappa,h})$	%	$MB(V_{\kappa,h})$	%	$MB(C_{\kappa,h})$	%
2264	1134	$1.0 \cdot 10^{-3}$	13.81	35.3	8.29	10.6	4.54	23.1
9056	4530	$1.0 \cdot 10^{-4}$	124.86	20.0	72.14	5.8	40.33	12.9
36224	18114	$1.0 \cdot 10^{-5}$	998.19	9.9	580.32	2.9	354.61	7.1

trix $C_{\kappa,h}$ can be seen in columns eight and nine of the Table 4.28. The third

Table 4.29. Accuracy of the Galerkin method, Neumann problem

N	M	$Iter$	$Error_1$	CF_1	$Error_2$	CF_2
2264	1134	177	$2.26 \cdot 10^{-2}$	-	$1.40 \cdot 10^{-3}$	-
9056	4530	201	$5.20 \cdot 10^{-3}$	4.3	$2.92 \cdot 10^{-4}$	4.8
36224	18114	244	$1.21 \cdot 10^{-3}$	4.3	$4.90 \cdot 10^{-5}$	5.9

column shows the number of iterations required by the GMRES method with diagonal preconditioning (4.21). The fourth column displays the L_2 error of the computed Dirichlet datum, and the next column shows its quadratic convergence. Column six of Table 4.29 displays the absolute error (cf. (2.67)) in a prescribed inner point $x^* \in \Omega$,

$$Error_2 = |u(x^*) - \tilde{u}(x^*)|, \ x^* = (0.0740705, 0.1, -0.05)^\top, \qquad (4.43)$$

for the value $\tilde{u}(x^*)$ obtained using an approximate representation formula
(2.65). Finally, the last column of this table indicates the quadratic (or even
better) convergence of this quantity.

The Cauchy data obtained when solving the Neumann boundary value
problem is optically the same as for the Dirichlet boundary value problem
and can be seen in Figs. 4.53– 4.54. The numerical curve obtained when using
an approximate representation formula in comparison with the curve of the
exact values (4.36) along the line (4.42) inside of the domain Ω is shown
in Figs. 4.62-4.63 for $N = 2264$, and, correspondingly, for $N = 9056$. The
values of the numerical solution $\tilde{u}$ and of the analytical solution u have been
computed in 512 points uniformly placed on the line (4.42). The thick dashed
line represents the course of the analytical solution (4.33) while the thin solid
line shows the course of the numerical solution $\tilde{u}$. The values of the variable
x_1 along the the line (4.16) are used for the axis of abscissas. Note that

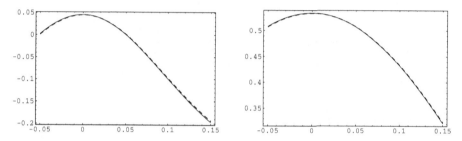

Fig. 4.62. Numerical and analytical curves for the Neumann Problem, $N = 2264$

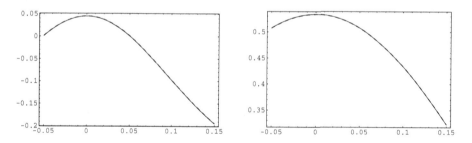

Fig. 4.63. Numerical and analytical curves for the Neumann Problem, $N = 9056$

the numerical solution for $N = 9056$ perfectly coinsides with the analytical
curves.

Exhaust Manifold. Multifrequency Analysis

Here we solve the Neumann boundary value problem for the Helmholtz equation on the surface, depicted in Fig. 4.11, for 17 uniformly distributed values of the wave number κ on the interval $[16, 24]$. In Fig. 4.64, we present the L_2 norm of the error for the computed Dirichlet datum (left plot) and the number of GMRES iterations (right plot) as functions of the wave number κ. The left plot in Fig. 4.64 clearly shows a total loss of accuracy close to the

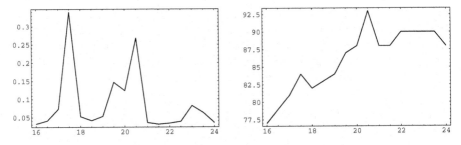

Fig. 4.64. Multifrequency computation for $N = 2264$, Neumann problem

wave numbers $\kappa = 17.5$ and $\kappa = 20.5$. If we plot the analytical and the numerical values of the solution of the boundary value problem for three subsequent points $17.0, 17.5$, and 18.0 for the parameter κ, we can see this loss of accuracy optically. The results are presented in Figs. 4.65–4.67. Thus, the numerical

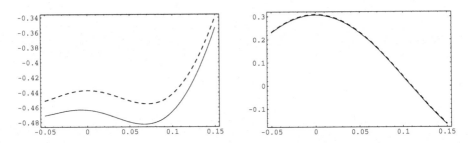

Fig. 4.65. Numerical and analytical curves for $\kappa = 17.0$ and $N = 2264$, Neumann problem

solution is quite wrong for $\kappa = 17.5$, while it is acceptable for $\kappa = 17.0$ and $\kappa = 18.0$.

The picture is similar, if we consider the numerical and the analytical solutions for $\kappa = 20.0, 20.5$, and $\kappa = 21.0$.

Note that the quality of the approximation of the boundary element matrices $K_{\kappa,h}$, $V_{\kappa,h}$, and $C_{\kappa,h}$ is more or less constant for all values of the wave number κ on the whole interval $[16, 24]$.

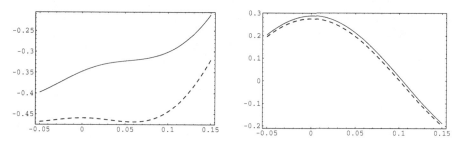

Fig. 4.66. Numerical and analytical curves for $\kappa = 17.5$ and $N = 2264$, Neumann problem

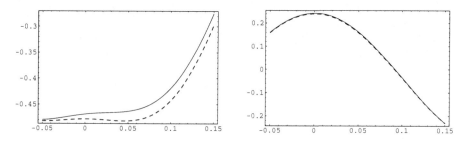

Fig. 4.67. Numerical and analytical curves for $\kappa = 18.0$ and $N = 2264$, Neumann problem

4.4.6 Exterior Dirichlet Problem

Here, we solve the Helmholtz equation (4.31) in $\Omega^e = \mathbb{R}^3 \setminus \overline{\Omega}$ together with the boundary condition $\gamma_0^{\text{ext}} u(x) = g(x)$ for $x \in \Gamma$, where $\Gamma = \partial \Omega$ is the boundary of the surface Ω. The variational problem (1.116)

$$\left\langle V_\kappa t, w \right\rangle_\Gamma = \left\langle \left(-\frac{1}{2}I + K_\kappa \right) g, w \right\rangle_\Gamma \quad \text{for all} \quad w \in H^{-1/2}(\Gamma)$$

is discretised and leads to a system of linear equations (cf. (2.57))

$$V_{\kappa,h} \widetilde{\underline{t}} = \left(-\frac{1}{2} M_h + K_{\kappa,h} \right) \underline{g}.$$

The matrix $V_{\kappa,h}$ of this system is identical with the corresponding matrix of the interior boundary value problem. Thus, in this subsection, we will choose different values of the wave number κ compared with those used in Subsection 4.4.6.

Unit Sphere

The analytical solution is taken in the form (4.33) for $\tilde{y} = (0.9, 0, 0)^\top \in \Omega$, i.e. close to the boundary of the domain Ω. The results of the computations

Table 4.30. ACA approximation of the Galerkin matrices $K_{\kappa,h}$ and $V_{\kappa,h}$

N	M	ε_1	MByte($K_{\kappa,h}$)	%	MByte($V_{\kappa,h}$)	%
80	42	$1.0 \cdot 10^{-2}$	0.05	100.0	0.05	50.6
320	162	$1.0 \cdot 10^{-3}$	0.75	94.5	0.64	40.8
1280	642	$1.0 \cdot 10^{-4}$	6.88	54.9	5.66	22.7
5120	2562	$1.0 \cdot 10^{-5}$	53.85	26.9	42.91	10.7
20480	10242	$1.0 \cdot 10^{-6}$	379.09	11.8	302.35	4.72

for the wave number $\kappa = 4$, which is still moderate, are shown in Tables 4.30 and 4.31. The number of boundary elements is listed in the first column of these tables. The second column contains the number of nodes, while in the third column of Table 4.30, the prescribed accuracy for the ACA algorithm for the approximation of both matrices $K_{\kappa,h} \in \mathbb{C}^{N \times M}$ and $V_{\kappa,h} \in \mathbb{C}^{N \times N}$ is given. The difference in the ACA approximation of these matrices for $\kappa = 4$ (Table 4.30) and for $\kappa = 2$ (Table 4.22) can be clearly seen. In Table 4.31, the

Table 4.31. Accuracy of the Galerkin method, Dirichlet problem

N	M	$Iter$	$Error_1$	CF_1	$Error_2$	CF_2
80	42	28	$9.43 \cdot 10^{-1}$	–	$1.88 \cdot 10^{-1}$	–
320	162	41	$6.95 \cdot 10^{-1}$	1.36	$4.39 \cdot 10^{-2}$	4.28
1280	642	52	$3.68 \cdot 10^{-1}$	1.89	$7.48 \cdot 10^{-3}$	5.87
5120	2562	63	$1.64 \cdot 10^{-1}$	2.24	$8.04 \cdot 10^{-4}$	9.30
20480	10242	75	$7.79 \cdot 10^{-2}$	2.11	$6.89 \cdot 10^{-5}$	11.68

third column shows the number of GMRES iterations without preconditioning needed to reach the prescribed accuracy $\varepsilon_2 = 10^{-8}$. The relative L_2 error for the Neumann datum,

$$Error_1 = \frac{\|\gamma_1^{int} u - \tilde{t}_h\|_{L_2(\Gamma)}}{\|\gamma_1^{int} u\|_{L_2(\Gamma)}},$$

is given in the fourth column. The next column represents the rate of convergence for the Neumann datum, i.e. the quotient of the errors in two consecutive lines of column four. We can see that linear convergence can be observed asymptotically. Finally, the last two columns show the absolute error (cf. (2.69)) in the point $x^* \in \Omega^e$,

$$Error_2 = |u(x^*) - \tilde{u}(x^*)|, \quad x^* = (1.1, 0, 0)^\top, \tag{4.44}$$

for the value $\tilde{u}(x^*)$ obtained using the approximate representation formula (2.68). Again, a rather high accuracy of the ACA approximation is necessary in order to be able to observe a third order (asymptotically even better) pointwise convergence rate within the domain Ω^e, shown in the last two columns of Table 4.31.

In Figs. 4.68–4.69, the given Dirichlet datum (real and imaginary parts) for $N = 320$ boundary elements is presented. The computed Neumann datum

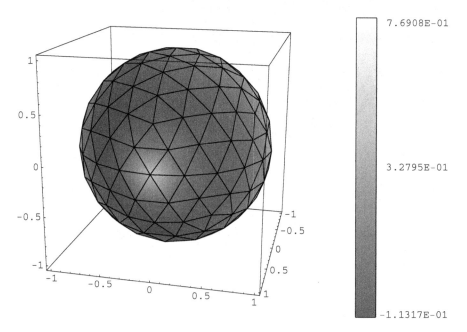

Fig. 4.68. Given Dirichlet datum (real part) for the unit sphere, $N = 320$

is shown in Figs. 4.70 (real part) and 4.71 (imaginary part). The numerical curves obtained when using an approximate representation formula in comparison with the curve of the exact values (4.33) along the line

$$x(t) = \begin{pmatrix} 1.1 \\ 0.0 \\ -4.0 \end{pmatrix} + t \begin{pmatrix} 0.0 \\ 0.0 \\ 8.0 \end{pmatrix} , \quad 0 \le t \le 1 \qquad (4.45)$$

inside of the domain Ω^e is shown in Fig. 4.72 for $N = 80$ and in Fig. 4.73 for $N = 320$. The values of the numerical solution $\tilde{u}$ and of the analytical solution u have been computed in 512 points uniformly placed on the line (4.45). In these figures, the thick dashed line represents the course of the analytical solution (4.36), while the thin solid line shows the course of the numerical solution $\tilde{u}$. The values of the variable x_3 along the line (4.38) are used for the axis of abscissas. The left plots in these figures correspond to the real parts of the solutions, while the imaginary parts are shown on the right. The next Fig. 4.74 shows these curves for $N = 1280$, but on the zoomed interval

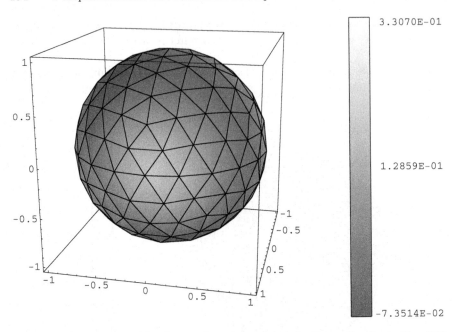

Fig. 4.69. Given Dirichlet datum (imaginary part) for the unit sphere, $N = 320$

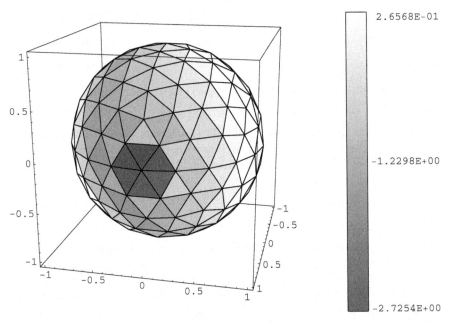

Fig. 4.70. Computed Neumann datum (real part) for the unit sphere, $N = 320$

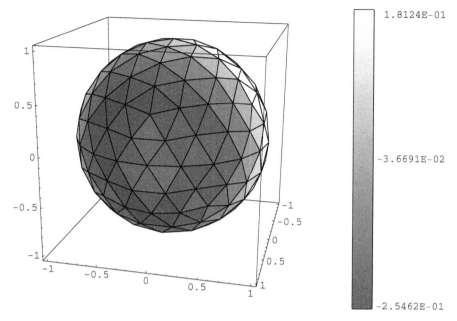

1.8124E-01

-3.6691E-02

-2.5462E-01

Fig. 4.71. Computed Neumann datum (imaginary part) for the unit sphere, $N = 320$

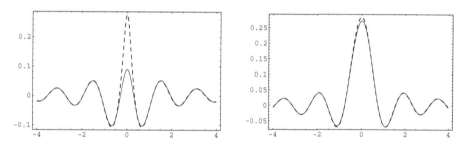

Fig. 4.72. Numerical and analytical curves for $N = 80$, Dirichlet problem

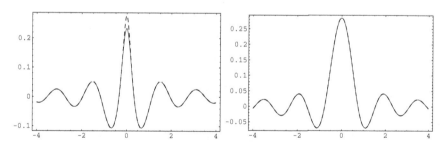

Fig. 4.73. Numerical and analytical curves for $N = 320$, Dirichlet problem

$[-1, 1]$ with respect to the variable x_3, in order to see the very small difference between them. It is almost impossible to see any optical difference between

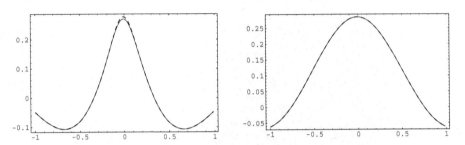

Fig. 4.74. Numerical and analytical curves for $N = 1280$, Dirichlet problem

the numerical and the analytical curves for higher values of N. Note that the point x^* in (4.44) is chosen close to the maximum of the function $\mathrm{Re}\, u$ along the line where the error seems to reach its maximum.

Thus, for the moderate value of the wave number $\kappa = 4$, the quality of the numerical results on the unit sphere is perfect. The ACA approximation is good, the number of GMRES iterations is low without any preconditioning and it grows corresponding to the theory, and, finally, the theoretical linear convergence order of the Neumann datum on the surface Γ as well as the cubic convergence order in the inner points of the domain Ω^e is perfectly illustrated.

4.4.7 Exterior Neumann Problem

We consider the exterior Neumann boundary value problem for the Helmholtz equation in $\Omega^e = \mathbb{R}^3 \setminus \overline{\Omega}$ with the boundary condition $\gamma_1^{\mathrm{ext}} u(x) = g(x)$ for $x \in \Gamma$. The variational problem (1.122)

$$\left\langle D_\kappa \bar{u}, v \right\rangle_\Gamma = -\left\langle \left(\frac{1}{2} I + K_\kappa' \right) g, v \right\rangle_\Gamma \quad \text{for all } v \in H^{1/2}(\Gamma)$$

is discretised and leads to a system of linear equations (cf. (2.61))

$$D_{\kappa,h} \widetilde{\underline{u}} = -\left(\frac{1}{2} M_h^\top - K_{\kappa,h}^\top \right) \underline{g}.$$

This symmetric, but complex valued system is then solved using the GMRES method up to the relative accuracy $\varepsilon_2 = 10^{-8}$.

Unit Sphere

We consider again the analytical solution in the form (4.33) for an interior point $\tilde{y} = (0.9, 0, 0)^\top \in \Omega$ and $\kappa = 4$ as the exact solution. The results for the

Table 4.32. ACA approximation of the Galerkin matrix $C_{\kappa,h}$, Neumann problem

N	M	ε_1	$MByte(C_{\kappa,h})$	%
80	42	$1.0 \cdot 10^{-2}$	0.01	51.2
320	162	$1.0 \cdot 10^{-3}$	0.20	50.3
1280	642	$1.0 \cdot 10^{-4}$	2.41	38.3
5120	2562	$1.0 \cdot 10^{-5}$	19.17	19.1
20480	10242	$1.0 \cdot 10^{-6}$	135.47	8.46

Table 4.33. Accuracy of the Galerkin method, Neumann problem

N	M	$Iter$	$Error_1$	CF_1	$Error_2$	CF_2
80	42	16	$6.78 \cdot 10^{-1}$	–	$3.24 \cdot 10^{-2}$	–
320	162	23	$1.91 \cdot 10^{-1}$	3.55	$4.99 \cdot 10^{-3}$	6.49
1280	642	31	$5.81 \cdot 10^{-2}$	3.29	$1.24 \cdot 10^{-3}$	4.02
5120	2562	44	$1.42 \cdot 10^{-2}$	4.09	$2.90 \cdot 10^{-4}$	4.28
20480	10242	62	$3.27 \cdot 10^{-3}$	4.34	$7.16 \cdot 10^{-5}$	4.04

ACA approximation of the matrices $K_{\kappa,h}$ and $V_{\kappa,h}$ are the same as for the exterior Dirichlet boundary value problems and can be seen in Table 4.30. The approximation results for the matrix $C_{\kappa,h}$ are shown in Table 4.32. In Table 4.33, the accuracy obtained for the whole numerical procedure is presented and the numbers in this table have the usual meaning. The third column shows the number of iterations required by the GMRES method without preconditioning. Note that the convergence of the Galerkin method for the unknown Dirichlet datum in the L_2 norm,

$$Error_1 = \frac{\|\gamma_0^{int} u - \widetilde{u}_h\|_{L_2(\Gamma)}}{\|\gamma_0^{int} u\|_{L_2(\Gamma)}},$$

is close to quadratic. In the inner point $x^* = (1.1, 0.0, 1.0978)^\top \in \Omega^e$ (close to a local minimum of the exact solution), we now observe quadratic convergence (7th column) as it was predicted in (2.70) instead of the cubic order obtained for the Dirichlet problem (cf. Table 4.31). This fact is clearly illustrated in Figs. 4.75–4.77, where the convergence of the boundary element method can be seen optically. The results obtained for $N = 80$ are plotted in Fig. 4.75, where the left plot shows the course of the real part of the solution, while the imaginary part is presented on the right. Thus, these curves are rather rough approximations of the exact solution. The numerical curves in Fig. 4.76 are notedly better than the previous ones. However, their quality is not as high as of the corresponding curves obtained while solving the Dirichlet problem (cf. Fig. 4.73). The numerical curves for $N = 1280$ are acceptable.

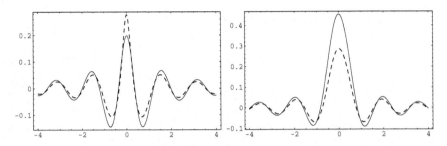

Fig. 4.75. Numerical and analytical curves for $N = 80$, exterior Neumann problem

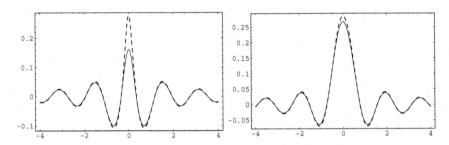

Fig. 4.76. Numerical and analytical curves for $N = 320$, exterior Neumann problem

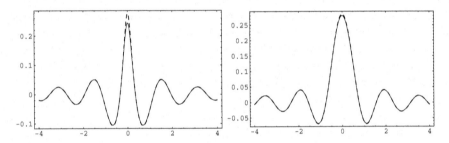

Fig. 4.77. Numerical and analytical curves for $N = 1280$, exterior Neumann problem

A

Mathematical Foundations

A.1 Function Spaces

Let $\alpha = (\alpha_1, \ldots, \alpha_d) \in \mathbb{N}_0^d$ be a multiindex with $|\alpha| = \alpha_1 + \ldots + \alpha_d$ and $\alpha! = \alpha_1! \ldots \alpha_d!$, $d = 2, 3$. Moreover, for $x \in \mathbb{R}^d$ we define $x^\alpha = x_1^{\alpha_1} \ldots x_d^{\alpha_d}$ as well as the partial derivatives

$$D^\alpha u(x) = \frac{\partial^{|\alpha|}}{\partial x_1^{\alpha_1} \ldots \partial x_d^{\alpha_d}} u(x_1, \ldots, x_d).$$

For an open bounded domain $\Omega \subset \mathbb{R}^d$ and $k \in \mathbb{N}_0$, we denote by $C^k(\Omega)$ the space of k times continuously differentiable functions equipped with the norm

$$\|u\|_{C^k(\Omega)} = \sum_{|\alpha| \le k} \sup_{x \in \Omega} |D^\alpha u(x)|;$$

$C^\infty(\Omega)$ is defined accordingly. The support of a given function is the closed set

$$\operatorname{supp} u = \overline{\{x \in \Omega : u(x) \neq 0\}}.$$

Then, $C_0^\infty(\Omega)$ is the space of infinite times continuously differentiable functions with compact support,

$$C_0^\infty(\Omega) = \{u \in C^\infty(\Omega) : \operatorname{supp} u \subset \Omega\}.$$

For $k \in \mathbb{N}_0$ and $\kappa \in (0, 1]$, we define the space $C^{k,\kappa}(\Omega)$ of Hölder continuous functions equipped with the norm

$$\|u\|_{C^{k,\kappa}(\Omega)} = \|u\|_{C^k(\Omega)} + \sum_{|\alpha|=k} \sup_{x,y \in \Omega} \frac{|D^\alpha u(x) - D^\alpha u(y)|}{|x-y|^\kappa}.$$

In particular, for $k = 0$ and $\kappa = 1$, we obtain the space $C^{0,1}(\Omega)$ of Lipschitz continuous functions.

The boundary Γ of an open and bounded set $\Omega \subset \mathbb{R}^3$ is given as

$$\Gamma = \partial\Omega = \overline{\Omega} \cap (\mathbb{R}^3 \backslash \Omega).$$

We assume that the boundary Γ can be represented by a certain decomposition

$$\Gamma = \bigcup_{i=1}^{p} \Gamma_i, \tag{A.1}$$

where each boundary segment Γ_i is described via a local parametrisation,

$$\Gamma_i = \left\{ x \in \mathbb{R}^3 : x = \chi_i(\xi) \text{ for } \xi \in T_i \subset \mathbb{R}^2 \right\}, \tag{A.2}$$

with respect to some open parameter domain T_i. A domain Ω is said to be a Lipschitz domain, when all functions χ_i in (A.2) are Lipschitz continuous for any arbitrary decomposition (A.1).

For $1 \le p < \infty$, we define $L_p(\Omega)$ as the space of all measurable functions u on Ω with $\|u\|_{L_p(\Omega)} < \infty$, where $\| \cdot \|_{L_p(\Omega)}$ denotes the norm:

$$\|u\|_{L_p(\Omega)} = \left(\int_{\Omega} |u(x)|^p dx \right)^{1/p}.$$

In fact, two functions $u, v \in L_p(\Omega)$ are identified, if they differ only on a set K with Lebesgue zero measure $\mu(K) = 0$. $L_\infty(\Omega)$ is the space of all measurable functions which are bounded almost everywhere,

$$\|u\|_{L_\infty(\Omega)} = \operatorname*{ess\,sup}_{x \in \Omega} |u(x)| = \inf_{K \subset \Omega, \mu(K)=0} \sup_{x \in \Omega \backslash K} |u(x)|.$$

Note that for $u \in L_p(\Omega)$ and $v \in L_q(\Omega)$ with adjoint parameters p, q satisfying

$$\frac{1}{p} + \frac{1}{q} = 1,$$

the Hölder inequality holds:

$$\int_{\Omega} |u(x)v(x)| \, dx \le \|u\|_{L_p(\Omega)} \|v\|_{L_q(\Omega)}.$$

Defining the duality pairing

$$\langle u, v \rangle_{\Omega} = \int_{\Omega} u(x)v(x)dx,$$

we then obtain

$$\|v\|_{L_q(\Omega)} = \sup_{0 \ne u \in L_p(\Omega)} \frac{|\langle u, v \rangle_{\Omega}|}{\|u\|_{L_p(\Omega)}} \quad \text{for } 1 \le p < \infty, \ \frac{1}{p} + \frac{1}{q} = 1.$$

In particular, for $p = 2$, the Hilbert space $L_2(\Omega)$ is the function space of square integrable functions. Finally, $L_1^{loc}(\Omega)$ is the space of locally integrable functions.

A function $u \in L_1^{loc}(\Omega)$ is said to have a generalised partial derivative

$$v = \frac{\partial}{\partial x_i} u \in L_1^{loc}(\Omega),$$

if it satisfies the equality

$$\int_\Omega v(x)\varphi(x)dx = -\int_\Omega u(x)\frac{\partial}{\partial x_i}\varphi(x)dx \quad \text{for all } \varphi \in C_0^\infty(\Omega). \tag{A.3}$$

In the same way, we may define the generalised derivative $D^\alpha u \in L_1^{loc}(\Omega)$ by

$$\int_\Omega D^\alpha u(x)\varphi(x)dx = (-1)^{|\alpha|}\int_\Omega u(x)D^\alpha\varphi(x)dx \quad \text{for all } \varphi \in C_0^\infty(\Omega).$$

Then, for $k \in \mathbb{N}_0$ and $1 \le p < \infty$,

$$\|v\|_{W_p^k(\Omega)} = \left(\sum_{|\alpha|\le k} \|D^\alpha v\|_{L_p(\Omega)}^p\right)^{1/p}$$

defines a norm, while for $p = \infty$ we set

$$\|v\|_{W_\infty^k(\Omega)} = \max_{|\alpha|\le k} \|D^\alpha v\|_{L_\infty(\Omega)}.$$

Now we are able to define the Sobolev spaces $W_p^k(\Omega)$ as the closure of $C^\infty(\Omega)$ with respect to the Sobolev norms as introduced above,

$$W_p^k(\Omega) = \overline{C^\infty(\Omega)}^{\|\cdot\|_{W_p^k(\Omega)}}.$$

In particular, for any $v \in W_p^k(\Omega)$, there exists a sequence $\{\varphi_j\}_{j\in\mathbb{N}} \subset C^\infty(\Omega)$ such that

$$\lim_{j\to\infty} \|v - \varphi_j\|_{W_p^k(\Omega)} = 0.$$

In the same way, we may also define the Sobolev spaces

$$\overset{\circ}{W}_p^k(\Omega) = \overline{C_0^\infty(\Omega)}^{\|\cdot\|_{W_p^k(\Omega)}}.$$

Up to now, the above Sobolev spaces are defined only for $k \in \mathbb{N}_0$. However, the definition of the Sobolev norms $\|\cdot\|_{W_p^k(\Omega)}$, and, therefore, of the Sobolev spaces, can be generalised for arbitrary $s \in \mathbb{R}$. For $s > 0$ with $s = k + \kappa$, $k \in \mathbb{N}_0$, $\kappa \in (0,1)$, we define the Sobolev–Slobodeckii norm

$$\|v\|_{W_p^s(\Omega)} = \left(\|v\|^p_{W_p^k(\Omega)} + |v|^p_{W_p^s(\Omega)} \right)^{1/p}$$

with the semi-norm

$$|v|^p_{W_p^s(\Omega)} = \sum_{|\alpha|=k} \int_\Omega \int_\Omega \frac{|D^\alpha v(x) - D^\alpha v(y)|^p}{|x-y|^{d+p\kappa}} \, dx dy.$$

For $s < 0$ and $1 < p < \infty$, the Sobolev space $W_p^s(\Omega)$ is the dual space of $\overset{\circ}{W}_q^{-s}(\Omega)$, where $1/p + 1/q = 1$. The corresponding norm is given by

$$\|f\|_{W_p^s(\Omega)} = \sup_{v \in \overset{\circ}{W}_q^{-s}(\Omega)} \frac{|\langle f, v \rangle_\Omega|}{\|v\|_{W_q^{-s}(\Omega)}}.$$

Accordingly, $\overset{\circ}{W}_p^s(\Omega)$ is the dual space of $W_q^{-s}(\Omega)$.

Next we will collect some properties of the Sobolev spaces needed later.

Theorem A.1 (Sobolev's imbedding theorem). *Let* $\Omega \subset \mathbb{R}^3$ *be a bounded domain with Lipschitz boundary* $\Gamma = \partial\Omega$, *and let* $s \geq 3$ *for* $p = 1$, *and* $s > 3/p$ *for* $p > 1$. *Then every function* $v \in W_p^s(\Omega)$ *is continuous,* $v \in C(\Omega)$, *satisfying*

$$\|v\|_{L_\infty(\Omega)} \leq c \|v\|_{W_p^s(\Omega)} \quad \text{for all } v \in W_p^s(\Omega).$$

In particular, we are interested in the Sobolev spaces $W_2^s(\Omega)$, i.e. for $p = 2$. For $s = 1$, the norm in $W_2^1(\Omega)$ is given by

$$\|v\|_{W_2^1(\Omega)} = \left(\|v\|^2_{L_2(\Omega)} + \|\nabla v\|^2_{L_2(\Omega)} \right)^{1/2}.$$

Now, we will derive equivalent norms in $W_2^1(\Omega)$. A norm $\|\cdot\|_{W_2^1(\Omega),f}$ is called equivalent to the norm $\|\cdot\|_{W_2^1(\Omega)}$, if there are some positive constants c_1 and c_2 such that

$$c_1 \|v\|_{W_2^1(\Omega)} \leq \|v\|_{W_2^1(\Omega),f} \leq c_2 \|v\|_{W_2^1(\Omega)} \quad \text{for all } v \in W_2^1(\Omega).$$

Let $f : W_2^1(\Omega) \to \mathbb{R}$ be a bounded linear functional satisfying

$$|f(v)| \leq c_f \|v\|_{W_2^1(\Omega)} \quad \text{for all } v \in W_2^1(\Omega).$$

Let $c \in \mathbb{R}$ be an arbitrary constant. If we can always conclude $c = 0$ from $f(c) = 0$, then

$$\|v\|_{W_2^1(\Omega),f} = \left(|f(v)|^2 + \|\nabla v\|^2_{L_2(\Omega)} \right)^{1/2}$$

defines an equivalent norm in $W_2^1(\Omega)$. Examples are the norms

$$\|v\|^2_{W_2^1(\Omega),\Omega} = \left(\int_\Omega v(x) dx \right)^2 + \|\nabla v\|^2_{L_2(\Omega)}$$

and

$$\|v\|_{W_2^1(\Omega),\Gamma}^2 = \left(\int_\Gamma v(x)ds_x\right)^2 + \|\nabla v\|_{L_2(\Omega)}^2 \,.$$

As a first consequence, the Sobolev norms $\|\nabla \cdot \|_{L_2(\Omega)}$ and $\| \cdot \|_{W_2^1(\Omega)}$ are equivalent norms in $\overset{\circ}{W}_p^k(\Omega)$. Secondly, there holds the Poincaré inequality

$$\int_\Omega |v(x)|^2 dx \leq c_P \left(\left(\int_\Omega v(x)dx\right)^2 + \int_\Omega |\nabla v(x)|^2 dx\right) \quad \text{for all } v \in W_2^1(\Omega).$$

Theorem A.2 (Bramble–Hilbert–Lemma). *Let $f : W_2^{p+1}(\Omega) \to \mathbb{R}$ for $p = 0, 1$ be a bounded linear functional satisfying*

$$|f(v)| \leq c_f \|v\|_{W_2^{p+1}(\Omega)} \quad \text{for all } v \in W_2^{p+1}(\Omega).$$

Let

$$P_0(\Omega) = \{q(x) = q_0 : x \in \Omega \subset \mathbb{R}^d\}$$

be the space of constant polynomials and let

$$P_1(\Omega) = \{q(x) = q_0 + q_1 x_1 + \ldots + q_d x_d : x \in \Omega \subset \mathbb{R}^d\}$$

be the space of linear polynomials defined on Ω. If $f(q) = 0$ is satisfied for all $q \in P_p(\Omega)$, then it follows that

$$|f(v)| \leq c_p\, c_f\, |v|_{W_2^{p+1}(\Omega)} \quad \text{for all } v \in W_2^{p+1}(\Omega).$$

Recall that the definition of the Sobolev spaces $W_2^s(\Omega)$ is based on the generalised derivatives (cf. (A.3)) in the sense of $L_1^{loc}(\Omega)$. In what follows, we will give a second definition of Sobolev spaces, which is based on derivatives of distributions. A distribution $T \in D'(\Omega)$ is a complex continuous linear functional with respect to $D(\Omega) = C_0^\infty(\Omega)$. For $v \in L_1^{loc}(\Omega)$, the equality

$$T_v(\varphi) = \int_\Omega v(x)\varphi(x)dx \quad \text{for } \varphi \in D(\Omega)$$

defines a regular distribution $T_v \in D'(\Omega)$. The most famous distribution, which is not regular, is the Dirac δ−distribution satisfying

$$\delta_0(\varphi) = \varphi(0)\,, \quad \text{for all } \varphi \in D(\Omega)\,.$$

The equality

$$(D^\alpha T_v)(\varphi) = (-1)^{|\alpha|} T_v(D^\alpha \varphi) \quad \text{for all } \varphi \in D(\Omega)$$

defines the derivative $D^\alpha T_v \in D'(\Omega)$ of a distribution $T_v \in D'(\Omega)$.

Next, we consider the Schwartz space $S(\mathbb{R}^d)$ of smooth fast decreasing functions and its dual space $S'(\mathbb{R}^d)$ of tempered distributions. For $\varphi \in S(\mathbb{R}^d)$, we define the Fourier transform $\mathcal{F} : S(\mathbb{R}^d) \to S(\mathbb{R}^d)$ as

$$\widehat{\varphi}(\xi) = (\mathcal{F}\varphi)(\xi) = (2\pi)^{-d/2} \int_{\mathbb{R}^d} e^{-i\,(x,\xi)} \varphi(x)dx \quad \text{for } \xi \in \mathbb{R}^d\,,$$

as well as the inverse Fourier transform

$$\varphi(x) = (\mathcal{F}^{-1}\widehat{\varphi})(x) = (2\pi)^{-d/2} \int_{\mathbb{R}^d} e^{i\,(x,\xi)} \widehat{\varphi}(\xi)d\xi \quad \text{for } x \in \mathbb{R}^d.$$

Note that for $\varphi \in S(\mathbb{R}^d)$, we have

$$D^\alpha(\mathcal{F}\varphi)(\xi) = (-i)^{|\alpha|}\mathcal{F}(x^\alpha \varphi)(\xi), \quad \xi^\alpha(\mathcal{F}\varphi)(\xi) = (-i)^{|\alpha|}\mathcal{F}(D^\alpha \varphi)(\xi).$$

For a distribution $T \in S'(\mathbb{R}^d)$, the Fourier transformation $\widehat{T} \in S'(\mathbb{R}^d)$ is defined via

$$\widehat{T}(\varphi) = T(\widehat{\varphi}) \quad \text{for all } \varphi \in S(\mathbb{R}^d).$$

For $s \in \mathbb{R}$ and $v \in S(\mathbb{R}^d)$ we define the Bessel potential operator $\mathcal{J}^s : S(\mathbb{R}^d) \to S(\mathbb{R}^d)$ as

$$(\mathcal{J}^s v)(x) = \int_{\mathbb{R}^d} (1 + |\xi|^2)^{s/2}\widehat{v}(\xi)e^{i\,(x,\xi)}d\xi \quad \text{for } x \in \mathbb{R}^d.$$

The application of the Fourier transform yields

$$(\mathcal{F}\mathcal{J}^s v)(\xi) = (1 + |\xi|^2)^{s/2}(\mathcal{F}v)(\xi)\,.$$

Thus, $\mathcal{J}^s$ acts similar to a differential operator of order s. For a distribution $T \in S'(\mathbb{R}^d)$, we can define $\mathcal{J}^s T \in S'(\mathbb{R}^d)$ via

$$(\mathcal{J}^s T)(\varphi) = T(\mathcal{J}^s \varphi) \quad \text{for all } \varphi \in S(\mathbb{R}^d).$$

Now we are in a position to define the Sobolev space $H^s(\mathbb{R}^d)$ as a space of all distributions $v \in S'(\mathbb{R}^d)$ with $\mathcal{J}^s v \in L_2(\mathbb{R}^d)$, and with the inner product

$$\langle u, v \rangle_{H^s(\mathbb{R}^d)} = \langle \mathcal{J}^s u, \mathcal{J}^s v \rangle_{L_2(\mathbb{R}^d)}\,,$$

and the induced norm

$$\|v\|^2_{H^s(\mathbb{R}^d)} = \|\mathcal{J}^s v\|^2_{L_2(\mathbb{R}^d)} = \int_{\mathbb{R}^d} (1 + |\xi|^2)^s |\widehat{v}(\xi)|^2 d\xi.$$

It turns out that $H^s(\mathbb{R}^d) = W_2^s(\mathbb{R}^d)$ for all $s \in \mathbb{R}$. For a bounded domain $\Omega \subset \mathbb{R}^d$, the Sobolev space $H^s(\Omega)$ is defined by restriction

$$H^s(\Omega) = \{v = \tilde{v}_{|\Omega} : \tilde{v} \in H^s(\mathbb{R}^d)\}$$

with the norm

$$\|v\|_{H^s(\Omega)} = \inf_{\tilde{v} \in H^s(\mathbb{R}^d) : \tilde{v}_{|\Omega}=v} \|\tilde{v}\|_{H^s(\mathbb{R}^d)} \,.$$

Moreover, we introduce the Sobolev spaces

$$\widetilde{H}^s(\Omega) = \overline{C_0^\infty(\Omega)}^{\|\cdot\|_{H^s(\mathbb{R}^d)}}, \quad H_0^s(\Omega) = \overline{C_0^\infty(\Omega)}^{\|\cdot\|_{H^s(\Omega)}}$$

and state the following result:

Theorem A.3. *Let* $\Omega \subset \mathbb{R}^3$ *be a Lipschitz domain and* $s \geq 0$. *Then* $\widetilde{H}^s(\Omega) \subset H_0^s(\Omega)$. *In particular, there holds*

$$\widetilde{H}^s(\Omega) = H_0^s(\Omega) \quad \text{for } s \neq \frac{1}{2}, \frac{3}{2}, \frac{5}{2}, \cdots .$$

Moreover,

$$\widetilde{H}^s(\Omega) = \left(H^{-s}(\Omega)\right)', \quad H^s(\Omega) = \left(\widetilde{H}^{-s}(\Omega)\right)' \quad \text{for all } s \in \mathbb{R}.$$

Finally, we comment on the equivalence of the Sobolev spaces $W_2^s(\Omega)$ and $H^s(\Omega)$, where we have to impose additional restrictions on the bounded domain Ω. In particular, let us assume that there is given a bounded linear extension operator

$$E_\Omega : W_2^s(\Omega) \to W_2^s(\mathbb{R}^d).$$

Note that this condition is satisfied, if an uniform cone condition holds for Ω. In fact, if Ω is a Lipschitz domain, we conclude $H^s(\Omega) = W_2^s(\Omega)$ for all $s > 0$.

To define Sobolev spaces on the boundary $\Gamma = \partial\Omega$ of a bounded domain $\Omega \subset \mathbb{R}^3$ we start with an arbitrary overlapping parametrisation

$$\Gamma = \bigcup_{i=1}^J \Gamma_i, \quad \Gamma_i = \left\{x \in \mathbb{R}^3 : x = \chi_i(\xi), \, \xi \in \mathcal{T}_i \subset \mathbb{R}^2\right\}.$$

We consider a partition of unity subordinated to the above decomposition, i.e. non–negative functions $\varphi_i \in C_0^\infty(\mathbb{R}^3)$ satisfying

$$\sum_{i=1}^J \varphi_i(x) = 1 \quad \text{for } x \in \Gamma, \quad \varphi_i(x) = 0 \quad \text{for } x \in \Gamma \backslash \Gamma_i \,.$$

Any function v on Γ can then be written as

$$v(x) = \sum_{i=1}^J \varphi_i(x)v(x) = \sum_{i=1}^J v_i(x) \quad \text{for } x \in \Gamma,$$

with

$$v_i(x) = \varphi_i(x)v(x) \quad \text{for } x \in \Gamma_i.$$

Inserting the local parametrisation, this gives

$$v_i(x) = \varphi_i(x)v(x) = \varphi_i(\chi_i(\xi))v(\chi_i(\xi)) = \tilde{v}_i(\xi) \quad \text{for } \xi \in T_i \subset \mathbb{R}^2.$$

For the above defined functions $\tilde{v}_i(\xi)$, for $\xi \in T_i \subset \mathbb{R}^2$, and $s \geq 0$, we now consider the Sobolev space $H^s(T_i)$ associated with the norms $\|\tilde{v}_i\|_{H^s(T_i)}$. Hence, we can define the Sobolev spaces $H^s(\Gamma)$ equipped with the norm

$$\|v\|_{H^s_\chi(\Gamma)} = \left(\sum_{i=1}^{J} \|\tilde{v}_i\|^2_{H^s(T_i)} \right)^{1/2}.$$

Note that derivatives $D^\alpha_\xi \tilde{v}_i(\xi)$ of the order $|\alpha| \leq k$ require the existence of derivatives $D^\alpha_\xi \chi_i(\xi)$ due to the chain rule. Hence, we assume $\chi_i \in C^{k-1,1}(T_i)$. Therefore, if Ω is a Lipschitz domain, we can define the Sobolev spaces $H^s(\Gamma)$ on the boundary only for $|s| \leq 1$.

Note that the definition of the above Sobolev norm $\| \cdot \|_{H^s_\chi(\Gamma)}$ depends on the parametrisation chosen, but all of the norms are equivalent. In particular, for $s = 0$

$$\|v\|_{L_2(\Gamma)} = \left(\int_\Gamma |v(x)|^2 ds_x \right)^{1/2}$$

is equivalent to $\|v\|_{H^0_\chi(\Gamma)}$. For $s \in (0,1)$, an equivalent norm is the Sobolev–Slobodeckii norm

$$\|v\|^2_{H^s(\Gamma)} = \left(\|v\|^2_{L_2(\Gamma)} + \int_\Gamma \int_\Gamma \frac{|v(x) - v(y)|^2}{|x - y|^{2+2s}} ds_x ds_y \right)^{1/2}.$$

Moreover,

$$\|v\|_{H^{1/2}(\Gamma),\Gamma} = \left(\left(\int_\Gamma v(x) ds_x \right)^2 + \int_\Gamma \int_\Gamma \frac{|v(x) - v(y)|^2}{|x - y|^3} ds_x ds_y \right)^{1/2}$$

defines an equivalent norm in $H^{1/2}(\Gamma)$.

For $s < 0$, Sobolev spaces $H^s(\Gamma) = \left(H^{-s}(\Gamma) \right)'$ are defined by duality,

$$\|w\|_{H^s(\Gamma)} = \sup_{0 \neq v \in H^{-s}(\Gamma)} \frac{\langle w, v \rangle_\Gamma}{\|v\|_{H^{-s}(\Gamma)}},$$

with respect to the duality pairing

$$\langle w, v \rangle_\Gamma = \int_\Gamma w(x)v(x) ds_x.$$

Next, we consider open boundary parts $\Gamma_0 \subset \Gamma = \partial\Omega$. For $s \geq 0$, we define

$$H^s(\Gamma_0) = \{v = \tilde{v}_{|\Gamma_0} : \tilde{v} \in H^s(\Gamma),\}$$

and

$$\|v\|_{H^s(\Gamma_0)} = \inf_{\tilde{v} \in H^s(\Gamma):\tilde{v}_{|\Gamma_0}=v} \|\tilde{v}\|_{H^s(\Gamma)}.$$

Correspondingly,

$$\tilde{H}^s(\Gamma_0) = \{v = \tilde{v}_{|\Gamma_0} : \tilde{v} \in H^s(\Gamma), \operatorname{supp}\tilde{v} \subset \Gamma_0\}.$$

If $s < 0$, then by duality

$$H^s(\Gamma_0) = \left(\tilde{H}^{-s}(\Gamma_0)\right)', \quad \tilde{H}^s(\Gamma_0) = \left(H^{-s}(\Gamma_0)\right)'.$$

Finally, we consider a piecewise smooth boundary

$$\Gamma = \bigcup_{i=1}^{J} \overline{\Gamma}_i, \quad \Gamma_i \cap \Gamma_j = \varnothing \quad \text{for } i \neq j,$$

and define for $s \geq 0$ the Sobolev spaces

$$H^s_{\mathrm{pw}}(\Gamma) = \{v \in L_2(\Gamma) : v_{|\Gamma_i} \in H^s(\Gamma_i), i = 1, \ldots, J\},$$

equipped with the norm

$$\|v\|_{H^s_{\mathrm{pw}}(\Gamma)} = \left(\sum_{i=1}^{J} \|v_{|\Gamma_i}\|^2_{H^s(\Gamma_i)}\right)^{1/2}.$$

For $s < 0$, the corresponding Sobolev spaces are given as a product space

$$H^s_{\mathrm{pw}}(\Gamma) = \prod_{i=1}^{J} \tilde{H}^s(\Gamma_i),$$

with the norm

$$\|w\|_{H^s_{\mathrm{pw}}(\Gamma)} = \sum_{i=1}^{J} \|w_{|\Gamma_i}\|_{\tilde{H}^s(\Gamma)}.$$

Note that for all $w \in H^s_{\mathrm{pw}}(\Gamma)$ and $s < 0$, we conclude $w \in H^s(\Gamma)$, satisfying

$$\|w\|_{H^s(\Gamma)} \leq \|w\|_{H^s_{\mathrm{pw}}(\Gamma)}.$$

At the end of this subsection, we state some relations between the Sobolev spaces $H^s(\Omega)$ in the domain Ω and $H^s(\Gamma)$ on the boundary $\Gamma = \partial\Omega$.

Theorem A.4 (Trace Theorem). *Let $\Omega \subset \mathbb{R}^3$ be a bounded domain with boundary $\Gamma \in C^{k-1,1}$. The trace operator*

$$\gamma_0^{int} : H^s(\Omega) \to H^{s-1/2}(\Gamma)$$

is continuous for $1/2 < s \leq k$, i.e.

$$\|\gamma_0^{int} v\|_{H^{s-1/2}(\Gamma)} \leq c_T \|v\|_{H^s(\Omega)} \quad \text{for all } v \in H^s(\Omega).$$

Theorem A.5 (Inverse Trace Theorem). *Let $\Omega \subset \mathbb{R}^3$ be a bounded domain with boundary $\Gamma \in C^{k-1,1}$. For $1/2 < s \leq k$, there exists a continuous extension operator*

$$\mathcal{E} : H^{s-1/2}(\Gamma) \to H^s(\Omega),$$

i.e.

$$\|\mathcal{E}v\|_{H^s(\Omega)} \leq c_{IT} \|v\|_{H^{s-1/2}(\Gamma)} \quad \text{for all } v \in H^{s-1/2}(\Gamma),$$

satisfying $\gamma_0^{int} \mathcal{E}v = v$.

A.2 Fundamental Solutions

A.2.1 Laplace Equation

The fundamental solution (1.7) of the Laplace equation can be found from the relation (1.5), which can be written as a partial differential equation in the distributional sense,

$$-\Delta_y u^*(x, y) = \delta_0(y - x) \quad \text{for } x, y \in \mathbb{R}^3,$$

where δ_0 denotes the Dirac δ–distribution. Since the Laplace operator is invariant with respect to translations and rotations, we may use the transformation $z = y - x$ to find v satisfying

$$-\Delta_z v(z) = \delta_0(z) \quad \text{for } z \in \mathbb{R}^3.$$

The application of the Fourier transform together with the transformation rules for derivatives gives

$$|\xi|^2 \hat{v}(\xi) = \frac{1}{(2\pi)^{3/2}},$$

and, therefore,

$$\hat{v}(\xi) = \frac{1}{(2\pi)^{3/2}} \frac{1}{|\xi|^2} \in \mathcal{S}'(\mathbb{R}^3).$$

The Fourier transform of the tempered distributions of the above form is well studied and can be found, for example, in [35, Chapter 2]. Thus, in d–dimensional space, it holds for $\lambda \neq -d, -d - 2, \ldots$

$$\left(\mathcal{F}^{-1}|\xi|^\lambda\right)(z) = 2^{\lambda+d/2}\frac{\Gamma\left(\frac{\lambda+d}{2}\right)}{\Gamma\left(-\frac{\lambda}{2}\right)}|z|^{-\lambda-d}.$$

For $d = 3$ and $\lambda = -2$, the above formula leads to

$$\left(\mathcal{F}^{-1}|\xi|^{-2}\right)(z) = \sqrt{\frac{\pi}{2}}|z|^{-1},$$

where the the values

$$\Gamma(1/2) = \pi/2, \quad \Gamma(1) = 1$$

of the Gamma function Γ have been used. Therefore, with $z = x - y$, the fundamental solution of the Laplace operator is

$$u^*(x, y) = \frac{1}{4\pi}\frac{1}{|x - y|}.$$

A.2.2 Lame System

The fundamental solution of linear elastostatics, given by the Kelvin tensor (1.66), can be found from (1.64), which can be written as a system (for $\ell = 1, 2, 3$) of partial differential equations in the distributional sense:

$$-\mu\Delta\underline{U}_\ell^*(x, y) - (\lambda + \mu)\operatorname{grad}\operatorname{div}\underline{U}_\ell^*(x, y) = \delta_0(y - x)\underline{e}_\ell \quad \text{for } x, y \in \mathbb{R}^3.$$

Defining

$$\underline{U}_\ell^*(x, y) = \Delta\underline{w}_\ell(x, y) - \frac{\lambda + \mu}{\lambda + 2\mu}\operatorname{grad}\operatorname{div}\underline{w}_\ell(x, y) \quad \text{for } x, y \in \mathbb{R}^3,$$

the solution of the inhomogeneous system of linear elastostatics is equivalent to a system of scalar Bi–Laplace equations

$$-\mu\Delta^2\underline{w}_\ell(x, y) = \delta_0(y - x)\underline{e}_\ell \quad \text{for } x, y \in \mathbb{R}^3.$$

In particular, for $\ell = 1$, we set $w_{\ell,2}(x, y) = w_{\ell,3}(x, y) = 0$, and it remains to solve

$$-\mu\Delta^2 w_{\ell,1}(x, y) = \delta_0(y - x) \quad \text{for } x, y \in \mathbb{R}^3.$$

Using the transformation $z = y - x$, we have to find v satisfying

$$-\mu\Delta^2 v(z) = \delta_0(z) \quad \text{for } z \in \mathbb{R}^3$$

or

$$-\mu\Delta\varphi(z) = \delta_0(z), \quad \Delta v(z) = \varphi(z).$$

From this, we first get

$$\varphi(z) = \frac{1}{\mu}\frac{1}{4\pi}\frac{1}{|z|}.$$

To determine v, we rewrite the Laplace equation in spherical coordinates as

$$\frac{1}{r^2} \frac{\partial}{\partial r} \left(r^2 \frac{\partial}{\partial r} \widetilde{v}(r) \right) = \frac{1}{\mu} \frac{1}{4\pi} \frac{1}{r} \quad \text{for } r > 0.$$

Thus, we obtain the general solution

$$\widetilde{v}(r) = \frac{1}{\mu} \frac{1}{4\pi} \left(\frac{1}{2} r + \frac{a}{r} + b \right) \quad \text{for } r > 0, \quad a, b \in \mathbb{R}.$$

Choosing $a = b = 0$, we, therefore, get

$$v(z) = \frac{1}{\mu} \frac{1}{8\pi} |z|.$$

From this, we find

$$U_{1,1}^*(z) = \Delta v(z) - \frac{\lambda + \mu}{\lambda + 2\mu} \frac{\partial^2}{\partial z_1^2} v(z),$$

$$U_{2,1}^*(z) = -\frac{\lambda + \mu}{\lambda + 2\mu} \frac{\partial^2}{\partial z_1 \partial z_2} v(z),$$

$$U_{3,1}^*(z) = -\frac{\lambda + \mu}{\lambda + 2\mu} \frac{\partial}{\partial z_1 \partial z_3} v(z),$$

and, hence,

$$U_{1,1}^*(z) = \frac{1}{8\pi} \frac{\lambda + 3\mu}{\mu(\lambda + 2\mu)} \frac{1}{|z|} + \frac{1}{8\pi} \frac{\lambda + \mu}{\mu(\lambda + 2\mu)} \frac{z_1^2}{|z|^3},$$

$$U_{2,1}^*(z) = \frac{1}{8\pi} \frac{\lambda + \mu}{\mu(\lambda + 2\mu)} \frac{z_1 z_2}{|z|^3},$$

$$U_{3,1}^*(z) = \frac{1}{8\pi} \frac{\lambda + \mu}{\mu(\lambda + 2\mu)} \frac{z_1 z_3}{|z|^3}.$$

Doing the same computations for $\ell = 2, 3$, and inserting the Lamé constants, this gives the Kelvin tensor as the fundamental solution of linear elastostatics,

$$U_{k\ell}^*(x, y) = \frac{1}{8\pi} \frac{1}{E} \frac{1 + \nu}{1 - \nu} \left((3 - 4\nu) \frac{\delta_{k\ell}}{|x - y|} + \frac{(y_k - x_k)(y_\ell - x_\ell)}{|x - y|^3} \right)$$

for $k, \ell = 1, 2, 3$.

A.2.3 Stokes System

To find the fundamental solution for the Stokes system, we have to solve the following system of partial differential equations

$$-\varrho\Delta\underline{U}_\ell^*(x,y) + \nabla q_\ell^*(x,y) = \delta_0(y-x)\underline{e}_\ell, \quad \mathrm{div}\,\underline{U}_\ell^*(x,y) = 0 \quad \text{for } x,y \in \mathbb{R}^3,$$

for $\ell = 1,2,3$, and

$$-\varrho\Delta\underline{U}_4^*(x,y) + \nabla q_4^*(x,y) = \underline{0}, \quad \mathrm{div}\,\underline{U}_4^*(x,y) = \delta_0(y-x) \quad \text{for } x,y \in \mathbb{R}^3.$$

As in the case of linear elastostatics, we set

$$\underline{U}_\ell^*(x,y) = \Delta\underline{w}_\ell(x,y) - \mathrm{grad}\,\mathrm{div}\,\underline{w}_\ell(x,y), \quad q_\ell^*(x,y) = \varrho\,\Delta\,\mathrm{div}\,\underline{w}_\ell(x,y)$$

for $x,y \in \mathbb{R}^3$ and for $\ell = 1,2,3$. This implies $\mathrm{div}\underline{U}_\ell^*(x,y) = 0$, and

$$-\varrho\Delta^2\underline{w}_\ell(x,y) = \delta_0(y-x)\underline{e}_\ell.$$

For $\ell = 1$, we find

$$\Delta w_{1,1}(x,y) = \frac{1}{\varrho}\frac{1}{4\pi}\frac{1}{|x-y|}$$

and

$$w_{1,1}(x,y) = \frac{1}{\varrho}\frac{1}{8\pi}|x-y|, \quad w_{1,2}(x,y) = w_{1,3}(x,y) = 0.$$

Using

$$\mathrm{div}\,\underline{w}_1(x,y) = \frac{1}{\varrho}\frac{1}{8\pi}\frac{\partial}{\partial y_1}|x-y| = \frac{1}{\varrho}\frac{1}{8\pi}\frac{y_1-x_1}{|x-y|},$$

we obtain

$$U_{11}^*(x,y) = \Delta w_{1,1}(x,y) - \frac{\partial}{\partial y_1}\mathrm{div}\,\underline{w}_1(x,y) =$$

$$= \frac{1}{\varrho}\frac{1}{4\pi}\frac{1}{|x-y|} - \frac{1}{\varrho}\frac{1}{8\pi}\frac{\partial}{\partial y_1}\frac{y_1-x_1}{|x-y|} =$$

$$= \frac{1}{\varrho}\frac{1}{8\pi}\left(\frac{1}{|x-y|} + \frac{(x_1-y_1)^2}{|x-y|^3}\right),$$

$$U_{12}^*(x,y) = -\frac{\partial}{\partial y_2}\mathrm{div}\,\underline{w}_1(x,y) =$$

$$= -\frac{1}{\varrho}\frac{1}{8\pi}\frac{\partial}{\partial y_2}\frac{y_1-x_1}{|x-y|} = \frac{1}{\varrho}\frac{1}{8\pi}\frac{(x_1-y_1)(x_2-y_2)}{|x-y|^3},$$

$$U_{13}^*(x,y) = -\frac{\partial}{\partial y_3}\mathrm{div}\,\underline{w}_1(x,y) =$$

$$= -\frac{1}{\varrho}\frac{1}{8\pi}\frac{\partial}{\partial y_3}\frac{y_1-x_1}{|x-y|} = \frac{1}{\varrho}\frac{1}{8\pi}\frac{(x_1-y_1)(x_3-y_3)}{|x-y|^3},$$

and

$$q_1^*(x,y) = \varrho \operatorname{div} \Delta \underline{w}_1(x,y) = \frac{1}{4\pi} \frac{\partial}{\partial y_1} \frac{1}{|x-y|} = \frac{1}{4\pi} \frac{x_1-y_1}{|x-y|^3}.$$

Doing the same computations for $\ell = 2, 3$, we obtain the fundamental solution of the Stokes system as

$$U_{k\ell}^*(x,y) = \frac{1}{\varrho} \frac{1}{8\pi} \left(\frac{\delta_{k\ell}}{|x-y|} + \frac{(x_k-y_k)(x_\ell-y_\ell)}{|x-y|^3} \right),$$

$$q_\ell^*(x,y) = \frac{1}{4\pi} \frac{x_\ell - y_\ell}{|x-y|^3}$$

for $k, \ell = 1, 2, 3$.

To find the fundamental solution $\underline{U}_4^*(x,y)$ and $q_4^*(x,y)$, we have to solve the system

$$-\varrho \Delta \underline{U}_4^*(x,y) + \nabla q_4^*(x,y) = \underline{0}, \quad \operatorname{div} \underline{U}_4^*(x,y) = \delta_0(y-x) \quad \text{for } x, y \in \mathbb{R}^3$$

in the distributional sense. Setting

$$\underline{U}_4^*(x,y) = -\nabla \varphi(x,y) \quad \text{for } x, y \in \mathbb{R}^3,$$

we get

$$-\Delta \varphi(x,y) = \delta_0(y-x) \quad \text{for } x, y \in \mathbb{R}^3$$

and therefore

$$\varphi(x,y) = \frac{1}{4\pi} \frac{1}{|x-y|}.$$

Hence,

$$U_{4k}^*(x,y) = -\frac{\partial}{\partial y_k} \varphi(x,y) = \frac{1}{4\pi} \frac{x_k-y_k}{|x-y|^3}, \quad k = 1, 2, 3.$$

In addition, taking the divergence of the first equation, this gives

$$\Delta q_4^*(x,y) = \varrho \Delta \operatorname{div} \underline{U}_4^*(x,y) = \varrho \Delta \delta_0(y-x),$$

implying

$$q_4^*(x,y) = \varrho \, \delta_0(y-x).$$

A.2.4 Helmholtz Equation

The fundamental solution $u_\kappa^*(x,y)$ satisfies the relation (1.94),

$$\int_\Omega \left(-\Delta u_\kappa^*(x,y) - \kappa^2 u_\kappa^*(x,y) \right) u(y) dy = u(x) \quad \text{for } x \in \Omega,$$

which can be written as a partial differential equation in the distributional sense,

$$-\Delta u_\kappa^*(x,y) - \kappa^2 u_\kappa^*(x,y) = \delta_0(y - x) \quad \text{for } x, y \in \mathbb{R}^3.$$

Setting $z = y - x$, we have to find v as a solution of

$$-\Delta v(z) - \kappa^2 v(z) = \delta_0(z) \quad \text{for } z \in \mathbb{R}.$$

Using spherical coordinates as well as $v(z) = \widetilde{v}(r)$ with $r = |z|$, this is equivalent to

$$-\frac{1}{r^2}\frac{\partial}{\partial r}\left[r^2 \frac{\partial}{\partial r}\widetilde{v}(r)\right] - \kappa^2 \widetilde{v}(r) = 0 \quad \text{for } r > 0.$$

With the substitutions

$$\widetilde{v}(r) = \frac{w(r)}{r}, \quad \frac{\partial}{\partial r}\widetilde{v}(r) = \frac{\partial}{\partial r}\frac{w(r)}{r} = \frac{1}{r}\frac{\partial}{\partial r}w(r) - \frac{1}{r^2}w(r),$$

we get

$$\frac{\partial^2}{\partial r^2}w(r) + \kappa^2 w(r) = 0 \quad \text{for } r > 0,$$

with the general solution

$$w(r) = Ae^{i\kappa r} + Be^{-i\kappa r} \quad \text{for } r > 0.$$

Choosing $A = 1/4\pi$ and $B = 0$, this gives the fundamental solution for the Helmholtz equation,

$$u_\kappa^*(x,y) = \frac{1}{4\pi}\frac{e^{i\kappa|x-y|}}{|x-y|} \quad \text{for } x, y \in \mathbb{R}^3.$$

A.3 Mapping Properties

In this section, we will summarise the mapping properties of all boundary integral operators used. For simplicity, we will restrict ourselves to the case of the Laplace equation. For a more detailed presentation, we refer, for example, to [24, 71, 105].

The Newton potential

$$u(x) = (\widetilde{N}f)(x) = \int_\Omega u^*(x,y)f(y)dy \quad \text{for } x \in \Omega$$

is a generalised solution of the Poisson equation

$$-\Delta u(x) = f(x) \quad \text{for } x \in \Omega.$$

In particular, for a given $f \in \widetilde{H}^{-1}(\Omega)$, we have $\widetilde{N}f \in H^1(\Omega)$ satisfying

$$\|u\|_{H^1(\Omega)} = \|\widetilde{N}f\|_{H^1(\Omega)} \le c\|f\|_{\widetilde{H}^{-1}(\Omega)} \quad \text{for all } f \in \widetilde{H}^{-1}(\Omega).$$

Taking the interior trace, we obtain

$$(Nf)(x) = \gamma_0^{\text{int}}(\widetilde{N}f)(x) = \int_\Omega u^*(x,y)f(y)dy \quad \text{for } x \in \Gamma.$$

Combining the mapping properties of the Newton potential

$$\widetilde{N} : \widetilde{H}^{-1}(\Omega) \to H^1(\Omega)$$

with those of the interior trace operator

$$\gamma_0^{\text{int}} : H^1(\Omega) \to H^{1/2}(\Gamma),$$

this gives

$$N = \gamma_0^{\text{int}}\widetilde{N} : \widetilde{H}^{-1}(\Omega) \to H^{1/2}(\Gamma).$$

With this we can derive corresponding mapping properties of the single layer potential

$$(\widetilde{V}w)(x) = \int_\Gamma u^*(x,y)w(y)ds_y \quad \text{for } x \in \mathbb{R}^3 \backslash \Gamma.$$

For $f \in \widetilde{H}^{-1}(\Omega)$, we have, by exchanging the order of integration,

$$\langle \widetilde{V}w, f \rangle_\Omega = \int_\Omega f(x) \int_\Gamma u^*(x,y)w(y)ds_y dx$$

$$= \int_\Gamma w(y) \int_\Omega u^*(x,y)f(x)dx\, ds_y = \langle w, Nf \rangle_\Gamma.$$

Hence, due to $Nf \in H^{1/2}(\Gamma)$ for $f \in \widetilde{H}^{-1}(\Omega)$, we find $w \in H^{-1/2}(\Gamma)$ and $\widetilde{V}w \in H^1(\Omega)$. In fact, the single layer potential

$$\widetilde{V} : H^{-1/2}(\Gamma) \to H^1(\Omega)$$

is a bounded operator, and, combining this with the mapping property of the interior trace operator

$$\gamma_0^{\text{int}} : H^1(\Omega) \to H^{1/2}(\Gamma),$$

this gives

$$V = \gamma_0^{\text{int}}\widetilde{V} : H^{-1/2}(\Gamma) \to H^{1/2}(\Gamma).$$

Moreover, for $x \in \Omega$ we have

$$-\Delta(\widetilde{V}w)(x) = -\Delta \int_\Gamma u^*(x,y)w(y)ds_y = \int_\Gamma \left(-\Delta_x u^*(x,y)\right)w(y)ds_y = 0,$$

i.e. $\widetilde{V}w$ is a generalised solution of the Laplace equation for any $w \in H^{-1/2}(\Gamma)$.

The above considerations are essentially based on the symmetry of the fundamental solution, i.e.

$$u^*(x, y) = u^*(y, x).$$

Therefore, these considerations are valid only for self–adjoint partial differential operators. For more general partial differential operators, one has to incorporate also all the volume and surface potentials which are defined by the fundamental solution of the formally adjoint partial differential operator, see, for example, [71].

It remains to describe an explicit representation of the single layer potential operator

$$(Vw)(x) = \gamma_0^{\mathrm{int}}(\widetilde{V}w)(x)$$

for $x \in \Gamma$. For $w \in L_\infty(\Gamma)$, we obtain

$$(Vw)(x) = \gamma_0^{\mathrm{int}}(\widetilde{V}w)(x) = \int_\Gamma u^*(x, y)w(y)ds_y$$

as a weakly singular boundary integral (cf. Lemma 1.1). For $\varepsilon > 0$, we consider $\widetilde{x} \in \Omega$ and $x \in \Gamma$ with $|\widetilde{x} - x| < \varepsilon$. Then,

$$\lim_{\widetilde{x} \to x} \left| \int_\Gamma u^*(\widetilde{x}, y)w(y)ds_y - \int_{y \in \Gamma: |y-x| > \varepsilon} u^*(x, y)w(y)ds_y \right|$$

$$\leq \lim_{\widetilde{x} \to x} \left| \int_{y \in \Gamma: |y-x| > \varepsilon} \left(u^*(\widetilde{x}, y) - u^*(x, y) \right) w(y)ds_y \right|$$

$$+ \lim_{\widetilde{x} \to x} \left| \int_{y \in \Gamma: |y-x| < \varepsilon} u^*(\widetilde{x}, y)w(y)ds_y \right|$$

$$= \lim_{\widetilde{x} \to x} \left| \int_{y \in \Gamma: |y-x| < \varepsilon} u^*(\widetilde{x}, y)w(y)ds_y \right|$$

$$\leq \sup_{y \in \Gamma: |y-x| < \varepsilon} \left| w(y) \right| \lim_{\widetilde{x} \to x} \int_{y \in \Gamma: |y-x| < \varepsilon} \left| u^*(\widetilde{x}, y) \right| ds_y$$

$$\leq \frac{1}{4\pi} \|w\|_{L_\infty(\Gamma)} \lim_{\widetilde{x} \to x} \int_{y \in \Gamma: |y-x| < \varepsilon} \frac{1}{|\widetilde{x} - y|} ds_y$$

$$= \frac{1}{4\pi} \|w\|_{L_\infty(\Gamma)} \int_{y \in \Gamma: |x-y| < \varepsilon} \frac{1}{|x - y|} ds_y \leq c \|w\|_{L_\infty(\Gamma)} \varepsilon,$$

where we used polar coordinates to obtain the last inequality. Taking the limit $\varepsilon \to 0$, this gives the definition of the single layer potential as a weakly singular integral. In the same way we get for the exterior trace

$$\gamma_0^{\text{ext}}(\widetilde{V}w)(x) = (Vw)(x) \quad \text{for } x \in \Gamma.$$

Since $\widetilde{V}w \in H^1(\Omega)$ is a generalised solution of the Laplace equation for any $w \in H^{1/2}(\Gamma)$, we can compute the associated conormal derivative of $\widetilde{V}w$ as $\gamma_1^{\text{int}}\widetilde{V}w \in H^{-1/2}(\Gamma)$. For $v \in H^1(\Omega)$, and using Green's first formula as well as interpreting the surface integral as weakly singular integral, we obtain

$$
\int_\Gamma \gamma_1^{\text{int}}(\widetilde{V}w)(x)\gamma_0^{\text{int}}v(x)ds_x = \int_\Omega \Big(\nabla(\widetilde{V}w)(x), \nabla v(x)\Big)dx
$$

$$
= \int_\Omega \Big(\nabla \int_\Gamma u^*(x,y)w(y)ds_y, \nabla v(x)\Big)dx
$$

$$
= \int_\Omega \Big(\nabla \lim_{\varepsilon \to 0} \int_{y \in \Gamma : |y-x| \geq \varepsilon} u^*(x,y)w(y)ds_y, \nabla v(x)\Big)dx
$$

$$
= \int_\Gamma w(y) \lim_{\varepsilon \to 0} \int_{x \in \Omega : |x-y| \geq \varepsilon} \Big(\nabla_x u^*(x,y), \nabla v(x)\Big)dx\, ds_y.
$$

Again, Green's first formula gives

$$
\int_{x \in \Omega : |x-y| \geq \varepsilon} \Big(\nabla_x u^*(x,y), \nabla_x v(x)\Big)dx = \int_{x \in \Gamma : |x-y| \geq \varepsilon} \gamma_{1,x}^{\text{int}}u^*(x,y)\gamma_0^{\text{int}}v(x)ds_x
$$

$$
+ \int_{x \in \Omega : |x-y| = \varepsilon} \gamma_{1,x}^{\text{int}}u^*(x,y)\gamma_0^{\text{int}}v(x)ds_x\,,
$$

and, therefore,

$$
\int_\Gamma \gamma_1^{\text{int}}(\widetilde{V}w)(x)\gamma_0^{\text{int}}v(x)ds_x = \int_\Gamma w(y) \lim_{\varepsilon \to 0} \int_{x \in \Gamma : |x-y| \geq \varepsilon} \gamma_{1,x}^{\text{int}}u^*(x,y)\gamma_0^{\text{int}}v(x)ds_x ds_y
$$

$$
+ \int_\Gamma w(y) \lim_{\varepsilon \to 0} \int_{x \in \Omega : |x-y| = \varepsilon} \gamma_{1,x}^{\text{int}}u^*(x,y)\gamma_0^{\text{int}}v(x)ds_x ds_y
$$

$$
= \int_\Gamma \lim_{\varepsilon \to 0} \int_{y \in \Gamma : |y-x| \geq \varepsilon} \gamma_{1,x}^{\text{int}}u^*(x,y)w(y)ds_y\gamma_0^{\text{int}}v(x)ds_x
$$

$$
+ \int_\Gamma w(y) \lim_{\varepsilon \to 0} \int_{x \in \Omega : |x-y| = \varepsilon} \gamma_{1,x}^{\text{int}}u^*(x,y)v(x)ds_x ds_y
$$

$$
= \int_\Gamma (K'w)(x)\gamma_0^{\text{int}}v(x)ds_x + \int_\Gamma w(y)\sigma(y)v(y)ds_y\,.
$$

In the above, the adjoint double layer potential

$$(K'w)(x) = \lim_{\varepsilon \to 0} \int_{y \in \Gamma:|y-x| \geq \varepsilon} \gamma_{1,x}^{\text{int}} u^*(x,y)w(y) ds_y \quad \text{for } x \in \Gamma$$

has been introduced. Furthermore, we have used

$$\lim_{\varepsilon \to 0} \int_{x \in \Omega:|x-y|=\varepsilon} \gamma_{1,x}^{\text{int}} u^*(x,y)v(x) ds_x = \sigma(y)v(y),$$

with

$$\sigma(y) = \lim_{\varepsilon \to 0} \int_{x \in \Omega:|x-y|=\varepsilon} \gamma_{1,x}^{\text{int}} u^*(x,y) ds_x$$

$$= \lim_{\varepsilon \to 0} \frac{1}{4\pi} \int_{x \in \Omega:|x-y|=\varepsilon} \frac{(y-x, \underline{n}(x))}{|x-y|^3} ds_x = \lim_{\varepsilon \to 0} \frac{1}{4\pi} \frac{1}{\varepsilon^2} \int_{x \in \Omega:|x-y|=\varepsilon} ds_x,$$

which is related to the interior angle of Ω in $y \in \Gamma$. In particular, if Γ is smooth in $y \in \Gamma$, we find $\sigma(y) = 1/2$ (cf. Lemma 1.3).

Summarising the above, for $w \in H^{-1/2}(\Gamma)$, we have

$$\gamma_1^{\text{int}} \widetilde{V} w = \sigma(x)w(x) + (K'w)(x) \quad \text{for } x \in \Gamma$$

in the sense of $H^{-1/2}(\Gamma)$. Correspondingly, the application of the exterior conormal derivative gives

$$\gamma_1^{\text{ext}} \widetilde{V} w = (\sigma(x) - 1)w(x) + (K'w)(x) \quad \text{for } x \in \Gamma,$$

and, therefore, we obtain the jump relation of the adjoint double layer potential,

$$[\gamma_1 \widetilde{V} w] = \gamma_1^{\text{ext}}(\widetilde{V}w)(x) - \gamma_1^{\text{int}}(\widetilde{V}w)(x) = -w(x) \quad \text{for } x \in \Gamma.$$

Next, we consider the double layer potential

$$(Wv)(x) = \int_{\Gamma} \gamma_{1,y}^{\text{int}} u^*(x,y)v(y) ds_y \quad \text{for } x \in \mathbb{R}^3 \backslash \Gamma.$$

For $f \in \widetilde{H}^{-1}(\Omega)$, by exchanging the order of integration, we have,

$$\langle Wv, f \rangle_\Omega = \int_{\Omega} f(x) \int_{\Gamma} \gamma_{1,y}^{\text{int}} u^*(x,y)v(y) ds_y \, dx$$

$$= \int_{\Gamma} v(y) \gamma_{1,y}^{\text{int}} \int_{\Omega} u^*(x,y)f(x) dx \, ds_y$$

$$= \int_{\Gamma} v(y) \gamma_{1,y}^{\text{int}} (\widetilde{N} f)(y) ds_y = \langle v, \gamma_1^{\text{int}} \widetilde{N} f \rangle_\Gamma.$$

From this, we find $Wv \in H^1(\Omega)$ for any $v \in H^{1/2}(\Gamma)$. Moreover, $Wv \in H^1(\Omega)$ is a generalised solution of the Laplace equation for any $v \in H^{1/2}(\Gamma)$. The application of the interior trace operator

$$\gamma_0^{\text{int}} \; : \; H^1(\Omega) \to H^{1/2}(\Gamma)$$

gives

$$\gamma_0^{\text{int}} W \; : \; H^{1/2}(\Gamma) \to H^{1/2}(\Gamma).$$

It turns out (cf. Lemma 1.2) that

$$\gamma_0^{\text{int}}(Wv)(x) \; = \; (-1 + \sigma(x))v(x) + (Kv)(x) \quad \text{for } x \in \Gamma,$$

with the double layer potential

$$(Kv)(x) \; = \; \lim_{\varepsilon \to 0}(K_\varepsilon v)(x) \; = \; \lim_{\varepsilon \to 0} \int\limits_{y \in \Gamma:|y-x| \geq \varepsilon} \gamma_{1,y}^{\text{int}} u^*(x,y)v(y)ds_y \quad \text{for } x \in \Gamma.$$

This representation follows, when considering $\widetilde{x} \in \Omega$ and $x \in \Gamma$ satisfying $|\widetilde{x} - x| < \varepsilon$, from

$$(Wv)(\widetilde{x}) - (K_\varepsilon v)(x) = \int\limits_{\Gamma} \gamma_{1,y}^{\text{int}} u^*(\widetilde{x},y)v(y)ds_y - \int\limits_{y \in \Gamma:|y-x| \geq \varepsilon} \gamma_{1,y}^{\text{int}} u^*(x,y)v(y)ds_y$$

$$= \int\limits_{y \in \Gamma:|y-x| \geq \varepsilon} \left(\gamma_{1,y}^{\text{int}} u^*(\widetilde{x},y) - \gamma_{1,y}^{\text{int}} u^*(x,y)\right)v(y)ds_y$$

$$+ \int\limits_{y \in \Gamma:|y-x| \leq \varepsilon} \gamma_{1,y}^{\text{int}} u^*(\widetilde{x},y)\left(v(y) - v(x)\right)ds_y$$

$$+ v(x) \int\limits_{y \in \Gamma:|y-x| \leq \varepsilon} \gamma_{1,y}^{\text{int}} u^*(\widetilde{x},y)ds_y.$$

Note that for all $\varepsilon > 0$ we have

$$\lim_{\widetilde{x} \to x} \int\limits_{y \in \Gamma:|y-x| \geq \varepsilon} \left(\gamma_{1,y}^{\text{int}} u^*(\widetilde{x},y) - \gamma_{1,y}^{\text{int}} u^*(x,y)\right)v(y)ds_y = 0.$$

In addition,

$$\lim_{\varepsilon \to 0} \int\limits_{y \in \Gamma:|y-x| \leq \varepsilon} \gamma_{1,y}^{\text{int}} u^*(\widetilde{x},y)\left(v(y) - v(x)\right)ds_y = 0.$$

Moreover, for $\widetilde{x} \in B_\varepsilon(x) = \{z \in \Omega : |z - x| < \varepsilon\}$, we have

$$\int\limits_{y \in \Gamma:|y-x| \leq \varepsilon} \gamma_{1,y}^{\text{int}} u^*(\widetilde{x},y)ds_y = \int\limits_{\partial B_\varepsilon(x)} \gamma_{1,y}^{\text{int}} u^*(\widetilde{x},y)ds_y - \int\limits_{y \in \Omega:|y-x|=\varepsilon} \gamma_{1,y}^{\text{int}} u^*(\widetilde{x},y)ds_y.$$

Taking into account that

$$\int\limits_{\partial B_\varepsilon(x)} \gamma_{1,y}^{int} u^*(\widetilde{x}, y) ds_y = -1 \quad \text{for } \widetilde{x} \in B_\varepsilon(x)$$

as well as the direction of the normal vector along $y \in \Omega : |y - x| = \varepsilon$, we finally obtain the above representation for $\gamma_0^{int} W v$.

Correspondingly, the application of the exterior trace operator gives

$$\gamma_0^{ext}(Wv)(x) = \sigma(x)v(x) + (Kv)(x) \quad \text{for } x \in \Gamma,$$

and, therefore, the jump relation of the double layer potential reads

$$[\gamma_0 W v] = \gamma_0^{ext}(Wv)(x) - \gamma_0^{int}(Wv)(x) = v(x) \quad \text{for } x \in \Gamma.$$

Since the double layer potential $Wv \in H^1(\Omega)$ is a solution of the Laplace equation for any $v \in H^{1/2}(\Gamma)$, the application of the conormal derivative γ_1^{int} defines the bounded operator

$$D = -\gamma_1^{int} W : H^{1/2}(\Gamma) \to H^{-1/2}(\Gamma).$$

Note that for $x \in \Gamma$, we have

$$(Dv)(x) = -\gamma_1^{int}(Wv)(x) = \lim_{\Omega \ni \widetilde{x} \to x \in \Gamma} \left(\nabla_{\widetilde{x}}(Wv)(\widetilde{x}), \underline{n}(x) \right)$$

$$= \frac{1}{4\pi} \lim_{\varepsilon \to 0} \int\limits_{y \in \Gamma : |y-x| \geq \varepsilon} \left(\frac{(\underline{n}(x), \underline{n}(y))}{|x-y|^3} - 3\frac{(y-x, \underline{n}(y))(y-x, \underline{n}(x))}{|x-y|^5} \right) v(y) ds_y,$$

which does not exist. In what follows we will sketch the computations to obtain the alternative representation (1.9) (cf. Lemma 1.4). For the derivatives of the double layer potential for $\widetilde{x} \in \Omega$ we first have

$$\frac{\partial}{\partial \widetilde{x}_i}(Wv)(\widetilde{x}) = \frac{\partial}{\partial \widetilde{x}_i} \int\limits_\Gamma \gamma_1^{int} u^*(\widetilde{x}, y) v(y) ds_y$$

$$= \frac{1}{4\pi} \frac{\partial}{\partial \widetilde{x}_i} \int\limits_\Gamma \left(\underline{n}(y), \nabla_y \frac{1}{|\widetilde{x}-y|} \right) v(y) ds_y$$

$$= -\frac{1}{4\pi} \int\limits_\Gamma \left(\underline{n}(y), \nabla_y \frac{\partial}{\partial y_i} \frac{1}{|\widetilde{x}-y|} \right) v(y) ds_y$$

$$= \frac{1}{4\pi} \int\limits_\Gamma \left(\underline{n}(y), \underline{\text{curl}}_y \left(\underline{e}_i \times \nabla_y \frac{1}{|\widetilde{x}-y|} \right) \right) v(y) ds_y$$

$$= \frac{1}{4\pi} \int\limits_\Gamma v(y) \, \text{curl}_{\Gamma, y} \left(\underline{e}_i \times \nabla_y \frac{1}{|\widetilde{x}-y|} \right) ds_y.$$

Now, using integration by parts, i.e.

$$\int_\Gamma \mathrm{curl}_\Gamma \underline{\varphi}(y)\psi(y)ds_y = -\int_\Gamma \left(\underline{\varphi}(y), \mathrm{curl}_\Gamma \psi(y)\right) ds_y,$$

we obtain

$$\frac{\partial}{\partial \tilde{x}_i}(Wv)(\tilde{x}) = -\frac{1}{4\pi}\int_\Gamma \left(\underline{\mathrm{curl}}_\Gamma v(y), \underline{e}_i \times \nabla_y \frac{1}{|\tilde{x}-y|}\right)ds_y$$

$$= \frac{1}{4\pi}\int_\Gamma \left(\underline{e}_i, \underline{\mathrm{curl}}_\Gamma v(y) \times \nabla_y \frac{1}{|\tilde{x}-y|}\right)ds_y,$$

and, therefore,

$$\nabla_{\tilde{x}}(Wv)(\tilde{x}) = \frac{1}{4\pi}\int_\Gamma \left(\underline{\mathrm{curl}}_\Gamma v(y) \times \nabla_y \frac{1}{|\tilde{x}-y|}\right)ds_y$$

$$= -\frac{1}{4\pi}\int_\Gamma \left(\underline{\mathrm{curl}}_\Gamma v(y) \times \nabla_{\tilde{x}} \frac{1}{|\tilde{x}-y|}\right)ds_y.$$

Hence,

$$(\nabla_{\tilde{x}}(Wv)(\tilde{x}),\underline{n}(x)) = -\frac{1}{4\pi}\int_\Gamma \left(\underline{\mathrm{curl}}_\Gamma v(y) \times \nabla_{\tilde{x}} \frac{1}{|\tilde{x}-y|}, \underline{n}(x)\right)ds_y$$

$$= \frac{1}{4\pi}\int_\Gamma \left(\underline{\mathrm{curl}}_\Gamma v(y), \underline{n}(x) \times \nabla_{\tilde{x}} \frac{1}{|\tilde{x}-y|}\right)ds_y.$$

Taking the limes $\tilde{x} \to x$, we obtain

$$(Dv)(x) = -\frac{1}{4\pi}\int_\Gamma \left(\underline{\mathrm{curl}}_\Gamma v(y), \underline{n}(x) \times \nabla_x \frac{1}{|x-y|}\right)ds_y$$

$$= -\frac{1}{4\pi}\int_\Gamma \left(\underline{\mathrm{curl}}_\Gamma v(y), \underline{\mathrm{curl}}_{\Gamma,x} \frac{1}{|x-y|}\right)ds_y,$$

and, therefore,

$$\langle Dv, u\rangle_\Gamma = -\frac{1}{4\pi}\int_\Gamma u(x)\int_\Gamma \left(\underline{\mathrm{curl}}_\Gamma v(y), \underline{\mathrm{curl}}_{\Gamma,x} \frac{1}{|x-y|}\right)ds_y ds_x$$

$$= \frac{1}{4\pi}\int_\Gamma\int_\Gamma \mathrm{curl}_{\Gamma,x}\left(u(x)\underline{\mathrm{curl}}_{\Gamma,y} v(y)\right)\frac{1}{|x-y|}ds_x ds_y.$$

With

$$\mathrm{curl}_{\Gamma,x}\Big(u(x)\underline{\mathrm{curl}}_{\Gamma,y}v(y)\Big) = \Big(\underline{n}(x), \nabla_x \times \Big(u(x)\underline{\mathrm{curl}}_{\Gamma,y}v(y)\Big)\Big)$$

$$= \Big(\underline{n}(x), \nabla_x u(x) \times \underline{\mathrm{curl}}_{\Gamma,y}v(y)\Big)$$

$$= \Big(\underline{n}(x) \times \nabla_x u(x), \underline{\mathrm{curl}}_{\Gamma,y}v(y)\Big)$$

$$= \Big(\underline{\mathrm{curl}}_{\Gamma,x}u(x), \underline{\mathrm{curl}}_{\Gamma,y}v(y)\Big) \ ,$$

we finally obtain the alternative representation (1.9),

$$\langle Dv, u \rangle_{\Gamma} = \frac{1}{4\pi} \int_{\Gamma} \int_{\Gamma} \frac{(\underline{\mathrm{curl}}_{\Gamma}u(x), \underline{\mathrm{curl}}_{\Gamma}v(y))}{|x - y|} ds_x ds_y.$$

Since the ellipticity of boundary integral operators is essential to derive results on the unique solvability of boundary integral equations, we will sketch the proof of the ellipticity of the single layer potential V, cf. Lemma 1.1.

The single layer potential

$$u(x) = (\widetilde{V}w)(x) = \int_{\Gamma} u^*(x, y)w(y)ds_y \quad \text{for } x \in \mathbb{R}^3 \backslash \Gamma$$

is a solution of the interior Dirichlet boundary value problem

$$-\Delta u(x) = 0 \quad \text{for } x \in \Omega, \quad \gamma_0^{\mathrm{int}}u(x) = \gamma_0^{\mathrm{int}}(\widetilde{V}w)(x) = (Vw)(x) \quad \text{for } x \in \Gamma.$$

By applying Green's first formula (cf. (1.2)), we obtain

$$a_\Omega(u, u) = \int_\Omega |\nabla u(x)|^2 dx = \int_\Gamma \gamma_1^{\mathrm{int}}u(x)\gamma_0^{\mathrm{int}}u(x)ds_x$$

$$= \int_\Gamma \gamma_1^{\mathrm{int}}(\widetilde{V}w)(x)\gamma_0^{\mathrm{int}}(\widetilde{V}w)(x)ds_x$$

$$= \int_\Gamma \Big(\frac{1}{2}w(x) + (K'w)(x)\Big)(Vw)(x)ds_x.$$

From Green's first formula (1.2), we also find the estimate

$$c_1 \|\gamma_1^{\mathrm{int}}u\|^2_{H^{-1/2}(\Gamma)} \leq a_\Omega(u, u).$$

Since $u = \widetilde{V}w$ is also a solution of the exterior Dirichlet boundary value problem,

$$-\Delta u(x) = 0 \quad \text{for } x \in \Omega^e, \quad \gamma_0^{\mathrm{ext}}u(x) = \gamma_0^{\mathrm{ext}}(\widetilde{V}w)(x) = (Vw)(x) \text{ for } x \in \Gamma,$$

satisfying the radiation condition

$$|u(x)| = \mathcal{O}\left(\frac{1}{|x|}\right) \quad \text{as } |x| \to \infty,$$

we also find the relations

$$a_{\Omega^e}(u, u) = -\int_{\Gamma} \gamma_1^{\text{ext}} u(x) \gamma_0^{\text{ext}} u(x) ds_x$$

$$= -\int_{\Gamma} \left(-\frac{1}{2}w(x) + (K'w)(x)\right)(Vw)(x) ds_x,$$

and

$$c_2 \|\gamma_0^{\text{ext}} u\|_{H^{-1/2}(\Gamma)}^2 \le a_{\Omega^e}(u, u).$$

Hence, we obtain

$$\langle Vw, w \rangle_{\Gamma} = \int_{\Gamma} w(x)(Vw)(x) ds_x$$

$$= \int_{\Gamma} \left(\frac{1}{2}w(x) + (K'w)(x)\right)(Vw)(x) ds_x$$

$$- \int_{\Gamma} \left(-\frac{1}{2}w(x) + (K'w)(x)\right)(Vw)(x) ds_x$$

$$= a_{\Omega}(u, u) + a_{\Omega^e}(u, u)$$

$$\ge \min\{c_1, c_2\}\left(\|\gamma_1^{\text{int}} u\|_{H^{-1/2}(\Gamma)}^2 + \|\gamma_1^{\text{ext}} u\|_{H^{-1/2}(\Gamma)}^2\right).$$

Using

$$\|w\|_{H^{-1/2}(\Gamma)}^2 = \left\|\frac{1}{2}w + K'w - \left(-\frac{1}{2}w + K'w\right)\right\|_{H^{-1/2}(\Gamma)}^2$$

$$= \|\gamma_1^{\text{int}} u - \gamma_1^{\text{ext}} u\|_{H^{-1/2}(\Gamma)}^2$$

$$\le 2\left(\|\gamma_1^{\text{int}} u\|_{H^{-1/2}(\Gamma)}^2 + \|\gamma_1^{\text{ext}} u\|_{H^{-1/2}(\Gamma)}^2\right),$$

we finally obtain the ellipticity estimate

$$\langle Vw, w \rangle_{\Gamma} \ge c_1^V \|w\|_{H^{-1/2}(\Gamma)}^2 \quad \text{for all } w \in H^{-1/2}(\Gamma).$$

Note that the ellipticity estimate of the hypersingular boundary integral operator D (cf. Lemma 1.4) follows almost in the same manner when considering the double layer potential $u = Wv$.

If $\Gamma = \partial\Omega$ is the boundary of a Lipschitz domain $\Omega \subset \mathbb{R}^3$, we then can extend the mapping properties of all boundary integral operators, see [24]. In particular, the boundary integral operators

$$V \ : H^{-1/2+s}(\Gamma) \to H^{1/2+s}(\Gamma),$$
$$K \ : H^{1/2+s}(\Gamma) \ \to H^{1/2+s}(\Gamma),$$
$$K' : H^{-1/2+s}(\Gamma) \to H^{-1/2+s}(\Gamma),$$
$$D \ : H^{1/2+s}(\Gamma) \ \to H^{-1/2+s}(\Gamma)$$

are bounded for all $s \in [-1/2, 1/2]$.

B

Numerical Analysis

B.1 Variational Methods

Let X be a Hilbert space with an inner product $\langle \cdot, \cdot \rangle_X$, which induces a norm

$$\|v\|_X = \sqrt{\langle v, v \rangle_X} \quad \text{for all } v \in X.$$

Let X' be the dual space of X with respect to the duality pairing

$$\langle \cdot, \cdot \rangle \; : \; X' \times X \to \mathbb{R}.$$

Then,

$$\|f\|_{X'} \leq \sup_{0 \neq v \in X} \frac{\langle f, v \rangle}{\|v\|_X} \quad \text{for all } f \in X'.$$

Let $A : X \to X'$ be a bounded linear operator with

$$\|Av\|_{X'} \leq c_2^A \|v\|_X \quad \text{for all } v \in X, \tag{B.1}$$

which is symmetric, i.e.

$$\langle Au, v \rangle = \langle Av, u \rangle \quad \text{for all } u, v \in X.$$

For a given $f \in X'$, we consider an operator equation to find $u \in X$ such that

$$Au = f \tag{B.2}$$

is satisfied in X'. Then,

$$0 = \|Au - f\|_{X'} = \sup_{0 \neq v \in X} \frac{\langle Au - f, v \rangle}{\|v\|_X},$$

and, therefore, $u \in X$ is a solution of the variational problem

$$\langle Au, v \rangle = \langle f, v \rangle \quad \text{for all } v \in X. \tag{B.3}$$

For a given operator $A : X \to X'$, we define the functional

$$F(v) = \frac{1}{2}\langle Av, v \rangle - \langle f, v \rangle .$$

If the operator A is assumed to be positive semi–elliptic, i.e.

$$\langle Av, v \rangle \geq 0 \quad \text{for all } v \in X ,$$

then the solution of the variational problem (B.3) is equivalent to the min-imisation problem

$$F(u) = \min_{v \in X} F(v) . \tag{B.4}$$

Essential for the analysis of variational problems is the Riesz representation theorem [118], i.e. any bounded linear functional $f \in X'$ is of the form

$$\langle f, v \rangle = \langle u, v \rangle_X \quad \text{for all } v \in X. \tag{B.5}$$

The element $u \in X$ is hereby uniquely determined and satisfies

$$\|u\|_X = \|f\|_{X'} .$$

The Riesz representation theorem (see (B.5)) defines a linear and bounded operator $J : X' \to X$ with

$$\langle Jf, v \rangle_X = \langle f, v \rangle \quad \text{for all } v \in X, \quad \|Jf\|_X = \|f\|_{X'} .$$

Now, instead of the operator equation (B.2), $Au = f$ in X', we consider the equivalent operator equation

$$JAu = Jf \quad \text{in } X. \tag{B.6}$$

Since $A : X \to X'$ is a bounded operator satisfying (B.1), we obtain

$$\|JAv\|_X = \|Av\|_{X'} \leq c_2^A \|v\|_X \quad \text{for all } v \in X,$$

i.e. the operator $JA : X \to X$ is bounded with

$$\|JA\|_{X \to X} \leq c_2^A .$$

An operator $A : X \to X'$ is called X–elliptic, if

$$\langle Av, v \rangle \geq c_1^A \|v\|_X^2 \quad \text{for all } v \in X . \tag{B.7}$$

Then, if A is X–elliptic, the estimate

$$\langle JAv, v \rangle_X = \langle Av, v \rangle \geq c_1^A \|v\|_X^2 \quad \text{for all } v \in X$$

follows, i.e. the operator $JA : X \to X$ is also X–elliptic. Therefore, instead of (B.6), we consider the fix point equation

$$u = u - \varrho J(Au - f) = T_\varrho u + \varrho J f, \tag{B.8}$$

with

$$T_\varrho = I - \varrho J A : X \to X$$

for some positive $\varrho \in \mathbb{R}_+$. For

$$0 < \varrho < 2 \frac{c_1^A}{\left(c_2^A\right)^2}$$

the operator $T_\varrho : X \to X$ defines a contraction in X with $\|T_\varrho\|_{X \to X} < 1$. Thus, the unique solvability of the fix point equation (B.8) follows from Banach's fix point theorem. Hence, if $A : X \to X'$ is bounded and X–elliptic, we conclude the unique solvability of the operator equation (B.2), and, therefore, of the equivalent variational problem (B.3), and of the minimisation problem (B.4). This result is just the well known Lax–Milgram lemma. Moreover, for the solution $u \in X$ of the operator equation $Au = f$, we obtain from the ellipticity estimate (B.7)

$$c_1^A \|u\|_X^2 \leq \langle Au, u \rangle = \langle f, u \rangle \leq \|f\|_{X'} \|u\|_X,$$

and, therefore,

$$\|u\|_X \leq \frac{1}{c_1^A} \|f\|_{X'}.$$

Now we consider a sequence $\{X_M\}_{M \in \mathbb{N}}$ of conformal trial spaces

$$X_M = \operatorname{span}\left\{\varphi_\ell\right\}_{\ell=1}^M \subset X,$$

which is assumed to be dense in X. In particular, we assume the approximation property

$$\lim_{M \to \infty} \inf_{v_M \in X_M} \|v - v_M\|_X = 0 \quad \text{for all } v \in X. \tag{B.9}$$

To define an approximate solution of the operator equation (B.2), or of the equivalent minimisation problem (B.4), we consider the finite dimensional minimisation problem

$$F(u_M) = \min_{v_M \in X_M} F(v_M),$$

with

$$F(v_M) = \frac{1}{2} \langle Av_M, v_M \rangle - \langle f, v_M \rangle$$

$$= \frac{1}{2} \sum_{\ell=1}^M \sum_{j=1}^M v_\ell v_j \langle A\varphi_\ell, \varphi_j \rangle - \sum_{\ell=1}^M v_\ell \langle f, \varphi_\ell \rangle = \tilde{F}(\underline{v}).$$

From the necessary condition

$$\frac{d}{dv_k}\widetilde{F}(\underline{v}) = 0 \quad \text{for } k = 1, \ldots, M,$$

we then obtain

$$\sum_{\ell=1}^{M} v_\ell \langle A\varphi_\ell, \varphi_k \rangle - \langle f, \varphi_k \rangle = 0 \quad \text{for } k = 1, \ldots, M.$$

Thus, $u_M \in X_M$ is the solution of the Galerkin variational problem

$$\sum_{\ell=1}^{M} u_\ell \langle A\varphi_\ell, \varphi_k \rangle = \langle f, \varphi_k \rangle \quad \text{for } k = 1, \ldots, M, \tag{B.10}$$

or, in the equivalent formulation,

$$\langle Au_M, v_M \rangle = \langle f, v_M \rangle \quad \text{for all } v_M \in X_M. \tag{B.11}$$

The Galerkin variational problem (B.10) is obviously equivalent to a linear system

$$A_M \underline{u} = \underline{f},$$

with

$$A_M[k, \ell] = \langle A\varphi_\ell, \varphi_k \rangle, \quad f_k = \langle f, \varphi_k \rangle$$

for $k, \ell = 1, \ldots, M$. Note that

$$(A_M \underline{u}, \underline{v}) = \sum_{k=1}^{M}\sum_{\ell=1}^{M} A[k, \ell] u_\ell v_k = \sum_{k=1}^{M}\sum_{\ell=1}^{M} \langle A\varphi_\ell, \varphi_k \rangle u_\ell v_k = \langle Au_M, v_M \rangle$$

for all $\underline{u}, \underline{v} \in \mathbb{R}^M$, i.e. $u_M, v_M \in X_M$. Therefore,

$$(A_M \underline{v}, \underline{v}) = \langle Av_M, v_M \rangle \geq c_1^A \|v_M\|_X^2$$

for all $\underline{v} \in \mathbb{R}^M$, i.e. $v_M \in X_M \subset X$. The Galerkin stiffness matrix A_M is hence positive definite, and, therefore, invertible. In particular, the Galerkin variational problem (B.11) is uniquely solvable. Moreover, from

$$c_1^A \|u_M\|_X^2 \leq \langle Au_M, u_M \rangle = \langle f, u_M \rangle \leq \|f\|_{X'} \|u_M\|_X,$$

we conclude the stability estimate

$$\|u_M\|_X \leq \frac{1}{c_1^A} \|f\|_{X'}.$$

From (B.3), we obtain

$$\langle Au, v_M \rangle = \langle f, v_M \rangle \quad \text{for all } v_M \in X_M \subset X.$$

Subtracting from this the Galerkin variational problem (B.11), this gives the Galerkin orthogonality

$$\langle A(u - u_M), v_M \rangle = 0 \quad \text{for all } v_M \in X_M. \tag{B.12}$$

Using the ellipticity estimate (B.7), the Galerkin orthogonality (B.12), and the boundedness of A, we then obtain

$$
\begin{aligned}
c_1^A \|u - u_M\|_X^2 &\leq \langle A(u - u_M), u - u_M \rangle \\
&= \langle A(u - u_M), u - v_M \rangle + \langle A(u - u_M), v_M - u_M \rangle \\
&= \langle A(u - u_M), u - v_M \rangle \\
&\leq \|A(u - u_M)\|_{X'} \|u - v_M\|_X \\
&\leq c_2^A \|u - u_M\|_X \|u - v_M\|_X ,
\end{aligned}
$$

and, therefore,

$$\|u - u_M\|_X \leq \frac{c_2^A}{c_1^A} \|u - v_M\|_X \quad \text{for all } v_M \in X_M.$$

This results in Cea's lemma, i.e.

$$\|u - u_M\|_X \leq \frac{c_2^A}{c_1^A} \inf_{v_M \in X_M} \|u - v_M\|_X. \tag{B.13}$$

Hence, we obtain convergence $u_M \to u$ for $M \to \infty$ from the approximation property (B.9).

In many applications, e.g. for direct boundary integral formulations, the right hand side in (B.2) is given as $f = Bg$, where $B : Y \to X'$ is some bounded operator satisfying

$$\|Bg\|_{X'} \leq c_2^B \|g\|_Y \quad \text{for all } g \in Y.$$

Then the Galerkin formulation (B.11) requires the evaluation of

$$f_k = \langle Bg, \varphi_k \rangle \quad \text{for all } k = 1, \ldots, M,$$

which may be complicated for general given $g \in Y$. Hence, introducing an approximation

$$g_N = \sum_{j=1}^{N} g_j \psi_j \in Y_N \subset Y,$$

we may compute the perturbed vector values

$$\widetilde{f}_k = \langle Bg_N, \varphi_k \rangle = \sum_{j=1}^{N} g_j \langle B\psi_j, \varphi_k \rangle = \sum_{j=1}^{N} B_N[k, j] g_j,$$

leading to the linear system

$$A_M \underline{\widetilde{u}} = B_N \underline{g}.$$

Note that the solution vector $\underline{\widetilde{u}} \in \mathbb{R}^M$ corresponds to the unique solution $\widetilde{u}_M$ of the perturbed variational problem

$$\langle A \widetilde{u}_M, v_M \rangle = \langle B g_N, v_M \rangle \quad \text{for all } v_M \in X_M.$$

Now, instead of the Galerkin orthogonality (B.12), we obtain

$$\langle A(u - \widetilde{u}_M), v_M \rangle = \langle A(u_M - \widetilde{u}_M), v_M \rangle = \langle B(g - g_N), v_M \rangle \quad \text{for all } v_M \in X_M.$$

From the ellipticity estimate (B.7), we then conclude

$$c_1^A \|u_M - \widetilde{u}_M\|_X^2 \leq \langle A(u_M - \widetilde{u}_M), u_M - \widetilde{u}_M \rangle$$
$$= \langle B(g - g_N), u_M - \widetilde{u}_M \rangle \leq c_2^B \|g - g_N\|_Y \|u_M - \widetilde{u}_M\|_X,$$

and, therefore,

$$\|u_M - \widetilde{u}_M\|_X \leq \frac{c_2^B}{c_1^A} \|g - g_N\|_Y.$$

Hence, we find from the triangle inequality the error estimate

$$\|u - \widetilde{u}_M\|_X \leq \|u - u_M\|_X + \|u_M - \widetilde{u}_M\|_X$$
$$\leq \|u - u_M\|_X + \frac{c_2^B}{c_1^A} \|g - g_N\|_Y.$$

Besides an approximation of the given right hand side f, we also have to consider an approximation of the given operator A, e.g., when applying numerical integration schemes. Hence, instead of the Galerkin formulation (B.11), we have to solve a perturbed variational problem to find the solution $\widetilde{u}_M \in X_M$ satisfying

$$\langle \widetilde{A} \widetilde{u}_M, v_M \rangle = \langle f, v_M \rangle \quad \text{for all } v_M \in X_M. \tag{B.14}$$

We assume that $\widetilde{A} : X \to X'$ is a bounded linear operator satisfying

$$\|\widetilde{A}v\|_{X'} \leq c_2^{\widetilde{A}} \|v\|_X \quad \text{for all } v \in X.$$

To ensure the unique solvability of the Galerkin formulation (B.14), we have to assume the X_M–ellipticity of $\widetilde{A}$, i.e.

$$\langle \widetilde{A} v_M, v_M \rangle \geq c_1^{\widetilde{A}} \|v_M\|_X^2 \quad \text{for all } v_M \in X_M. \tag{B.15}$$

The perturbed stiffness matrix $\widetilde{A}_M$, defined by

$$\widetilde{A}_M[k, \ell] = \langle \widetilde{A} \varphi_\ell, \varphi_k \rangle$$

for $k, \ell = 1, \ldots, M$, is then positive definite, and, therefore, invertible. Subtracting the perturbed variational formulation (B.14) from the Galerkin variational problem (B.11), this gives the orthogonality

$$\langle Au_M - \widetilde{A}\widetilde{u}_M, v_M \rangle = 0 \quad \text{for all } v_M \in X_M.$$

From this, and by using the X_M–ellipticity of $\widetilde{A}$, we find

$$c_1^{\widetilde{A}} \|u_M - \widetilde{u}_M\|_X^2 \leq \langle \widetilde{A}(u_M - \widetilde{u}_M), u_M - \widetilde{u}_M \rangle$$
$$= \langle (\widetilde{A} - A)u_M, u_M - \widetilde{u}_M \rangle$$
$$\leq \|(\widetilde{A} - A)u_M\|_{X'}\|u_M - \widetilde{u}_M\|_X \,,$$

and, therefore,

$$\|u_M - \widetilde{u}_M\|_X \leq \frac{1}{c_1^{\widetilde{A}}} \|(A - \widetilde{A})u_M\|_{X'}.$$

Hence, applying the triangle inequality, we finally obtain the error estimate

$$\|u - \widetilde{u}_M\|_X \leq \|u - u_M\|_X + \|u_M - \widetilde{u}_M\|_X$$
$$\leq \|u - u_M\|_X + \frac{1}{c_1^{\widetilde{A}}} \|(A - \widetilde{A})u_M\|_{X'}$$
$$\leq \|u - u_M\|_X + \frac{1}{c_1^{\widetilde{A}}} \left(\|(A - \widetilde{A})u\|_{X'} + \|(A - \widetilde{A})(u - u_M)\|_{X'} \right)$$
$$\leq \left(1 + \frac{c_2^A + c_2^{\widetilde{A}}}{c_1^{\widetilde{A}}} \right) \|u - u_M\|_X + \frac{1}{c_1^{\widetilde{A}}} \|(A - \widetilde{A})u\|_{X'}.$$

B.2 Approximation Properties

In this section, we will prove the approximation properties of piecewise polynomial basis functions defined in Chapter 2.

Let $\Gamma = \partial\Omega$ be a piecewise smooth Lipschitz boundary which is represented by a non–overlapping decomposition

$$\Gamma = \bigcup_{i=1}^{J} \overline{\Gamma}_i, \quad \Gamma_i \cap \Gamma_j = \varnothing \quad \text{for } i \neq j,$$

where each boundary segment Γ_i is described via a local parametrisation (A.2),

$$\Gamma_i = \left\{ x \in \mathbb{R}^3 : x = \chi_i(\xi) \text{ for } \xi \subset \mathcal{T}_i \subset \mathbb{R}^2 \right\},$$

with respect to some parameter domain $\mathcal{T}_i$.

For a sequence of boundary element discretisations (cf. Chapter 2),

$$\Gamma = \bigcup_{\ell=1}^{N} \overline{\tau}_\ell,$$

we further assume that for each boundary element τ_ℓ there exists exactly one boundary segment Γ_i with $\tau_\ell \subset \Gamma_i$. Hence, there also exists an element q_ℓ^i such that $\tau_\ell = \chi_i(q_\ell^i)$. For the area Δ_ℓ of the boundary element τ_ℓ, we then obtain

$$\Delta_\ell = \int\limits_{\tau_\ell} ds_x = \int\limits_{q_\ell^i} \sqrt{EG - F^2}\, d\xi\,,$$

with

$$E = \sum_{i=1}^{3} \left(\frac{\partial}{\partial \xi_1} x_i(\xi)\right)^2, \quad G = \sum_{i=1}^{3} \left(\frac{\partial}{\partial \xi_2} x_i(\xi)\right)^2, \quad F = \sum_{i=1}^{3} \frac{\partial}{\partial \xi_1} x_i(\xi) \frac{\partial}{\partial \xi_2} x_i(\xi).$$

We assume that the parametrisation is uniformly bounded, i.e.

$$c_1^\chi \le \sqrt{EG - F^2} \le c_2^\chi \quad \text{for all } \xi \in \mathcal{T}_i, \quad i = 1, \dots, J.$$

Hence, we obtain

$$c_1^\chi \, \text{area}\, q_\ell^i \le \Delta_\ell \le c_2^\chi \, \text{area}\, q_\ell^i\,.$$

Then,

$$\|v\|_{L_2(\tau_\ell)}^2 = \int\limits_{\tau_\ell} |v(x)|^2 ds_x = \int\limits_{q_\ell^i} |v(\chi_i(\xi))|^2 \sqrt{EG - F^2}\, d\xi$$

$$\le c_2^\chi \int\limits_{q_\ell^i} |v(\chi_i(\xi))|^2 d\xi \le \frac{c_2^\chi}{c_1^\chi} \frac{\Delta_\ell}{\text{area}\, q_\ell^i} \int\limits_{q_\ell^i} |v(\chi_i(\xi))|^2 d\xi\,.$$

Now, using a parametrisation $q_\ell^i = \chi_\ell^i(\tau)$ with respect to the parameter domain τ (cf. Chapter 2), this gives

$$\int\limits_{q_\ell^i} |v(\chi_i(\xi))|^2 d\xi = \int\limits_{\tau} |v(\chi_i(\chi_\ell^i(\eta)))|^2 \, |\det \chi_\ell^i|\, d\eta\,,$$

and with

$$\text{area}\, q_\ell^i = \int\limits_{q_\ell^i} d\xi = \int\limits_{\tau} |\det \chi_\ell^i|\, d\eta = \frac{1}{2} |\det \chi_\ell^i|\,,$$

we finally obtain

$$\|v\|_{L_2(\tau_\ell)}^2 \le 2c\, \Delta_\ell\, \|\tilde{v}_\ell\|_{L_2(\tau)}^2, \quad \tilde{v}_\ell(\eta) = v(\chi_i(\chi_\ell^i(\eta)))\,.$$

Note that this result would follow directly when considering a parametrisation of the boundary element τ_ℓ with respect to the reference element τ. However, the above approach is needed when considering higher order Sobolev spaces, for example

$$|v|^2_{H^m(\tau_\ell)} = \sum_{|\alpha|=m} \int_{q^i_\ell} |D^\alpha_\xi v(\chi_i(\xi))|^2 d\xi, \quad m \in \mathbb{N}.$$

For the parametrisation $q^i_\ell = \chi^i_\ell(\tau)$ and for $|\alpha| = m$, we now obtain the norm equivalence inequalities,

$$\frac{1}{c_m} (\text{area } q^i_\ell)^{1-m} \int_\tau |D^\alpha_\eta \widetilde{v}_\ell(\eta)|^2 d\eta \le \int_{q^i_\ell} |D^\alpha_\xi v(\chi_i(\xi))|^2 d\xi,$$

and

$$\int_{q^i_\ell} |D^\alpha_\xi v(\chi_i(\xi))|^2 d\xi \le c_m (\text{area } q^i_\ell)^{1-m} \int_\tau |D^\alpha_\eta \widetilde{v}_\ell(\eta)|^2 d\eta.$$

Hence, we have

$$c_1 \Delta^{1-m}_\ell |\widetilde{v}_\ell|^2_{H^m(\tau)} \le |v|^2_{H^m(\tau_\ell)} \le c_2 \Delta^{1-m}_\ell |\widetilde{v}_\ell|^2_{H^m(\tau)} \quad \text{for } m \in \mathbb{N}.$$

Let $Q_h w \in S^0_h(\Gamma)$ be the L_2 projection satisfying the variational problem

$$\sum_{\ell=1}^N w_\ell \int_\Gamma \psi_\ell(x)\psi_k(x)ds_x = \int_\Gamma w(x)\psi_k(x)ds_x \quad \text{for } k = 1,\dots,N.$$

Due to the definition of the piecewise constant basis functions ψ_k, we obtain the Galerkin orthogonality

$$\int_{\tau_\ell} \Big(w(x) - Q_h w(x)\Big)ds_x = 0 \quad \text{for } \ell = 1,\dots,N$$

as well as

$$w_\ell = \frac{1}{\Delta_\ell} \int_{\tau_\ell} w(y)ds_y \quad \text{for } \ell = 1,\dots,N.$$

For the local error, we first find

$$\|w - Q_h w\|^2_{L_2(\tau_\ell)} \le 2c\,\Delta_\ell \|\widetilde{w}_\ell - Q_\tau \widetilde{w}_\ell\|^2_{L_2(\tau)},$$

where $Q_\tau \widetilde{w}_\ell(\xi) = Q_h w(\chi_i(\xi))$.

For an arbitrary but fixed $\mu \in L_2(\tau)$, we define the linear functional

$$f_\mu(\widetilde{w}_\ell) = \int_\tau \Big(\widetilde{w}_\ell(\eta) - Q_\tau \widetilde{w}_\ell(\eta)\Big)\mu(\eta)d\eta$$

satisfying

$$|f_\mu(\widetilde{w}_\ell)| \le \|\widetilde{w}_\ell - Q_\tau \widetilde{w}_\ell\|_{L_2(\tau)} \|\mu\|_{L_2(\tau)}$$

$$\le 2\|\widetilde{w}_\ell\|_{L_2(\tau)} \|\mu\|_{L_2(\tau)} \le 2\|\widetilde{w}_\ell\|_{H^1(\tau)} \|\mu\|_{L_2(\tau)},$$

since the L_2 projection $Q_\tau : L_2(\tau) \to L_2(\tau)$ is bounded. If q is given as a constant function, we obviously have $Q_\tau q = q$, and, therefore,

$$f_\mu(q) = 0 \quad \text{for all } q \in P_0(\tau).$$

The Bramble–Hilbert lemma then implies

$$|f_\mu(\widetilde{w}_\ell)| \le c\,|\widetilde{w}_\ell|_{H^1(\tau)}\|\mu\|_{L_2(\tau)}.$$

In particular for $\bar{\mu} = \widetilde{w}_\ell - Q_\tau\widetilde{w}_\ell \in L_2(\tau)$, we then obtain

$$\|\widetilde{w}_\ell - Q_\tau\widetilde{w}_\ell\|_{L_2(\tau)}^2 = \int_\tau \big(\widetilde{w}_\ell(\eta) - Q_\tau\widetilde{w}_\ell(\eta)\big)^2 d\eta = |f_{\bar{\mu}}(\widetilde{w}_\ell)|$$

$$\le c\,|\widetilde{w}_\ell|_{H^1(\tau)}\|\bar{\mu}\|_{L_2(\tau)} = c\,|\widetilde{w}_\ell|_{H^1(\tau)}\|\widetilde{w}_\ell - Q_\tau\widetilde{w}_\ell\|_{L_2(\tau)},$$

and, therefore,

$$\|\widetilde{w}_\ell - Q_\tau\widetilde{w}_\ell\|_{L_2(\tau)} \le c\,|\widetilde{w}_\ell|_{H^1(\tau)}.$$

Altogether, we finally obtain

$$\|w - Q_h w\|_{L_2(\tau_\ell)}^2 = 2\Delta_\ell\,\|\widetilde{w}_\ell - Q_\tau\widetilde{w}_\ell\|_{L_2(\tau)}^2$$

$$\le \widetilde{c}\,\Delta_\ell\,|\widetilde{w}_\ell|_{H^1(\tau)}^2 \le \bar{c}\,h_\ell^2\,|w|_{H^1(\tau_\ell)}^2,$$

as well as

$$\|w - Q_h w\|_{L_2(\Gamma)}^2 \le \bar{c}\sum_{\ell=1}^{N} h_\ell^2\,|w|_{H^1(\tau_\ell)}^2$$

and

$$\|w - Q_h w\|_{L_2(\Gamma)} \le c\,h\,|w|_{H^1_{\mathrm{pw}}(\Gamma)},$$

when assuming $w \in H^1_{\mathrm{pw}}(\Gamma)$. This is the error estimate (2.4) for $s = 1$.

To obtain the error estimate (2.4) for $s \in (0,1)$, we consider $x \in \tau_\ell$, where we find

$$w(x) - Q_h w(x) = w(x) - \frac{1}{\Delta_\ell}\int_{\tau_\ell} w(y)ds_y = \frac{1}{\Delta_\ell}\int_{\tau_\ell}(w(x) - w(y))ds_y$$

for the error, and, therefore, by applying the Cauchy–Schwarz inequality,

$$|w(x) - Q_h w(x)|^2 = \left|\frac{1}{\Delta_\ell}\int_{\tau_\ell}(w(x) - w(y))ds_y\right|^2$$

$$= \frac{1}{\Delta_\ell^2}\left|\int_{\tau_\ell} \frac{w(x) - w(y)}{|x - y|^{1+s}}|x - y|^{1+s}ds_y\right|^2$$

$$\leq \frac{1}{\Delta_\ell^2} \int\limits_{\tau_\ell} \frac{(w(x) - w(y))^2}{|x - y|^{2+2s}} ds_y \int\limits_{\tau_\ell} |x - y|^{2+2s} ds_y$$

$$\leq \frac{d_\ell^{2+2s}}{\Delta_\ell} \int\limits_{\tau_\ell} \frac{(w(x) - w(y))^2}{|x - y|^{2+2s}} ds_y \,,$$

where d_ℓ is the element diameter (cf. Chapter 2). Integrating over τ_ℓ, and since the boundary element τ_ℓ is shape regular, i.e. $d_\ell \leq c_B h_\ell$, this gives with $\Delta_\ell = h_\ell^2$

$$\int\limits_{\tau_\ell} (w(x) - Q_h w(x))^2 ds_x \leq c_B^{2+2s} h_\ell^{2s} \int\limits_{\tau_\ell} \int\limits_{\tau_\ell} \frac{(w(x) - w(y))^2}{|x - y|^{2+2s}} ds_y ds_x.$$

Taking the sum over all boundary elements τ_ℓ, we obtain

$$\int\limits_{\Gamma} (w(x) - Q_h w(x))^2 ds_x \leq c_B^{2+2s} \sum_{\ell=1}^{N} h_\ell^{2s} \int\limits_{\tau_\ell} \int\limits_{\tau_\ell} \frac{(w(x) - w(y))^2}{|x - y|^{2+2s}} ds_y ds_x \,,$$

which is the error estimate (2.4) for $s \in (0, 1)$.

Hence, we have

$$\|w - Q_h w\|_{L_2(\Gamma)} \leq c \, h^s \, |w|_{H_{\text{pw}}^s(\Gamma)} \,,$$

when assuming $w \in H_{\text{pw}}^s(\Gamma)$ for some $s \in [0, 1]$. For $\sigma \in [-1, 0)$, we then find by duality,

$$\|w - Q_h w\|_{H^\sigma(\Gamma)} = \sup_{0 \neq v \in H^{-\sigma}(\Gamma)} \frac{\langle w - Q_h w, v \rangle_\Gamma}{\|v\|_{H^{-\sigma}(\Gamma)}}$$

$$= \sup_{0 \neq v \in H^{-\sigma}(\Gamma)} \frac{\langle w - Q_h w, v - Q_h v \rangle_\Gamma}{\|v\|_{H^{-\sigma}(\Gamma)}}$$

$$= \sup_{0 \neq v \in H^{-\sigma}(\Gamma)} \frac{\|w - Q_h w\|_{L_2(\Gamma)} \|v - Q_h v\|_{L_2(\Gamma)}}{\|v\|_{H^{-\sigma}(\Gamma)}}$$

$$\leq c \, h^{-\sigma} \|w - Q_h w\|_{L_2(\Gamma)} \leq c \, h^{s-\sigma} \, |w|_{H_{\text{pw}}^s(\Gamma)} \,,$$

and, therefore, the approximation property (2.5).

It remains to derive the approximation property (2.8) for piecewise linear but globally discontinuous basis functions. For this, we define the L_2 projection $Q_h w \in S_h^{1,-1}(\Gamma)$ satisfying the variational problem

$$\sum_{\ell=1}^{N} \sum_{i=1}^{3} w_{\ell,i} \int\limits_{\Gamma} \psi_{\ell,i}(x) \psi_{k,j}(x) ds_x = \int\limits_{\Gamma} w(x) \psi_{k,j}(x) ds_x$$

for $k = 1, \dots, N, j = 1, 2, 3$. As for piecewise constant basis functions, we then obtain

$$\|w - Q_h w\|^2_{L_2(\tau_\ell)} \leq 2c\, \Delta_\ell \, \|\widetilde{w}_\ell - Q_\tau \widetilde{w}_\ell\|_{L_2(\tau)}$$

for the local error. The Bramble–Hilbert lemma now implies

$$\|\widetilde{w} - Q_\tau \widetilde{w}\|_{L_2(\tau)} \leq c\, |\widetilde{w}_\ell|_{H^2(\tau)}.$$

Hence, we have

$$\|w - Q_h w\|^2_{L_2(\tau_\ell)} \leq c\, \Delta_\ell \, |\widetilde{w}_\ell|^2_{H^2(\tau)} \leq \widetilde{c}\, \Delta^2_\ell \, |w|^2_{H^2(\tau_\ell)} = \widetilde{c}\, h^4_\ell \, |w|^2_{H^2(\tau_\ell)},$$

and, therefore,

$$\|w - Q_h w\|^2_{L_2(\Gamma)} \leq \widetilde{c} \sum_{\ell=1}^{N} h^4_\ell \, |w|^2_{H^2(\tau_\ell)}$$

as well as

$$\|w - Q_h w\|_{L_2(\Gamma)} \leq c\, h^2 \, |w|_{H^2_{\mathrm{pw}}(\Gamma)},$$

when assuming $w \in H^2_{\mathrm{pw}}(\Gamma)$. Since

$$Q_h : L_2(\Gamma) \to L_2(\Gamma)$$

is bounded, we obtain, by applying an interpolation argument, the error estimate

$$\|w - Q_h w\|_{L_2(\Gamma)} \leq c\, h^s_\ell \, |w|_{H^s_{\mathrm{pw}}(\Gamma)},$$

when assuming $w \in H^s_{\mathrm{pw}}(\Gamma)$ for some $s \in [0,2]$. By duality, we then find

$$\|w - Q_h w\|_{H^\sigma(\Gamma)} \leq c\, h^{s-\sigma} \, |w|_{H^s_{\mathrm{pw}}(\Gamma)}$$

for $\sigma \in [-2,0]$ and $s \in [0,2]$, which is the approximation property (2.8).

Finally, we consider the piecewise linear, globally continuous L_2 projection $Q_h w \in S^1_h(\Gamma)$ as the unique solution of the variational problem

$$\int_\Gamma Q_h w(x) \varphi_j(x) ds_x = \int_\Gamma w(x) \varphi_j(x) ds_x \quad \text{for } j = 1, \ldots, M.$$

Due to $S^1_h(\Gamma) \subset S^{1,-1}_h(\Gamma)$, we immediately find the approximation property (2.10) for $\sigma \in [-2,0]$ and $s \in [0,2]$. To derive (2.10) for $\sigma \in (0,1]$, we introduce the H^1 projection of a given function $u \in H^1(\Gamma)$,

$$P_h u = \sum_{j=1}^{M} u_j \varphi_j \in S^1_h(\Gamma),$$

which minimises the error $u - P_h u$ in the $H^1(\Gamma)$–norm:

$$P_h u = \arg \min_{v_h \in S^1_h(\Gamma)} \|u - v_h\|_{H^1(\Gamma)}.$$

Thus, $P_h u \in S_h^1(\Gamma)$ is the unique solution of the variational problem

$$\langle P_h u, v_h \rangle_{H^1(\Gamma)} = \langle u, v_h \rangle_{H^1(\Gamma)} \quad \text{for all } v_h \in S_h^1(\Gamma).$$

As for the L_2 projection, we find the error estimate

$$\|u - P_h u\|_{H^1(\Gamma)} \leq c\, h\, |u|_{H_{\mathrm{pw}}^2(\Gamma)},$$

when assuming $u \in H_{\mathrm{pw}}^2(\Gamma)$. Since

$$P_h : H^1(\Gamma) \to H^1(\Gamma)$$

is bounded, we also conclude the trivial estimate

$$\|u - P_h u\|_{H^1(\Gamma)} \leq \|u\|_{H^1(\Gamma)}.$$

Then, using an interpolation argument, we also obtain the error estimate

$$\|u - P_h u\|_{H^1(\Gamma)} \leq c\, h^{s-1}\, |u|_{H^s(\Gamma)},$$

when assuming $u \in H_{\mathrm{pw}}^s(\Gamma)$ for some $s \in [1, 2]$. Then, for $\sigma \in [0, 1)$, we get by duality,

$$\|u - P_h u\|_{H^\sigma(\Gamma)} = \sup_{0 \neq v \in H^{2-\sigma}(\Gamma)} \frac{\langle u - P_h u, v \rangle_{H^1(\Gamma)}}{\|v\|_{H^{2-\sigma}(\Gamma)}}$$

$$= \sup_{0 \neq v \in H^{2-\sigma}(\Gamma)} \frac{\langle u - P_h u, v - P_h v \rangle_{H^1(\Gamma)}}{\|v\|_{H^{2-\sigma}(\Gamma)}}$$

$$\leq \sup_{0 \neq v \in H^{2-\sigma}(\Gamma)} \frac{\|u - P_h u\|_{H^1(\Gamma)} \|v - P_h v\|_{H^1(\Gamma)}}{\|v\|_{H^{2-\sigma}(\Gamma)}}$$

$$\leq c\, h^{1-\sigma}\, \|u - P_h u\|_{H^1(\Gamma)} \leq c\, h^{s-\sigma}\, |u|_{H^s(\Gamma)},$$

when assuming $u \in H^s(\Gamma)$ for some $s \in [1, 2]$. This is the approximation property (2.10) for $\sigma \in (0, 1]$ and $s \in [1, 2]$.

C

Numerical Algorithms

C.1 Numerical Integration

For the boundary $\Gamma = \partial\Omega$ of a Lipschitz domain $\Omega \subset \mathbb{R}^3$, we consider a sequence of boundary element meshes (2.1),

$$\Gamma_N = \bigcup_{\ell=1}^{N} \bar{\tau}_\ell.$$

We assume that all boundary elements τ_ℓ are plane triangles. Using the reference element

$$\tau = \left\{ \xi \in \mathbb{R}^2 : 0 < \xi_1 < 1,\ 0 < \xi_2 < 1 - \xi_1 \right\},$$

the boundary element $\tau_\ell = \chi_\ell(\tau)$ with nodes x_{ℓ_i} for $i = 1, 2, 3$ can be described via the parametrisation

$$x(\xi) = \chi_\ell(\xi) = x_{\ell_1} + \xi_1(x_{\ell_2} - x_{\ell_1}) + \xi_2(x_{\ell_3} - x_{\ell_1}) = x_{\ell_1} + J_\ell \xi \in \tau_\ell$$

for $\xi \in \tau$. Using

$$\Delta_\ell = \int_{\tau_\ell} ds_x,$$

we obtain for the integral of a given function f on τ_ℓ,

$$I = \int_{\tau_\ell} f(x)ds_x = 2\Delta_\ell \int_{\tau} \widetilde{f}_\ell(\xi)\, d\xi, \quad \widetilde{f}_\ell(\xi) = f(\chi_\ell(\xi)).$$

Hence, it is sufficient to consider numerical integration schemes

$$\frac{1}{2}\sum_{i=1}^{M} \omega_i \widetilde{f}_\ell(\xi_i) \approx \int_{\tau} \widetilde{f}_\ell(\xi)\, d\xi.$$

with respect to the reference element τ, which are exact for polynomials of a certain order. From this, we may find $3M$ parameters $(\xi_{i,1}, \xi_{i,2}, \omega_i)$ for $i = 1, \ldots, M$. Let

$$P_k = \text{span}\{\xi_1^{\alpha_1} \xi_2^{\alpha_2}\}_{\alpha_1 + \alpha_2 \le k}$$

be the space of polynomials of an order less or equal k. The dimensions of P_k and the minimal number M of integration points are given in Table C.1. The last line of this table shows the number $P = 3M$ of free parameters. These parameters $(\xi_{i,1}, \xi_{i,2}, \omega_i)$ for $i = 1, \ldots, M$ are solutions of the nonlinear

Table C.1. Dimensions of P_k and required number of integration nodes

k	0	1	2	3	4	5
$\dim P_k$	1	3	6	10	15	21
M		1	3	4		7
P		3	9	12		21

equations

$$\int_\tau \xi_1^{\alpha_1} \xi_2^{\alpha_2} d\xi = \frac{1}{2} \sum_{i=1}^{M} \omega_i \xi_{i,1}^{\alpha_1} \xi_{i,2}^{\alpha_2} \quad \text{for } \alpha_1, \alpha_2 : \alpha_1 + \alpha_2 \le k . \tag{C.1}$$

For $k = 1$ and $M = 1$, the equations (C.1) read

$$\int_\tau 1 \, d\xi = \frac{1}{2} = \frac{1}{2}\omega_1 ,$$

$$\int_\tau \xi_1 \, d\xi = \frac{1}{6} = \frac{1}{2}\omega_1 \xi_{1,1} ,$$

$$\int_\tau \xi_2 \, d\xi = \frac{1}{6} = \frac{1}{2}\omega_1 \xi_{1,2} ,$$

and, therefore, we obtain

$$\omega_1 = 1, \quad \xi_{1,1} = \frac{1}{3}, \quad \xi_{1,2} = \frac{1}{3} .$$

The resulting formula is the midpoint quadrature rule

$$\int_{\tau_\ell} f(x) \, ds_x \approx \Delta_\ell f(x_\ell^*) ,$$

which integrates linear functions exactly. For $k = 2$ and $M = 3$, we have from (C.1)

$$\int_\tau 1\,d\xi \;=\; \frac{1}{2} \;=\; \frac{1}{2}\left(\omega_1 + \omega_2 + \omega_3\right),$$

$$\int_\tau \xi_1\,d\xi \;=\; \frac{1}{6} \;=\; \frac{1}{2}\left(\omega_1\xi_{1,1} + \omega_2\xi_{2,1} + \omega_3\xi_{3,1}\right),$$

$$\int_\tau \xi_2\,d\xi \;=\; \frac{1}{6} \;=\; \frac{1}{2}\left(\omega_1\xi_{1,2} + \omega_2\xi_{2,2} + \omega_3\xi_{3,2}\right),$$

$$\int_\tau \xi_1^2\,d\xi \;=\; \frac{1}{12} \;=\; \frac{1}{2}\left(\omega_1\xi_{1,1}^2 + \omega_2\xi_{2,1}^2 + \omega_3\xi_{3,1}^2\right),$$

$$\int_\tau \xi_2^2\,d\xi \;=\; \frac{1}{12} \;=\; \frac{1}{2}\left(\omega_1\xi_{1,2}^2 + \omega_2\xi_{2,2}^2 + \omega_3\xi_{3,2}^2\right),$$

$$\int_\tau \xi_1\xi_2\,d\xi \;=\; \frac{1}{24} \;=\; \frac{1}{2}\left(\omega_1\xi_{1,1}\xi_{1,2} + \omega_2\xi_{2,1}\xi_{2,2} + \omega_3\xi_{3,1}\xi_{3,2}\right).$$

To find a solution of the above system of nonlinear equations, we add some additional constraints with respect to the geometrical setting. Due to symmetry, we first set

$$\omega_1 = \omega_2 = \omega_3 = \frac{1}{3}.$$

Introducing some real parameter $s \in \mathbb{R}$, we may define the integration nodes on the straight lines connecting the corner nodes with the midpoint,

$$\xi_1 = \frac{s}{3}\begin{pmatrix} 1 \\ 1 \end{pmatrix}, \quad \xi_2 = \begin{pmatrix} 1 \\ 0 \end{pmatrix} + \frac{s}{3}\begin{pmatrix} -2 \\ 1 \end{pmatrix}, \quad \xi_3 = \begin{pmatrix} 0 \\ 1 \end{pmatrix} + \frac{s}{3}\begin{pmatrix} 1 \\ -2 \end{pmatrix}.$$

Note that with this choice, the second and third equations are satisfied. For the remaining three equations, we obtain

$$4s^2 - 8s + 3 = 0,$$

with the solutions

$$s_{1/2} = 1 \pm \frac{1}{2}.$$

In particular, for $s_1 = 1/2$, the integration nodes and the corresponding weights are

$$\xi_1 = \left(\frac{1}{6}, \frac{1}{6}\right), \quad \xi_2 = \left(\frac{2}{3}, \frac{1}{6}\right), \quad \xi_3 = \left(\frac{1}{6}, \frac{2}{3}\right), \quad \omega_1 = \omega_2 = \omega_3 = \frac{1}{3},$$

while for $s_2 = 3/2$ the solution is

$$\xi_1 = \left(\frac{1}{2}, \frac{1}{2}\right), \quad \xi_2 = \left(0, \frac{1}{2}\right), \quad \xi_3 = \left(\frac{1}{2}, 0\right), \quad \omega_1 = \omega_2 = \omega_3 = \frac{1}{3}.$$

Note that both integration rules are exact for quadratic polynomials.

For $k = 3$ and $M = 4$, the nonlinear equations (C.1) read

$$\int_\tau \xi_1^{\alpha_1} \xi_2^{\alpha_2} d\xi = \frac{1}{2} \sum_{i=1}^{4} w_i \xi_{i,1}^{\alpha_1} \xi_{i,2}^{\alpha_2} \quad \text{for } \alpha_1, \alpha_2 : \alpha_1 + \alpha_2 \le 3.$$

Hence, we have to find 12 parameters $(\xi_{i,1}, \xi_{i,2}, w_i)$ satisfying 10 nonlinear equations. As in the previous case, we may introduce additional constraints to construct particular solutions of the above nonlinear system. Due to symmetry, we define the integration nodes as

$$\xi_1 = \frac{1}{3}\begin{pmatrix} 1 \\ 1 \end{pmatrix}, \ \xi_2 = \frac{s}{3}\begin{pmatrix} 1 \\ 1 \end{pmatrix}, \ \xi_3 = \begin{pmatrix} 1 \\ 0 \end{pmatrix} + \frac{s}{3}\begin{pmatrix} -2 \\ 1 \end{pmatrix}, \ \xi_4 = \begin{pmatrix} 0 \\ 1 \end{pmatrix} + \frac{s}{3}\begin{pmatrix} 1 \\ -2 \end{pmatrix},$$

and the associated integration weights as

$$w_1 = w, \quad w_2 = w_3 = w_4.$$

From the equation for $\alpha_1 = \alpha_2 = 0$, we then find

$$w_1 = w, \quad w_2 = w_3 = w_4 = \frac{1}{3}(1 - w).$$

Note that both equations, for $\alpha_1, \alpha_2 : \alpha_1 + \alpha_2 = 1$, are then satisfied. It turns out that it is sufficient to consider only two additional equations, the remaining equations are then satisfied automatically. For $\alpha_1 = 2$ and $\alpha_2 = 0$, it follows

$$(1 - w)(s^2 - 2s + 1) = \frac{1}{4},$$

while for $\alpha_1 = 3$ and $\alpha_2 = 0$, we have

$$(1 - w)(8 - 18s + 12s^2 - 2s^3) = \frac{17}{10}.$$

Thus, there are two equations for the two remaining unknowns w and s. We obtain

$$1 - w = \frac{1}{4} \frac{1}{s^2 - 2s + 1} = \frac{17}{10} \frac{1}{8 - 18s + 12s^2 - 2s^3},$$

and, therefore,

$$(s - 1)^2 (5s - 3) = 0,$$

with the solutions

$$s_{1/2} = 1, \quad s_3 = \frac{3}{5}.$$

For $s = 3/5$, we then get the integration weights

$$w_1 = -\frac{9}{16}, \quad w_2 = w_3 = w_4 = \frac{25}{48}$$

and the integration points

$$\xi_1 = \frac{1}{3}\begin{pmatrix}1\\1\end{pmatrix}, \quad \xi_2 = \frac{1}{5}\begin{pmatrix}1\\1\end{pmatrix},$$

$$\xi_3 = \frac{1}{5}\begin{pmatrix}3\\1\end{pmatrix}, \quad \xi_4 = \frac{1}{5}\begin{pmatrix}1\\3\end{pmatrix}.$$

However, the first weight w_1 is negative, and, therefore, the above quadrature can not be used for practical computations.

For $k = 5$ and $M = 7$, the nonlinear equations (C.1) read

$$\int_\tau \xi_1^{\alpha_1}\xi_2^{\alpha_2}\,d\xi = \frac{1}{2}\sum_{i=1}^{7} w_i\xi_{i,1}^{\alpha_1}\xi_{i,2}^{\alpha_2} \quad \text{for all } \alpha_1,\alpha_2 : \alpha_1 + \alpha_2 \leq 5.$$

Hence, we have to find 21 parameters $(\xi_{i,1},\xi_{i,2},w_i)$ satisfying 21 nonlinear equations. As in the previous cases, we introduce additional constraints to construct a solution of the above nonlinear equations. Due to symmetry, we define the integration nodes as

$$\xi_1 = \frac{1}{3}\begin{pmatrix}1\\1\end{pmatrix}, \quad \xi_2 = \frac{s}{3}\begin{pmatrix}1\\1\end{pmatrix}, \quad \xi_3 = \begin{pmatrix}1\\0\end{pmatrix} + \frac{s}{3}\begin{pmatrix}-2\\1\end{pmatrix}, \quad \xi_4 = \begin{pmatrix}0\\1\end{pmatrix} + \frac{s}{3}\begin{pmatrix}1\\-2\end{pmatrix}$$

and

$$\xi_5 = \frac{1}{2}\begin{pmatrix}1\\0\end{pmatrix} + \frac{t}{6}\begin{pmatrix}-1\\2\end{pmatrix}, \quad \xi_6 = \frac{1}{2}\begin{pmatrix}1\\1\end{pmatrix} + \frac{t}{6}\begin{pmatrix}-1\\-1\end{pmatrix}, \quad \xi_7 = \frac{1}{2}\begin{pmatrix}0\\1\end{pmatrix} + \frac{t}{6}\begin{pmatrix}2\\-1\end{pmatrix},$$

as well as the associated integration weights as

$$w_1 = w, \quad w_2 = w_3 = w_4, \quad w_5 = w_6 = w_7.$$

From the equation for $\alpha_1 = \alpha_2 = 0$, we easily find

$$w_1 = 1 - 3w_2 - 3w_5.$$

With this, the equations for $\alpha_1,\alpha_2 : \alpha_1 + \alpha_2 = 1$ are satisfied automatically. Moreover, all the remaining equations can be reduced to the four equations for the four parameters $w_2, w_5, s,$ and t:

$$12w_2(s-1)^2 + 3w_5(t-1)^2 = 1,$$
$$120w_2(s-1)^2(4-s) + 15w_5(t-1)^2(t+5) - 34,$$
$$240w_2(s-1)^2(3s^2 - 10s + 13) + 15w_5(t-1)^2(3t^2 + 2t + 19) = 176,$$
$$3360w_2(s-1)^2(8 - 11s + 6s^2 - s^3) + 105w_5(13 - t + 3t^2 + t^3) = 1184.$$

From the first two equations, we find

$$w_2 = \frac{1 - 3w_5(t-1)^2}{12(s-1)^2}, \quad w_5 = \frac{10s - 6}{15(t-1)^2(2s+t-3)}.$$

Inserting this into the third equation, this gives

$$s(9 - 5t) + 3(t - 1) = 0\,,$$

and, therefore,

$$s = 3\,\frac{t - 1}{5t - 9}\,.$$

The fourth equation is finally equivalent to

$$7t^2 - 18t + 3 = 0\,.$$

Thus,

$$t_{1/2} = \frac{9 \pm 2\sqrt{15}}{7}\,.$$

Hence, we have

$$t = \frac{9 - 2\sqrt{15}}{7}\,, \quad s = \frac{6 - \sqrt{15}}{7}\,.$$

For the integration weights, we find

$$\omega_5 = \frac{155 + \sqrt{15}}{1200}\,, \quad \omega_2 = \frac{155 - \sqrt{15}}{1200}\,, \quad \omega_1 = \frac{9}{40}\,,$$

while for the integration nodes, we finally obtain

$$\xi_2 = \frac{1}{21}\begin{pmatrix} 6 - \sqrt{15} \\ 6 - \sqrt{15} \end{pmatrix}\,, \quad \xi_3 = \frac{1}{21}\begin{pmatrix} 9 + 2\sqrt{15} \\ 6 - \sqrt{15} \end{pmatrix}\,, \quad \xi_4 = \frac{1}{21}\begin{pmatrix} 6 - \sqrt{15} \\ 9 + 2\sqrt{15} \end{pmatrix}$$

and

$$\xi_5 = \frac{1}{21}\begin{pmatrix} 6 + \sqrt{15} \\ 6 + \sqrt{15} \end{pmatrix}\,, \quad \xi_6 = \frac{1}{21}\begin{pmatrix} 6 + \sqrt{15} \\ 9 - 2\sqrt{15} \end{pmatrix}\,, \quad \xi_7 = \frac{1}{21}\begin{pmatrix} 9 - 2\sqrt{15} \\ 6 + \sqrt{15} \end{pmatrix}\,.$$

C.2 Analytic Integration

For the boundary $\Gamma = \partial\Omega$ of a Lipschitz domain $\Omega \subset \mathbb{R}^3$, we consider a sequence of boundary element meshes (2.1),

$$\Gamma_N = \bigcup_{\ell=1}^{N} \overline{\tau}_\ell.$$

We assume that all boundary elements τ_ℓ are plane triangles. In order not to overload the subsequent formulae, we consider a plane triangle $\tau \subset \mathbb{R}^3$ given via its three corner points x_1, x_2, and x_3 (cf. Fig. C.2), having in mind that this triangle is one of the boundary elements τ_ℓ, $\ell = 1 \dots, N$.

We first define a suitable local coordinate system connected to the boundary element τ with the origin in x_1 and having the basis vectors

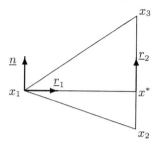

Fig. C.1. Boundary element τ

$$\left(\underline{r}_1\,,\,\underline{r}_2\,,\,\underline{n}\right).$$

The unit vector $\underline{r}_2$ directed from x_2 to x_3 is defined as follows

$$\underline{r}_2 \;=\; \frac{1}{t_\tau}(x_3 - x_2)\,, \quad t_\tau = |x_3 - x_2|\,.$$

Here, $t_\tau > 0$ denotes the length of the side $\overline{x_2\,x_3}$ of the triangle τ. This is one of the characteristic quantities, we will need for the analytical integration.

To find the unit vector $\underline{r}_1$ which is directed along the line through the origin in x_1 and perpendicular to $\underline{r}_2$, we first determine the intersection point x^*,

$$x^* \;=\; x_1 + s_\tau \underline{r}_1 \;=\; x_2 + t_* \underline{r}_2\,,$$

where $s_\tau > 0$, the height of the triangle τ, will be the second characteristic quantity. The parameter t_* can be easily computed as

$$t^* \;=\; (x_1 - x_2, \underline{r}_2)\,.$$

Then,

$$\underline{r}_1 \;=\; \frac{1}{s_\tau}(x^* - x_1)\,, \quad s_\tau = |x^* - x_1|\,.$$

Finally, the corresponding normal vector is defined by

$$\underline{n} \;=\; \underline{r}_1 \times \underline{r}_2\,.$$

Two further characteristic quantities, we will need for the analytical integration, are the tangents of the angles between the height $\overline{x_1\,x^*}$ and the sides $\overline{x_1\,x_2}$ and $\overline{x_1\,x_3}$, correspondingly. These quantities can be computed as

$$\alpha_1 \;=\; -\frac{t_*}{s_\tau}\,, \quad \alpha_2 \;=\; \frac{t_\tau - t_*}{s_\tau}\,.$$

Thus, an appropriate parametrisation of the boundary element τ would be

$$\tau \;=\; \left\{ y = y(s,t) = x_1 + s\,\underline{r}_1 + t\,\underline{r}_2 \,:\, 0 < s < s_\tau\,, \ \alpha_1\,s < t < \alpha_2\,s \right\}.$$

Any point $x \in \mathbb{R}^3$ can now be represented in the new coordinate system as

$$x = x_1 + s_x \, \underline{r}_1 + t_x \, \underline{r}_2 + u_x \, \underline{n}$$

with

$$s_x = (x - x_1, \underline{r}_1), \quad t_x = (x - x_1, \underline{r}_2), \quad u_x = (x - x_1, \underline{n}) \,.$$

From this, we find the distance between any arbitrary point $x \in \mathbb{R}^3$ and a point $y \in \tau$ as follows:

$$|x - y|^2 = (s - s_x)^2 + (t - t_x)^2 + u_x^2 \,.$$

C.2.1 Single Layer Potential for the Laplace operator

For $x \in \mathbb{R}^3$, we first consider the following function:

$$
\begin{aligned}
S(\tau, x) &= \frac{1}{4\pi} \int_\tau \frac{1}{|x-y|} ds_y \\
&= \frac{1}{4\pi} \int_0^{s_\tau} \int_{\alpha_1 s}^{\alpha_2 s} \frac{1}{\sqrt{(s-s_x)^2 + (t-t_x)^2 + u_x^2}} \, dt \, ds \\
&= \frac{1}{4\pi} \int_0^{s_\tau} \left[\log \left(t - t_x + \sqrt{(s-s_x)^2 + (t-t_x)^2 + u_x^2} \right) \right]_{\alpha_1 s}^{\alpha_2 s} ds \\
&= \frac{1}{4\pi} \Big(F(s_\tau, \alpha_2) - F(0, \alpha_2) - F(s_\tau, \alpha_1) + F(0, \alpha_1) \Big) \,.
\end{aligned}
$$

Hence, it is sufficient to compute the integral

$$
\begin{aligned}
F(s, \alpha) &= \int \log \left(\alpha s - t_x + \sqrt{(s-s_x)^2 + (\alpha s - t_x)^2 + u_x^2} \right) ds \\
&= \int \log \left(\alpha s - t_x + \sqrt{(1+\alpha^2)(s-p)^2 + q^2} \right) ds \,,
\end{aligned}
$$

with parameters

$$p = \frac{\alpha t_x + s_x}{1 + \alpha^2}, \quad q^2 = u_x^2 + \frac{(t_x - \alpha s_x)^2}{1 + \alpha^2} \,.$$

Integration by parts gives

$$F(s, \alpha) = (s - s_x) \log \left(\alpha s - t_x + \sqrt{(1+\alpha^2)(s-p)^2 + q^2} \right) - s$$

$$- \int \frac{(t_x - \alpha s_x)\sqrt{(1+\alpha^2)(s-p)^2 + q^2} + (1+\alpha^2)(s-p)(p-s_\ell) - q^2}{(1+\alpha)^2(s-p)^2 + q^2 + (\alpha s - t_x)\sqrt{(1+\alpha^2)(s-p)^2 + q^2}} \, ds \,.$$

With the substitution

$$s = p + \frac{q}{\sqrt{1+\alpha^2}} \sinh u, \qquad \frac{ds}{du} = \frac{q}{\sqrt{1+\alpha^2}} \cosh u,$$

we obtain for the remaining integral

$$I = \int \frac{(t_x - \alpha s_x)\sqrt{(1+\alpha^2)(s-p)^2 + q^2} + (1+\alpha^2)(s-p)(p-s_x) - q^2}{(1+\alpha)^2(s-p)^2 + q^2 + (\alpha s - t_x)\sqrt{(1+\alpha^2)(s-p)^2 + q^2}} \, ds$$

$$= q \int \frac{\sqrt{1+\alpha^2}(t_x - \alpha s_x)\cosh u + \alpha(t_x - \alpha s_x)\sinh u - \sqrt{1+\alpha^2}q}{q(1+\alpha^2)\cosh u + \alpha\sqrt{1+\alpha^2}q\sinh u + (\alpha s_x - t_x)} \, du.$$

The second substitution

$$v = \tanh\frac{u}{2}, \quad \sinh u = \frac{2v}{1-v^2}, \quad \cosh u = \frac{1+v^2}{1-v^2}, \quad \frac{du}{dv} = \frac{2}{1-v^2}$$

gives for the last integral

$$2q \int \frac{1}{1-v^2} \frac{\sqrt{1+\alpha^2}(t_x - \alpha s_x)(1+v^2) + 2\alpha(t_x - \alpha s_x)v - \sqrt{1+\alpha^2}q(1-v^2)}{q(1+\alpha^2)(1+v^2) + 2\alpha\sqrt{1+\alpha^2}qv + (\alpha s_x - t_x)(1-v^2)} \, dv,$$

and, therefore,

$$I = 2q \int \frac{1}{1-v^2} \frac{B_2 v^2 + B_1 v + B_0}{A_2 v^2 + A_1 v + A_0} \, dv,$$

with parameters

$$B_2 = \sqrt{1+\alpha^2}(t_x - \alpha s_x + q),$$
$$B_1 = 2\alpha(t_x - \alpha s_x),$$
$$B_0 = \sqrt{1+\alpha^2}(t_x - \alpha s_x - q),$$
$$A_2 = (1+\alpha^2)q - (\alpha s_x - t_x),$$
$$A_1 = 2\alpha\sqrt{1+\alpha^2}q,$$
$$A_0 = (1+\alpha^2)q + (\alpha s_x - t_x).$$

Due to

$$A_1^2 - 4A_0 A_2 = -4(1+\alpha^2)u_x^2 \leq 0,$$

we decompose

$$\frac{1}{1-v^2} \frac{B_2 v^2 + B_1 v + B_0}{A_2 v^2 + A_1 v + A_0} = \frac{C_1}{1-v} + \frac{C_2}{1+v} + \frac{C_3 v + C_4}{A_2 v^2 + A_1 v + A_0}$$

to obtain

$$C_1 = \frac{1}{2} \frac{B_0 + B_1 + B_2}{A_0 + A_1 + A_2},$$

$$C_2 = \frac{1}{2} \frac{B_0 - B_1 + B_2}{A_0 - A_1 + A_2},$$

$$C_3 = A_2(C_1 - C_2),$$

$$C_4 = B_0 - (C_1 + C_2)A_0,$$

and, therefore,

$$C_1 = C_2 = \frac{t_x - \alpha s_x}{2q\sqrt{1+\alpha^2}}, \quad C_3 = 0, \quad C_4 = -\frac{\sqrt{1+\alpha^2}u_x^2}{q}.$$

Hence,

$$I = \int \left(\frac{t_x - \alpha s_x}{\sqrt{1+\alpha^2}} \left(\frac{1}{1-v} + \frac{1}{1+v} \right) - 2\sqrt{1+\alpha^2}u_x^2 \frac{1}{A_2 v^2 + A_1 v + A_0} \right) dv.$$

Integrating the first part and resubstituting, this gives

$$\int \left(\frac{1}{1+v} - \frac{1}{v-1} \right) dv = \log|v+1| - \log|v-1| = \log\left| \frac{1+v}{1-v} \right|$$

$$= 2\operatorname{arctanh} v = u = \operatorname{arcsinh} \frac{\sqrt{1+\alpha^2}(s-p)}{q}$$

$$= \log \left(\frac{\sqrt{1+\alpha^2}(s-p)}{q} + \sqrt{\frac{(1+\alpha^2)(s-p)^2}{q^2} + 1} \right)$$

$$= \log \left(\sqrt{1+\alpha^2}(s-p) + \sqrt{(1+\alpha^2)(s-p)^2 + q^2} \right) - \log q.$$

For the remaining integral, we obtain

$$2\sqrt{1+\alpha^2}u_x^2 \int \frac{1}{A_2 v^2 + A_1 v + A_0} dv = 2u_x \arctan \frac{2A_2 v + A_1}{2\sqrt{1+\alpha^2}u_x}.$$

From $\sinh u = 2v/(1-v^2)$, we find

$$v = \frac{\sqrt{1 + \sinh^2 u} - 1}{\sinh u} = \frac{\sqrt{(1+\alpha^2)(s-p)^2 + q^2} - q}{\sqrt{1+\alpha^2}(s-p)},$$

and, therefore,

$$\frac{2A_2 v + A_1}{2\sqrt{1+\alpha^2}u_x} = \frac{A_2\sqrt{(1+\alpha^2)(s-p)^2 + q^2} + \left(\alpha(1+\alpha^2)(s-p) - A_2\right)q}{(1+\alpha^2)(s-p)\tau_\ell}$$

$$= \frac{\left(q - \frac{\alpha s_x - t_x}{1+\alpha^2}\right)\sqrt{(1+\alpha^2)(s-p)^2 + q^2} + (\alpha s - t_x - q)q}{(s-p)u_x}.$$

Collecting all terms together and ignoring constant parts, we finally obtain

$$F(s,\alpha) = (s - s_x)\log\left(\alpha s - t_x + \sqrt{(s - s_x)^2 + (\alpha s - t_x)^2 + u_x^2}\right) - s$$

$$+ \frac{\alpha s_x - t_x}{\sqrt{1 + \alpha^2}}\log\left(\sqrt{1 + \alpha^2}(s - p) + \sqrt{(1 + \alpha^2)(s - p)^2 + q^2}\right)$$

$$+ 2u_x\arctan\frac{\left(q - \frac{\alpha s_x - t_x}{1 + \alpha^2}\right)\sqrt{(1 + \alpha^2)(s - p)^2 + q^2} + (\alpha s - t_x - q)q}{(s - p)u_x}.$$

C.2.2 Double Layer Potential for the Laplace operator

For piecewise linear basis functions we have to compute the local contributions
of the integrals

$$D_i(\tau, x) = \frac{1}{4\pi}\int_\tau \frac{(x - y, \underline{n}(y))}{|x - y|^3}\psi_{\tau,i}(y)ds_y$$

$$= \frac{1}{4\pi}\int_0^{s_\tau}\frac{s_\tau - s}{s_\tau}\int_{\alpha_1 s}^{\alpha_2 s}\frac{u_x}{\left((t - t_x)^2 + (s - s_x)^2 + u_x^2\right)^{3/2}}dtds$$

$$= \frac{u_x}{4\pi}\int_0^{s_\tau}\frac{s_\tau - s}{s_\tau}\frac{1}{(s - s_x)^2 + u_x^2}\left[\frac{t - t_x}{\sqrt{(t - t_x)^2 + (s - s_x)^2 + u_x^2}}\right]_{\alpha_1 s}^{\alpha_2 s}ds$$

$$= \frac{1}{4\pi s_\tau}\left(F(s_\tau, \alpha_2) - F(0, \alpha_2) - F(s_\tau, \alpha_1) + F(0, \alpha_1)\right).$$

Hence, it is sufficient to compute the integral

$$F(s,\alpha) = u_x\int\frac{(s_\tau - s)(\alpha s - t_x)}{\left((s - s_x)^2 + u_x^2\right)\sqrt{(1 + \alpha^2)(s - p)^2 + q^2}}ds$$

$$= u_x\int\frac{(\alpha(s_\tau - s_x) - (\alpha s_x - t_x))(s - s_x) + (s_\tau - s_x)(\alpha s_x - t_x) + \alpha u_x^2}{\left((s - s_x)^2 + u_x^2\right)\sqrt{(1 + \alpha^2)(s - p)^2 + q^2}}ds$$

$$-\alpha u_x\int\frac{1}{\sqrt{(1 + \alpha^2)(s - p)^2 + q^2}}ds$$

$$= u_x\int\frac{(\alpha(s_\tau - s_x) - (\alpha s_x - t_x))(s - s_x) + (s_\tau - s_x)(\alpha s_x - t_x) + \alpha u_x^2}{\left((s - s_x)^2 + u_x^2\right)\sqrt{(1 + \alpha^2)(s - p)^2 + q^2}}ds$$

$$-u_x\frac{\alpha}{\sqrt{1 + \alpha^2}}\log\left(\sqrt{1 + \alpha^2}(s - p) + \sqrt{(1 + \alpha^2)(s - p)^2 + q^2}\right),$$

with parameters p and q already used for the single layer potential:

$$p = \frac{\alpha t_x + s_x}{1 + \alpha^2}, \quad q^2 = u_x^2 + \frac{(t_x - \alpha s_x)^2}{1 + \alpha^2}.$$

With the substitution

$$s = p + \frac{q}{\sqrt{1+\alpha^2}}\sinh u, \qquad \frac{ds}{du} = \frac{q}{\sqrt{1+\alpha^2}}\cosh u,$$

we obtain for the remaining integral

$$I = u_x \int \frac{\left(\alpha(s_\tau - s_x) - (\alpha s_x - t_x)\right)(s - s_x) + (s_\tau - s_x)(\alpha s_x - t_x) + \alpha u_x^2}{\left((s - s_x)^2 + u_x^2\right)\sqrt{(1+\alpha^2)(s-p)^2 + q^2}}\,ds =$$

$$u_x \int \frac{q\left(\alpha s_\tau + t_x - 2\alpha s_x\right)\sinh u + \frac{1}{\sqrt{1+\alpha^2}}\left(\alpha(1+\alpha^2)q^2 + (s_x - s_\tau)(t_x - \alpha s_x)\right)}{q^2\sinh^2 u + 2\sqrt{1+\alpha^2}q(p - s_x)\sinh u + (1+\alpha^2)\left((p - s_x)^2 + u_x^2\right)}\,du.$$

The second substitution

$$v = \tanh\frac{u}{2}, \qquad \sinh u = \frac{2v}{1-v^2}, \qquad \cosh u = \frac{1+v^2}{1-v^2}, \qquad \frac{du}{dv} = \frac{2}{1-v^2}$$

gives for the last integral

$$2u_x \int \frac{2q\left(\alpha s_\tau + t_x - 2\alpha s_x\right)v + \frac{1}{\sqrt{1+\alpha^2}}\left(\alpha(1+\alpha^2)q^2 + (s_x - s_\tau)(t_x - \alpha s_x)\right)(1-v^2)}{4q^2v^2 + 4\sqrt{1+\alpha^2}q(p - s_x)v(1-v^2) + (1+\alpha^2)\left((p - s_x)^2 + u_x^2\right)(1-v^2)^2}\,dv,$$

and, thereforfe,

$$I = 2u_x \int \frac{C_2 v^2 + C_1 v - C_2}{a_1 v^4 - a_2 v^3 + a_3 v^2 + a_2 v + a_1}\,dv,$$

with parameters

$$C_2 = \frac{1}{\sqrt{1+\alpha^2}}\left((s_\tau - s_x)(t_x - \alpha s_x) - \alpha(1+\alpha^2)q^2\right),$$

$$C_1 = 2q\left(\alpha s_\tau + t_x - 2\alpha s_x\right),$$

$$a_1 = (1+\alpha^2)\left(u_x^2 + (p - s_x)^2\right) = u_x^2 + \alpha^2 q^2 > 0,$$

$$a_2 = 4q\sqrt{1+\alpha^2}(p - s_s) = \frac{4\alpha q(t_x - \alpha s_x)}{\sqrt{1+\alpha^2}},$$

$$a_3 = 4q^2 - 2a_1,$$

or

$$I = \frac{2u_x}{u_x^2 + \alpha^2 q^2} \int \frac{C_2 v^2 + C_1 v - C_2}{v^4 - av^3 + bv^2 + av + 1}\,dv,$$

with

$$a = \frac{4\alpha q(t_x - \alpha s_x)}{\sqrt{1+\alpha^2}\left(u_x^2 + \alpha^2 q^2\right)}, \qquad b = \frac{4q^2}{u_x^2 + \alpha^2 q^2} - 2.$$

For the decomposition

$$\frac{C_2 v^2 + C_1 v - C_2}{v^4 - av^3 + bv^2 + av + 1} = \frac{D_1 v + E_1}{v^2 + A_1 v + B_1} + \frac{D_2 v + E_2}{v^2 + A_2 v + B_2},$$

we find the coefficients

$$A_{1/2} = -\frac{2\alpha\sqrt{1+\alpha^2}q}{u_x^2 + \alpha^2 q^2}\left(\frac{t_x - \alpha s_x}{1+\alpha^2} \pm q\right),$$

$$B_{1/2} = \frac{1+\alpha^2}{u_x^2 + \alpha^2 q^2}\left(\frac{t_x - \alpha s_x}{1+\alpha^2} \pm q\right)^2,$$

as well as

$$D_1 = -\frac{1}{2}\left(u_x^2 + \alpha^2 q^2\right),$$

$$D_2 = \frac{1}{2}\left(u_x^2 + \alpha^2 q^2\right),$$

$$E_1 = \frac{1}{2\sqrt{1+\alpha^2}}\left(s_\tau - s_x + \alpha q\right)\left(t_x - \alpha s_x + (1+\alpha^2)q\right),$$

$$E_2 = \frac{1}{2\sqrt{1+\alpha^2}}\left(s_\tau - s_x - \alpha q\right)\left(t_x - \alpha s_x - (1+\alpha^2)q\right).$$

Hence, we have

$$I = \frac{2u_x}{u_x^2 + \alpha^2 q^2}\left(\int \frac{D_1 v + E_1}{v^2 + A_1 v + B_1}dv + \int \frac{D_2 v + E_2}{v^2 + A_2 v + B_2}dv\right),$$

and it remains to integrate

$$I_{1/2} = \frac{2u_x}{u_x^2 + \alpha^2 q^2}\int \frac{D_{1/2}v + E_{1/2}}{v^2 + A_{1/2}v + B_{1/2}}dv$$

$$= \frac{2u_x}{u_x^2 + \alpha^2 q^2}\left(\frac{1}{2}D_{1/2}\log\left(v^2 + A_{1/2}v + B_{1/2}\right)\right.$$

$$\left. +\left(E_{1/2} - \frac{1}{2}A_{1/2}B_{1/2}\right)\int \frac{1}{v^2 + A_{1/2}v + B_{1/2}}dv\right)$$

$$= \mp\frac{1}{2}u_x\log\left(v^2 + A_{1/2}v + B_{1/2}\right)$$

$$+\frac{2u_x}{u_x^2 + \alpha^2 q^2}\left(E_{1/2} - \frac{1}{2}A_{1/2}B_{1/2}\right)\int \frac{1}{v^2 + A_{1/2}v + B_{1/2}}dv.$$

Due to

$$G_{1/2}^2 = B_{1/2} - \frac{1}{4}A_{1/2}^2 = \frac{(1+\alpha^2)u_x^2}{\left(u_x^2 + \alpha^2 q^2\right)^2}\left(\frac{t_x - \alpha s_x}{1+\alpha^2} \pm q\right)^2,$$

we find

$$G_{1/2} = \frac{\sqrt{1+\alpha^2}|u_x|}{u_x^2 + \alpha^2 q^2}\left(q \pm \frac{t_x - \alpha s_x}{1+\alpha^2}\right) > 0.$$

Hence, we obtain

$$
\left(E_{1/2} - \frac{1}{2}A_{1/2}D_{1/2}\right) \int \frac{1}{v^2 + A_{1/2}v + B_{1/2}}\, dv
$$

$$
= \left(E_{1/2} - \frac{1}{2}A_{1/2}D_{1/2}\right) \int \frac{1}{(v + \frac{1}{2}A_{1/2})^2 + G_{1/2}^2}\, dv
$$

$$
= \frac{1}{G_{1/2}}\left(E_{1/2} - \frac{1}{2}A_{1/2}D_{1/2}\right) \arctan \frac{2v + A_{1/2}}{2G_{1/2}}
$$

$$
= \pm \frac{u_x^2 + \alpha^2 q^2}{2|u_x|}(s_\tau - s_x) \arctan \frac{2v + A_{1/2}}{2G_{1/2}}\,.
$$

Therefore,

$$
I_{1/2} = \mp \frac{1}{2}u_x \log\left(v^2 + A_{1/2}v + B_{1/2}\right) \pm (s_\tau - s_x)\frac{u_x}{|u_x|} \arctan \frac{2v + A_{1/2}}{2G_{1/2}}\,.
$$

Recall that

$$
v = \frac{\sqrt{1 + \sinh^2 u} - 1}{\sinh u} = \frac{\sqrt{(1 + \alpha^2)(s - p)^2 + q^2} - q}{\sqrt{1 + \alpha^2}(s - p)}\,.
$$

C.2.3 Linear Elasticity Single Layer Potential

By the use of Lemma 1.15, it is sufficient to describe the computation of

$$
S_{ij}(\tau, x) = \frac{1}{4\pi} \int_\tau \frac{(x_i - y_i)(x_j - y_j)}{|x - y|^3}\, ds_y\,.
$$

With the local parametrisations

$$
y = x_1 + s\,\underline{r}_1 + t\,\underline{r}_2, \quad x = x_1 + s_x\underline{r}_1 + t_x\underline{r}_2 + u_x\underline{n}\,,
$$

the kernel function can be written as

$$
\frac{(x_i - y_i)(x_j - y_j)}{|x - y|^3} = \frac{a_{ij}}{\left((s - s_x)^2 + (t - t_x)^2 + u_x^2\right)^{1/2}} +
$$

$$
\frac{\left(b_{ij}(s - s_x) - c_{ij}u_x\right)(t - t_x)}{\left((s - s_x)^2 + (t - t_x)^2 + u_x^2\right)^{3/2}} + \frac{d_{ij}(s - s_x)^2 - e_{ij}(s - s_x)u_x + f_{ij}u_x^2}{\left((s - s_x)^2 + (t - t_x)^2 + u_x^2\right)^{3/2}}
$$

with the parameters

$$a_{ij} = r_{2,i}r_{2,j},$$
$$b_{ij} = r_{1,i}r_{2,j} + r_{2,i}r_{1,j},$$
$$c_{ij} = r_{2,i}n_j + r_{2,j}n_i,$$
$$d_{ij} = r_{1,i}r_{1,j} - a_{ij},$$
$$e_{ij} = r_{1,i}n_j + r_{1,j}n_i,$$
$$f_{ij} = n_i n_j - a_{ij}.$$

Hence, we have to compute

$$S_{ij}(\tau, x) = S_{ij}^1(\tau, x) + S_{ij}^2(\tau, x) + S_{ij}^3(\tau, x)$$

with

$$S_{ij}^1(\tau, x) = a_{ij}\frac{1}{4\pi}\int_0^{s_\tau}\int_{\alpha_1 s}^{\alpha_2 s} \frac{1}{\left((s - s_x)^2 + (t - t_x)^2 + u_x^2\right)^{1/2}}\,dt\,ds,$$

$$S_{ij}^2(\tau, x) = \frac{1}{4\pi}\int_0^{s_\tau}\int_{\alpha_1 s}^{\alpha_2 s} \frac{\left(b_{ij}(s - s_x) - c_{ij}u_x\right)(t - t_x)}{\left((s - s_x)^2 + (t - t_x)^2 + u_x^2\right)^{3/2}}\,dt\,ds,$$

$$S_{ij}^3(\tau, x) = \frac{1}{4\pi}\int_0^{s_\tau}\int_{\alpha_1 s}^{\alpha_2 s} \frac{d_{ij}(s - s_x)^2 - e_{ij}(s - s_x)u_x + f_{ij}u_x^2}{\left((s - s_x)^2 + (t - t_x)^2 + u_x^2\right)^{3/2}}\,dt\,ds.$$

Note that $S_{ij}^1(\tau, x)$ corresponds to the single layer potential function for the Laplace equation.

For the second entry we obtain

$$S_{ij}^2(\tau, x) = \frac{1}{4\pi}\int_0^{s_\tau}\int_{\alpha_1 s}^{\alpha_2 s} \frac{\left(b_{ij}(s - s_x) - c_{ij}u_x\right)(t - t_x)}{\left((s - s_x)^2 + (t - t_x)^2 + u_x^2\right)^{3/2}}$$

$$= -\frac{1}{4\pi}\int_0^{s_\tau}\left[\frac{b_{ij}(s - s_x) - c_{ij}u_x}{\left((s - s_x)^2 + (t - t_x)^2 + u_x^2\right)^{1/2}}\right]_{\alpha_1 s}^{\alpha_2 s}\,ds$$

$$= -\frac{1}{4\pi}\left(F_{ij}^2(s_\tau, \alpha_2) - F_{ij}^2(s_\tau, \alpha_1) - F_{ij}^2(0, \alpha_2) + F_{ij}^2(0, \alpha_1)\right),$$

with the integral

$$F_{ij}^2(s, \alpha) = \int \frac{b_{ij}(s - s_x) - c_{ij}u_x}{\left((s - s_x)^2 + (\alpha s - t_x)^2 + u_x^2\right)^{1/2}}\,ds$$

$$= \int \frac{b_{ij}(s - s_x) - c_{ij}u_x}{\sqrt{(1 + \alpha^2)(s - p)^2 + q^2}}\,ds$$

$$= b_{ij} \int \frac{s-p}{\sqrt{(1+\alpha^2)(s-p)^2+q^2}} ds + \int \frac{b_{ij}(p-s_x)-c_{ij}u_x}{\sqrt{(1+\alpha^2)(s-p)^2+q^2}} ds$$

$$= \frac{b_{ij}}{1+\alpha^2}\left((1+\alpha^2)(s-p)^2+q^2\right)^{1/2}$$

$$+ \frac{b_{ij}(p-s_x)-c_{ij}u_x}{\sqrt{1+\alpha^2}} \log\left(\sqrt{1+\alpha^2}(s-p)+\sqrt{(1+\alpha^2)(s-p)^2+q^2}\right).$$

For the third integral, we get

$$S_{ij}^3(\tau,x) = \frac{1}{4\pi} \int_0^{s_\tau} \int_{\alpha_1 s}^{\alpha_2 s} \frac{d_{ij}(s-s_x)^2 - e_{ij}(s-s_x)u_x + f_{ij}u_x^2}{\left((s-s_x)^2+(t-t_x)^2+u_x^2\right)^{3/2}} \, dt \, ds$$

$$= \int_0^{s_\tau} \left[\frac{t-t_x}{(s-s_x)^2+u_x^2} \frac{d_{ij}(s-s_x)^2 - e_{ij}(s-s_x)u_x + f_{ij}u_x^2}{\sqrt{(s-s_x)^2+(t-t_x)^2+u_x^2}}\right]_{\alpha_1 s}^{\alpha_2 s} ds$$

$$= \frac{1}{4\pi}\left(F_{ij}^3(s_\tau,\alpha_2) - F_{ij}^3(s_\tau,\alpha_1) - F_{ij}^3(0,\alpha_2) + F_{ij}^3(0,\alpha_1)\right)$$

with

$$F_{ij}^3(s,\alpha) = \int \frac{\alpha s - t_x}{(s-s_x)^2+u_x^2} \frac{d_{ij}(s-s_x)^2 - e_{ij}(s-s_x)u_x + f_{ij}u_x^2}{\sqrt{(s-s_x)^2+(\alpha s-t_x)^2+u_x^2}} ds$$

$$= \int \frac{\alpha s - t_x}{(s-s_x)^2+u_x^2} \frac{d_{ij}\left((s-s_x)^2+u_x^2\right) - e_{ij}(s-s_x)u_x + \left(f_{ij}-d_{ij}\right)u_x^2}{\sqrt{(s-s_x)^2+(\alpha s-t_x)^2+u_x^2}} ds$$

$$= d_{ij} \int \frac{\alpha s - t_x}{\sqrt{(1+\alpha^2)(s-p)^2+q^2}} ds$$

$$-e_{ij} \int \frac{(s-s_x)u_x}{(s-s_x)^2+u_x^2} \frac{\alpha s - t_x}{\sqrt{(1+\alpha^2)(s-p)^2+q^2}} ds$$

$$+\left(f_{ij}-d_{ij}\right) \int \frac{u_x^2}{(s-s_x)^2+u_x^2} \frac{\alpha s - t_x}{\sqrt{(1+\alpha^2)(s-p)^2+q^2}} ds$$

$$= d_{ij} F^{3,1}(s,\alpha) - e_{ij} F^{3,2}(s,\alpha) + \left(f_{ij}-d_{ij}\right) F^{3,3}(s,\alpha)$$

and

$$F^{3,1}(s,\alpha) = \int \frac{\alpha s - t_\ell}{\sqrt{(1+\alpha^2)(s-p)^2+q^2}} ds$$

$$= \alpha \int \frac{s-p}{\sqrt{(1+\alpha^2)(s-p)^2+q^2}} ds + \int \frac{\alpha p - t_\ell}{\sqrt{(1+\alpha^2)(s-p)^2+q^2}} ds$$

$$= \frac{\alpha}{\sqrt{1+\alpha^2}}\sqrt{(1+\alpha^2)(s-p)^2+q^2}$$

$$+ \frac{\alpha p - t_\ell}{\sqrt{1+\alpha^2}}\log\left(\sqrt{1+\alpha^2}(s-p) + \sqrt{(1+\alpha^2)(s-p)^2+q^2}\right).$$

The remaining integrals can be computed in the same way as for the double layer potential of the Laplace operator. In particular, up to the sign and the choice $s_\tau = s_\ell$, the integral

$$F^{3,2}(s,\alpha) = \int \frac{(s-s_x)u_x}{(s-s_x)^2+u_x^2}\frac{\alpha s - t_x}{\sqrt{(1+\alpha^2)(s-p)^2+q^2}}ds$$

coincides with the integral of the double layer potential function of the Laplace operator.

Furthermore, using the transformations as for the Laplace operator, we have

$$F^{3,3}(s,\alpha) = \int \frac{u_x^2}{(s-s_x)^2+u_x^2}\frac{\alpha s - t_x}{\sqrt{(1+\alpha^2)(s-p)^2+q^2}}ds$$

$$= u_x^2 \int \frac{\alpha q \sinh u + \sqrt{1+\alpha^2}(\alpha p - t_x)}{q^2\sinh^2 u + 2\sqrt{1+\alpha^2}q(p-s_x)\sinh u + (1+\alpha^2)[u_x^2+(p-s_x)^2]}du$$

$$= 2u_x^2 \int \frac{\alpha q2v + \sqrt{1+\alpha^2}(\alpha p - t_x)(1-v^2)}{4q^2v^2 + 4\sqrt{1+\alpha^2}q(p-s_x)v(1-v^2) + (1+\alpha^2)(u_x^2+(p-s_x)^2)(1-v^2)^2}dv$$

$$= 2u_x^2 \int \frac{C_2 v^2 + C_1 v - C_2}{a_1 v^4 - a_2 v^3 + a_3 v^2 + a_2 v + a_1}dv,$$

with the parameters

$$C_2 = -\sqrt{1+\alpha^2}(\alpha p - t_x), \quad C_1 = 2\alpha q,$$

and

$$a_1 = u_x^2 + \alpha^2 q^2, \quad a_2 = \frac{4\alpha q(t_x - \alpha s_x)}{\sqrt{1+\alpha^2}}, \quad a_3 = 4q^2 - 2a_1.$$

Hence, we get

$$F^{3,3}(s,\alpha) = \frac{2u_x^2}{u_x^2+\alpha^2 q^2}\int \frac{C_2 v^2 + C_1 v - C_2}{(v^2 + A_1 v + B_1)(v^2 + A_2 v + B_2)}dv$$

with

$$A_{1/2} = -\frac{2\alpha\sqrt{1+\alpha^2}q}{u_x^2+\alpha^2 q^2}\left(\frac{t_x - \alpha s_x}{1+\alpha^2} \pm q\right), \quad B_{1/2} = \frac{1+\alpha^2}{u_x^2+\alpha^2 q^2}\left(\frac{t_x - \alpha s_x}{1+\alpha^2} \pm q\right)^2.$$

For the decomposition

$$\frac{C_2v^2 + C_1v - C_2}{(v^2 + A_1v + B_1)(v^2 + A_2v + B_2)} = \frac{D_1v + E_1}{v^2 + A_1v + B_1} + \frac{D_2v + E_2}{v^2 + A_2v + B_2},$$

we find the coefficients

$$D_1 = D_2 = 0, \ E_1 = \frac{(t_x - \alpha s_x) + (1 + \alpha^2)q}{2\sqrt{1 + \alpha^2}}, \ E_2 = \frac{(t_x - \alpha s_x) - (1 + \alpha^2)q}{2\sqrt{1 + \alpha^2}}.$$

Therefore,

$$F^{3,3}(s, \alpha) = \frac{2\tau_\ell^2}{\tau_\ell^2 + \alpha^2 q^2} \left(\int \frac{E_1}{v^2 + A_1v + B_1} dv + \int \frac{E_2}{v^2 + A_2v + B_2} dv \right),$$

and it remains to integrate

$$I_{1/2} = \frac{u_x^2}{u_x^2 + \alpha^2 q^2} \frac{(t_x - \alpha s_x) \pm (1 + \alpha^2)q}{\sqrt{1 + \alpha^2}} \int \frac{1}{v^2 + A_{1/2}v + B_{1/2}} dv$$

$$= \frac{u_x^2}{u_x^2 + \alpha^2 q^2} \frac{(t_x - \alpha s_x) \pm (1 + \alpha^2)q}{\sqrt{1 + \alpha^2}} \int \frac{1}{(v + \frac{1}{2}A_{1/2})^2 + G_{1/2}^2} dv$$

$$= \frac{1}{G_{1/2}} \frac{u_x^2}{u_x^2 + \alpha^2 q^2} \frac{(t_x - \alpha s_x) \pm (1 + \alpha^2)q}{\sqrt{1 + \alpha^2}} \arctan \frac{2v + A_{1/2}}{2G_{1/2}},$$

with

$$G_{1/2} = \frac{\sqrt{1 + \alpha^2}|\tau_\ell|}{u_x^2 + \alpha^2 q^2} \left(q \pm \frac{t_x - \alpha s_x}{1 + \alpha^2} \right) > 0.$$

Finally,

$$I_{1/2} = \pm |u_x| \arctan \frac{2v + A_{1/2}}{2G_{1/2}}.$$

C.3 Iterative Solution Methods

In this section we consider the iterative solution of a linear system

$$A\underline{x} = \underline{f}, \quad A \in \mathbb{R}^{N \times N}, \quad \underline{x}, \underline{f} \in \mathbb{R}^N. \tag{C.2}$$

Note that the dimension N of the system is usually connected to the discretisation parameter h (cf. (2.2)), and, therefore, we have a sequence of linear systems to be solved for $N \to \infty$. We first consider symmetric and positive definite systems and later general systems with regular matrices.

C.3.1 Conjugate Gradient Method (CG)

For a symmetric and positive definite matrix $A \in \mathbb{R}^{N \times N}$, we may define an inner product

$$(\underline{u}, \underline{v})_A = (A\underline{u}, \underline{v}) \quad \text{for all } \underline{u}, \underline{v} \in \mathbb{R}^N.$$

A system of vectors

$$P = \{\underline{p}^k\}_{k=0}^{N-1}, \quad \underline{p}^k \neq 0, \ k = 0, \ldots, N-1 \tag{C.3}$$

is called conjugate, or A–orthogonal, if there holds

$$(A\underline{p}^\ell, \underline{p}^k) = 0 \quad \text{for } k \neq \ell.$$

Note that $(A\underline{p}^k, \underline{p}^k) > 0$ since A is assumed to be positive definite. The vector system P forms an A–orthogonal basis of the vector space $\mathbb{R}^N$. Hence, the solution vector $\underline{x} \in \mathbb{R}^N$ of the linear system (C.2) can be written as

$$\underline{x} = \underline{x}^0 - \sum_{\ell=0}^{N-1} \alpha_\ell\, \underline{p}^\ell$$

with an arbitrary given vector $\underline{x}^0 \in \mathbb{R}^N$. To find the yet unknown coefficients α_ℓ, we consider the linear system

$$A\underline{x} = A\underline{x}^0 - \sum_{\ell=0}^{N-1} \alpha_\ell\, A\underline{p}^\ell = \underline{f}.$$

Taking the Euclidean inner product with $\underline{p}^k$, this gives

$$\sum_{\ell=0}^{N-1} \alpha_\ell\, (A\underline{p}^\ell, \underline{p}^k) = (A\underline{x}^0 - \underline{f}, \underline{p}^k), \quad \text{for } k = 0, \ldots, N-1,$$

and, therefore, by using the A–orthogonality of the system P

$$\alpha_k = \frac{(A\underline{x}^0 - \underline{f}, \underline{p}^k)}{(A\underline{p}^k, \underline{p}^k)} \quad \text{for } k = 0, \ldots, N-1. \tag{C.4}$$

Thus, if a system P of A–orthogonal vectors $\underline{p}^k$ is given, we can compute the coefficients α_k from (C.4), and, therefore, the solution $\underline{x}$ of the linear system (C.2). With it, this method can be seen as a direct solution algorithm. To define an iterative process, we introduce approximate solutions as

$$\underline{x}^{k+1} = \underline{x}^0 - \sum_{\ell=0}^{k} \alpha_\ell\, \underline{p}^\ell = \underline{x}^k - \alpha_k \underline{p}^k \quad \text{for } k = 0, \ldots, N-1.$$

For the computation of the coefficient α_k, we obtain from (C.4)

$$\alpha_k = \frac{(A\underline{x}^0 - \underline{f}, \underline{p}^k)}{(A\underline{p}^k, \underline{p}^k)} = \frac{1}{(A\underline{p}^k, \underline{p}^k)} \left(A\underline{x}^0 - \underline{f} - \sum_{\ell=0}^{k-1} \alpha_\ell A\underline{p}^\ell, \underline{p}^k \right) = \frac{(A\underline{x}^k - \underline{f}, \underline{p}^k)}{(A\underline{p}^k, \underline{p}^k)},$$

when using the A–orthogonality of the system (C.3). Therefore,

$$\alpha_k = \frac{(\underline{r}^k, \underline{p}^k)}{(A\underline{p}^k, \underline{p}^k)} \quad \text{for } k = 0, \ldots, N-1, \tag{C.5}$$

where

$$\underline{r}^k = A\underline{x}^k - \underline{f}$$

is the residual vector induced by the approximate solution $\underline{x}^k$. Note that the following recursion holds:

$$\underline{r}^{k+1} = A\underline{x}^{k+1} - \underline{f} = A(\underline{x}^k - \alpha_k \underline{p}^k) - \underline{f} = \underline{r}^k - \alpha_k A\underline{p}^k$$

for $k = 0, \ldots, N-2$. Moreover, we have

$$(\underline{r}^{k+1}, \underline{p}^k) = (\underline{r}^k - \alpha_k A\underline{p}^k, \underline{p}^k) = (\underline{r}^k, \underline{p}^k) - \frac{(\underline{r}^k, \underline{p}^k)}{(A\underline{p}^k, \underline{p}^k)} (A\underline{p}^k, \underline{p}^k) = 0$$

for $k = 0, \ldots, N-2$. From

$$(\underline{r}^{k+1}, \underline{p}^\ell) = (\underline{r}^k, \underline{p}^\ell) - \alpha_k (A\underline{p}^k, \underline{p}^\ell),$$

it follows by induction that

$$(\underline{r}^{k+1}, \underline{p}^\ell) = 0 \quad \text{for } \ell = 0, \ldots, k, \quad k = 0, \ldots, N-2. \tag{C.6}$$

It remains to construct an A–orthogonal vector system P via the orthogonalisation method of Gram–Schmidt. Let W be any system of linear independent vectors $\underline{w}^k$. Setting $\underline{p}^0 = \underline{w}^0$, we can compute A–orthogonal vectors $\underline{p}^k$ for $k = 0, \ldots, N-2$ as follows:

$$\underline{p}^{k+1} = \underline{w}^{k+1} - \sum_{\ell=0}^{k} \beta_{k\ell} \underline{p}^\ell, \quad \beta_{k\ell} = \frac{(A\underline{w}^{k+1}, \underline{p}^\ell)}{(A\underline{p}^\ell, \underline{p}^\ell)}, \quad \text{for } \ell = 0, \ldots, k. \tag{C.7}$$

From (C.7) and (C.6), we further conclude

$$(\underline{r}^{k+1}, \underline{w}^\ell) = (\underline{r}^{k+1}, \underline{p}^\ell) + \sum_{j=0}^{l-1} \beta_{\ell j} (\underline{r}^{k+1}, \underline{p}^j) = 0 \tag{C.8}$$

for $\ell = 0, \ldots, k$. In particular, if the residual vector $\underline{r}^{k+1}$ does not vanish, the vectors

$$\{\underline{w}^0, \underline{w}^1, \ldots, \underline{w}^k, \underline{r}^{k+1}\}$$

are linear independent and we can choose $\underline{w}^{k+1} = \underline{r}^{k+1}$. If $\underline{r}^{k+1} = 0$, then the system (C.2) is solved. In general, this will not happen and we can choose

$$\underline{w}^k = \underline{r}^k \quad \text{for } k = 0, \ldots, N-1.$$

From (C.8), we then conclude

$$(\underline{r}^{k+1}, \underline{r}^{\ell}) = 0 \quad \text{for all } \ell = 0, \ldots, k.$$

For the enumerator of the coefficient α_k defined in (C.5), we obtain

$$(\underline{r}^k, \underline{p}^k) = (\underline{r}^k, \underline{r}^k) - \sum_{\ell=0}^{k-1} \beta_{k-1,\ell} (\underline{r}^k, \underline{p}^{\ell}) = (\underline{r}^k, \underline{r}^k) = \varrho_k.$$

by using the recursion (C.7) and the orthogonality relation (C.6). From $\varrho_\ell > 0$, it follows $\alpha_\ell > 0$ for $\ell = 0, \ldots, k$, and, therefore, we can write

$$A\underline{p}^{\ell} = \frac{1}{\alpha_\ell} \left(\underline{r}^{\ell} - \underline{r}^{\ell+1} \right).$$

For the enumerator of the coefficients β_{kj} in (C.7), we then obtain

$$(A\underline{w}^{k+1}, \underline{p}^{\ell}) = (\underline{r}^{k+1}, A\underline{p}^{\ell}) = \frac{1}{\alpha_\ell}(\underline{r}^{k+1}, \underline{r}^{\ell} - \underline{r}^{\ell+1}) = 0$$

for $\ell = 0, \ldots, k-1$ and

$$(A\underline{w}^{k+1}, \underline{p}^k) = -\frac{1}{\alpha_k}(\underline{r}^{k+1}, \underline{r}^{k+1}) = -\frac{\varrho_{k+1}}{\alpha_k}$$

for $\ell = k$. Hence, we have $\beta_{k\ell} = 0$ for $\ell = 0, \ldots, k-1$ and

$$\beta_{kk} = \beta_k = -\frac{\varrho_{k+1}}{\alpha_k(A\underline{p}^k, \underline{p}^k)}.$$

Using $\underline{r}^{k+1} = \underline{r}^k - \alpha_k A\underline{p}^k$, we finally obtain

$$\alpha_k(A\underline{p}^k, \underline{p}^k) = (\underline{r}^k - \underline{r}^{k+1}, \underline{p}^k) = (\underline{r}^k, \underline{p}^k) = \varrho_k,$$

and, therefore,

$$\underline{p}^{k+1} = \underline{r}^{k+1} + \beta_k \underline{p}^k, \quad \beta_k = \frac{\varrho_{k+1}}{\varrho_k}.$$

Hence, we end up with the conjugate gradient method as summarised in Algorithm C.1.

Algorithm C.1

1. Compute for an arbitrary given initial solution $\underline{x}^0 \in \mathbb{R}^N$

$$\underline{r}^0 = A\underline{x}^0 - \underline{f}, \quad \underline{p}^0 = \underline{r}^0, \quad \varrho_0 = (\underline{r}^0, \underline{r}^0).$$

2. Iterate for $k = 0, \ldots, N-2$

$$\underline{s}^k = A\underline{p}^k, \quad \sigma_k = (\underline{s}^k, \underline{p}^k), \quad \alpha_k = \frac{\varrho_k}{\sigma_k}, \quad \underline{x}^{k+1} = \underline{x}^k - \alpha_k \underline{p}^k$$

and compute the new residual

$$\underline{r}^{k+1} = \underline{r}^k - \alpha_k \underline{s}^k, \quad \varrho_{k+1} = (\underline{r}^{k+1}, \underline{r}^{k+1}).$$

Stop, if

$$\varrho_{k+1} \leq \varepsilon^2 \varrho_0$$

is satisfied with some prescribed accuracy ε. Otherwise compute

$$\beta_k = \frac{\varrho_{k+1}}{\varrho_k}, \quad \underline{p}^{k+1} = \underline{r}^{k+1} + \beta_k \underline{p}^k.$$

For the error of the computed approximate solution $\underline{x}^{k+1}$, we obtain

$$\underline{x}^{k+1} - \underline{x} = \sum_{\ell=k+1}^{N-1} \alpha_\ell A\underline{p}^\ell,$$

and using the A–orthogonality of the vector system P, we further conclude

$$\left\| \underline{x}^{k+1} - \underline{x} \right\|_A^2 = \left(A(\underline{x}^{k+1} - \underline{x}), \underline{x}^{k+1} - \underline{x} \right) = \sum_{\ell=k+1}^{N-1} \alpha_\ell^2 \left(A\underline{p}^\ell, \underline{p}^\ell \right).$$

For an arbitrary given vector

$$\underline{u}^{k+1} = \underline{x}^0 - \sum_{\ell=0}^{k} \gamma_\ell \, \underline{p}^\ell, \tag{C.9}$$

we obtain in the same way

$$\left\| \underline{u}^{k+1} - \underline{x} \right\|_A^2 = \sum_{\ell=0}^{k} (\alpha_\ell - \gamma_\ell)^2 (A\underline{p}^\ell, \underline{p}^\ell) + \sum_{\ell=k+1}^{N-1} \alpha_\ell^2 (A\underline{p}^\ell, \underline{p}^\ell)$$

Hence, we find the approximate solution $\underline{x}^{k+1}$ as the solution of the minimisation problem

$$\left\| \underline{x}^{k+1} - \underline{x} \right\|_A = \min_{\underline{u}^{k+1}} \left\| \underline{u}^{k+1} - \underline{x} \right\|_A,$$

where the minimum is taken over all vectors $\underline{u}^{k+1}$ of the form (C.9). From the recursions

$$\underline{p}^0 = \underline{r}^0, \quad \underline{p}^{k+1} = \underline{r}^{k+1} + \beta_k \underline{p}^k, \quad \underline{r}^{k+1} = \underline{r}^k - \alpha_k A\underline{p}^k,$$

we find representations

$$\underline{p}^\ell = \psi_\ell(A)\underline{r}^0$$

with a matrix polynomials $\psi_\ell(A)$ of degree ℓ. Hence, we conclude with $\underline{e}^0 = \underline{x}^0 - \underline{x}$

$$\underline{u}^{k+1} - \underline{x} = \underline{x}^0 - \underline{x} - \sum_{\ell=0}^{k} \gamma_\ell \underline{p}^\ell = \underline{e}^0 - \sum_{\ell=0}^{k} \gamma_\ell \psi_\ell(A)\underline{r}^0 = \underline{e}^0 - \sum_{\ell=0}^{k} \gamma_\ell \psi_\ell(A) A\underline{e}^0,$$

and, therefore,

$$\underline{u}^{k+1} - \underline{x} = p_{k+1}(A)\underline{e}^0$$

with some matrix polynomial $p_{k+1}(A)$ of degree $k + 1$ and having the property $p_{k+1}(0) = 1$. The polynomial $p_{k+1}(A)$ obtained for the vector $\underline{x}^{k+1}$ is, therefore, the solution of the minimisation problem

$$\left\| \underline{x}^{k+1} - \underline{x} \right\|_A = \min_{p_{k+1}(A)} \left\| p_{k+1}(A)\underline{e}^0 \right\|_A, \tag{C.10}$$

where the minimum is taken over all polynomials $p_{k+1}(A)$ having the property $p_{k+1}(0) = 1$. The space

$$S_k(A, \underline{r}^0) = \mathrm{span}\{\underline{p}^0, \underline{p}^1, \dots, \underline{p}^k\} = \mathrm{span}\{\underline{r}^0, A\underline{r}, \dots, A^k\underline{r}^0\}$$

is called a Krylov space of the matrix A induced by the residual vector $\underline{r}^0$.

From the minimisation problem (C.10), we further conclude the estimate

$$\left\| \underline{e}^{k+1} \right\|_A \leq \min_{p_{k+1}(A)} \max_\lambda \left| p_{k+1}(\lambda) \right| \left\| \underline{e}^0 \right\|_A,$$

where the minimum is again taken over all polynomials $p_{k+1}(A)$ having the property $p_{k+1}(0) = 1$, and the maximum over the spectrum of the matrix A, i.e. $\lambda \in [\lambda_{\min}(A), \lambda_{\max}(A)]$. The above min − max–problem will be solved by the scaled Tschebyscheff polynomials $\widetilde{T}_{k+1}(\lambda)$, and we find

$$\min_{p_{k+1}(A)} \max_\lambda \left| p_{k+1}(\lambda) \right| = \max_\lambda \left| \widetilde{T}_{k+1}(\lambda) \right| = \frac{2q^{k+1}}{1 + q^{2(k+1)}},$$

with

$$q = \frac{\sqrt{\lambda_{\max}(A)} + \sqrt{\lambda_{\min}(A)}}{\sqrt{\lambda_{\max}(A)} - \sqrt{\lambda_{\min}(A)}} = \frac{\sqrt{\kappa_2(A)} + 1}{\sqrt{\kappa_2(A)} - 1},$$

where

$$\kappa_2(A) = \frac{\lambda_{\max}(A)}{\lambda_{\min}(A)}$$

is the spectral condition number of the symmetric and positive definite matrix A. Using the Raleigh quotient

$$\lambda_{\min}(A) = \min_{\underline{x} \neq 0} \frac{(A\underline{x}, \underline{x})}{(\underline{x}, \underline{x})} \leq \max_{\underline{x} \neq 0} \frac{(A\underline{x}, \underline{x})}{(\underline{x}, \underline{x})} = \lambda_{\max}(A),$$

we find from the spectral equivalence inequalities

$$c_1^A (\underline{x}, \underline{x}) \leq (A\underline{x}, \underline{x}) \leq c_2^A (\underline{x}, \underline{x}) \quad \text{for } \underline{x} \in \mathbb{R}^N$$

an upper bound for the spectral condition number

$$\kappa_2(A) \leq \frac{c_2^A}{c_1^A}.$$

Since the spectral condition number of the boundary element stiffness matrices may depend on mesh parameters such as the mesh size h or the mesh ratio h_{max}/h_{min}, an appropriate preconditioning is mandatory in many cases.

Hence, we assume that there is a symmetric and positive definite matrix $C_A \in \mathbb{R}^{N \times N}$ which can be factorised as $C_A = C_A^{1/2} C_A^{1/2}$, where $C_A^{1/2}$ is again symmetric and positive definite. Instead of the linear system (C.2), we now consider the transformed system

$$\widetilde{A}\widetilde{\underline{x}} = C_A^{-1/2} A C_A^{-1/2} C_A^{1/2} \underline{x} = C_A^{-1/2} \underline{f} = \widetilde{\underline{f}},$$

where the transformed system matrix $\widetilde{A} = C_A^{-1/2} A C_A^{-1/2}$ is again symmetric and positive definite. For the solution of the linear system $\widetilde{A}\widetilde{\underline{x}} = \widetilde{\underline{f}}$, we can apply Algorithm C.1 to obtain, by substituting the transformations, the precondioned conjugate gradient method as described in Algorithm C.2.

Algorithm C.2

1. Compute for an arbitrary given initial solution $\underline{x}^0 \in \mathbb{R}^N$

$$\underline{r}^0 = A\underline{x}^0 - \underline{f}, \quad \underline{v}^0 = C_A^{-1}\underline{r}^0, \quad \underline{p}^0 = \underline{v}^0, \quad \varrho_0 = (\underline{v}^0, \underline{r}^0).$$

2. Iterate for $k = 0, \ldots, N-2$

$$\underline{s}^k = A\underline{p}^k, \quad \sigma_k = (\underline{s}^k, \underline{p}^k), \quad \alpha_k = \frac{\varrho_k}{\sigma_k}, \quad \underline{x}^{k+1} = \underline{x}^k - \alpha_k \underline{p}^k$$

and compute the new residual

$$\underline{r}^{k+1} = \underline{r}^k - \alpha_k \underline{s}^k, \quad \underline{v}^{k+1} = C_A^{-1} \underline{r}^{k+1}, \quad \varrho_{k+1} = (\underline{v}^{k+1}, \underline{r}^{k+1}).$$

Stop, if

$$\varrho_{k+1} \leq \varepsilon^2 \varrho_0$$

is satisfied with some prescribed accuracy ε.
Otherwise compute

$$\beta_k = \frac{\varrho_{k+1}}{\varrho_k}, \quad \underline{p}^{k+1} = \underline{v}^{k+1} + \beta_k \underline{p}^k.$$

For the preconditioned conjugate gradient scheme, we then obtain the error estimate

$$\|\underline{e}^{k+1}\|_{\widetilde{A}} \leq \frac{2q^{k+1}}{1 + q^{2(k+1)}} \|\underline{e}^0\|_{\widetilde{A}},$$

where

$$q = \sqrt{\frac{\kappa_2(\tilde{A}) + 1}{\kappa_2(\tilde{A}) - 1}}, \quad \kappa_2(\tilde{A}) = \frac{\lambda_{\max}(\tilde{A})}{\lambda_{\min}(\tilde{A})} \le \frac{c_2^{\tilde{A}}}{c_1^{\tilde{A}}},$$

and $c_1^{\tilde{A}}, c_2^{\tilde{A}}$ are the positive constants from the spectral equivalence inequalities

$$c_1^{\tilde{A}}(\tilde{x}, \tilde{x}) \le (\tilde{A}\tilde{x}, \tilde{x}) \le c_2^{\tilde{A}}(\tilde{x}, \tilde{x}) \quad \text{for all } \tilde{x} \in \mathbb{R}^N.$$

Inserting $\tilde{x} = C_A^{1/2} x$, this is equivalent to the spectral equivalence inequalities

$$c_1^{\tilde{A}}(C_A x, x) \le (A x, x) \le c_2^{\tilde{A}}(C_A x, x) \quad \text{for } x \in \mathbb{R}^N. \tag{C.11}$$

Hence, the quite often challenging problem is to find a preconditioning matrix C_A satisfying the spectral equivalence inequalities (C.11) and allowing a simple and efficient application of the inverse matrix C_A^{-1} as needed in Algorithm C.2.

C.3.2 Generalised Minimal Residual Method (GMRES)

For a symmetric and positive definite matrix A, we have used the Krylov space

$$S_k(A, r^0) = \text{span}\{r^0, A r^0, \ldots, A^k r^0\}$$

to construct an A–orthogonal vector system P (cf. (C.3)). Formally, such a vector system can be defined for any arbitrary matrix $A \in \mathbb{R}^{N \times N}$. However, a nonsymmetric and possibly indefinite matrix A does not induce an inner product. Instead of an A–orthogonal vector system, we therefore define an orthonormal vector system

$$V = \{v^k\}_{k=0}^{N-1}$$

satisfying

$$(v^k, v^\ell) = \begin{cases} 1 & \text{for } k = \ell, \\ 0 & \text{for } k \ne \ell \end{cases}$$

using the method of Arnoldi as described in Algorithm C.3.

Algorithm C.3

1. Compute for an arbitrary given initial solution $x^0 \in \mathbb{R}^N$

$$r^0 = A x^0 - f, \quad v^0 = \frac{r^0}{\|r^0\|_2}.$$

2. Iterate for $k = 0, \ldots, N - 1$

$$\hat{v}^{k+1} = A v^k - \sum_{\ell=0}^{k} \beta_{k\ell} v^\ell, \quad \beta_{k\ell} = (A v^k, v^\ell).$$

```
Stop, if ‖v̂^{k+1}‖_2 = 0 is satisfied.
Otherwise, compute
```

$$\underline{v}^{k+1} = \frac{\underline{\hat{v}}^{k+1}}{\|\underline{\hat{v}}^{k+1}\|_2}.$$

Note that the method of Arnoldi (Algorithm C.3) may fail if $\|\underline{\hat{v}}^{k+1}\|_2 = 0$ is satisfied. We will comment this break down situation later.

Using the orthonormal basis vectors from the system V, we may define an approximate solution of the linear system $A\underline{x} = \underline{f}$ as

$$\underline{x}^{k+1} = \underline{x}^0 - \sum_{\ell=0}^{k} \alpha_\ell \, \underline{v}^\ell \,,$$

where we have to find the yet unknown coefficients α_ℓ. To this end, we may require to minimise the residual $\underline{r}^{k+1} = A\underline{x}^{k+1} - \underline{f}$ with respect to the Euclidean vector norm,

$$\left\|\underline{r}^{k+1}\right\|_2 = \left\|A\underline{x}^{k+1} - \underline{f}\right\|_2 = \left\|A\underline{x}^0 - \underline{f} - \sum_{\ell=0}^{k} \alpha_\ell A\underline{v}^\ell\right\|_2 \longrightarrow \min,$$

using the parameters $\alpha_0, \ldots, \alpha_k$. From the method of Arnoldi (Algorithm C.3), we obtain

$$A\underline{v}^\ell = \underline{\hat{v}}^{\ell+1} + \sum_{j=0}^{\ell} \beta_{\ell j} \, \underline{v}^j = \sum_{j=0}^{\ell+1} \beta_{\ell j} \, \underline{v}^j, \quad \beta_{\ell\ell+1} = \left\|\underline{\hat{v}}^{\ell+1}\right\|_2.$$

Hence, we have

$$\underline{r}^{k+1} = \underline{r}^0 - \sum_{\ell=0}^{k} \alpha_\ell \sum_{j=0}^{\ell+1} \beta_{\ell j} \, \underline{v}^j = \underline{r}^0 - V_{k+1} H_k \underline{\alpha}$$

with the orthogonal matrix

$$V_{k+1} = \left(\underline{v}^0, \underline{v}^1, \ldots, \underline{v}^{k+1}\right) \in \mathbb{R}^{N \times (k+2)},$$

and with a upper Hessenberg matrix $H_k \in \mathbb{R}^{(k+2) \times (k+1)}$ defined by

$$H_k[j, \ell] = \begin{cases} \beta_{\ell j} & \text{for } j \leq \ell+1, \\ 0 & \text{for } j > \ell+1. \end{cases}$$

Moreover, we can write

$$\underline{r}^0 = \|\underline{r}^0\|_2 \underline{v}^0 = \|\underline{r}^0\|_2 V_{k+1} \underline{e}^0,$$

where the notation $\underline{e}^0 = (1, 0, \ldots, 0)^\top \in \mathbb{R}^{k+2}$ has been used. Since V_{k+1} is orthogonal, we deduce

$$\left\|\underline{r}^{k+1}\right\|_2 = \left\|V_{k+1}\left(\|\underline{r}^0\|_2\underline{e}^0 - H_k\underline{\alpha}\right)\right\|_2$$
$$= \left\|\|\underline{r}^0\|_2\underline{e}^0 - H_k\underline{\alpha}\right\|_2 = \left\|\|\underline{r}^0\|_2 Q_k\underline{e}^0 - Q_kH_k\underline{\alpha}\right\|_2,$$

where $Q_k \in \mathbb{R}^{(k+2)\times(k+2)}$ is an orthogonal matrix such that $R_k = Q_kH_k \in \mathbb{R}^{(k+2)\times(k+1)}$ is an upper triangular matrix. Then, we obtain

$$\left\|\underline{r}^{k+1}\right\|_2^2 = \left\|\|\underline{r}^0\|_2 Q_k\underline{e}^0 - R_k\underline{\alpha}\right\|_2^2$$

$$= \sum_{\ell=0}^{k+1}\left(\|\underline{r}^0\|_2 Q_k\underline{e}^0 - R_k\underline{\alpha}\right)_\ell^2$$

$$= \sum_{\ell=0}^{k}\left(\|\underline{r}^0\|_2 Q_k\underline{e}^0 - R_k\underline{\alpha}\right)_\ell^2 + \left(\|\underline{r}^0\|_2 Q_k\underline{e}^0\right)_{k+1}^2 = \left(\|\underline{r}^0\|_2 Q_k\underline{e}^0\right)_{k+1}^2,$$

if the coefficient vector $\underline{\alpha} \in \mathbb{R}^{k+1}$ is found from the upper triangular linear system

$$R_k\underline{\alpha} = \|\underline{r}^0\|_2 Q_k\underline{e}^0.$$

It remains to find an orthogonal matrix $Q_k \in \mathbb{R}^{(k+2)\times(k+2)}$ transforming the upper Hessenberg matrix

$$H_k = \begin{pmatrix} \beta_{0,0} & \beta_{1,0} & \cdots & \beta_{k,0} \\ \beta_{0,1} & \beta_{1,1} & & \vdots \\ 0 & \beta_{1,2} & \ddots & \vdots \\ & 0 & \ddots & \beta_{k,k} \\ & & & \beta_{k,k+1} \end{pmatrix} \in \mathbb{R}^{(k+2)\times(k+1)}$$

into an upper triangular matrix

$$R_k = Q_kH_k = \begin{pmatrix} r_{0,0} & r_{0,1} & \cdots & r_{0,k} \\ 0 & r_{1,1} & & \vdots \\ 0 & 0 & \ddots & \vdots \\ & & \ddots & r_{k,k} \\ & & & 0 \end{pmatrix} \in \mathbb{R}^{(k+2)\times(k+1)}.$$

This can be done by the use of the Givens rotations. Let us first consider the column vector

$$\underline{h}^j = (\beta_{j,0}, \ldots, \beta_{j,j-1}, \beta_{j,j}, \beta_{j,j+1}, 0, \ldots, 0)^\top \in \mathbb{R}^{k+2},$$

where we have to find an orthogonal matrix G_j such that

$$G_j\underline{h}^j = \left(\beta_{j,0}, \ldots, \beta_{j,j-1}, \tilde{\beta}_{j,j}, 0, 0, \ldots, 0\right)^\top$$

is satisfied. For this, it is sufficient to consider the orthogonal matrix $\tilde{G}_j \in \mathbb{R}^{2\times 2}$ such that

$$\tilde{G}_j \begin{pmatrix} \beta_{j,j} \\ \beta_{j,j+1} \end{pmatrix} = \begin{pmatrix} \tilde{\beta}_{j,j} \\ 0 \end{pmatrix}$$

is fulfilled. The orthogonal matrix $\tilde{G}_j \in \mathbb{R}^{2\times 2}$ allows the general representation

$$\tilde{G}_j = \begin{pmatrix} a_j & b_j \\ -b_j & a_j \end{pmatrix}, \quad a_j^2 + b_j^2 = 1,$$

where the coefficients a_j and b_j can be found from the condition

$$-b_j \beta_{j,j} + a_j \beta_{j,j+1} = 0$$

as

$$a_j = \frac{\beta_{j,j}}{\sqrt{\beta_{j,j}^2 + \beta_{j,j+1}^2}}, \quad b_j = \frac{\beta_{j,j+1}}{\sqrt{\beta_{j,j}^2 + \beta_{j,j+1}^2}},$$

and, therefore, when assuming $\beta_{j,j+1} > 0$,

$$\tilde{\beta}_{j,j} = a_j \beta_{j,j} + b_j \beta_{j,j+1} = \sqrt{\beta_{j,j}^2 + \beta_{j,j+1}^2} > 0. \tag{C.12}$$

For $j = 0, \ldots, k$ the resulting orthogonal matrices G_j are of the form

$$G_j = \begin{pmatrix} 1 & & & & & & \\ & \ddots & & & & & \\ & & 1 & & & & \\ & & & a_j & b_j & & \\ & & & -b_j & a_j & & \\ & & & & 1 & & \\ & & & & & \ddots & \\ & & & & & & 1 \end{pmatrix} \in \mathbb{R}^{(k+2)\times(k+2)}$$

with $G_j[j,j] = G_j[j+1,j+1] = a_j$. Their recursive application gives

$$G_k G_{k-1} \ldots G_2 G_1 G_0 H_k = G_k G_{k-1} \ldots G_2 G_1 G_0 \begin{pmatrix} \beta_{0,0} & \beta_{1,0} & \cdots & \beta_{k,0} \\ \beta_{0,1} & \beta_{1,1} & \cdots & \beta_{k,1} \\ 0 & \beta_{1,2} & \ddots & \vdots \\ & 0 & \ddots & \beta_{k,k} \\ & & & \beta_{k,k+1} \end{pmatrix}$$

$$= G_k G_{k-1} \ldots G_2 G_1 \begin{pmatrix} \tilde{\beta}_{0,0} & \tilde{\beta}_{1,0} & \cdots & \tilde{\beta}_{k,0} \\ 0 & \bar{\beta}_{1,1} & \cdots & \bar{\beta}_{k,1} \\ & \beta_{1,2} & \ddots & \vdots \\ & & \ddots & \beta_{k,k} \\ & & & \beta_{k,k+1} \end{pmatrix}$$

$$= G_k G_{k-1} \ldots G_2 \begin{pmatrix} \tilde{\beta}_{0,0} & \tilde{\beta}_{1,0} & \cdots & \tilde{\beta}_{k,0} \\ 0 & \tilde{\beta}_{1,1} & \cdots & \tilde{\beta}_{k,1} \\ & 0 & \ddots & \vdots \\ & & \ddots & \tilde{\beta}_{k,k} \\ & & & \tilde{\beta}_{k,k+1} \end{pmatrix}$$

$$= \begin{pmatrix} \tilde{\beta}_{0,0} & \tilde{\beta}_{1,0} & \cdots & \tilde{\beta}_{k,0} \\ 0 & \tilde{\beta}_{1,1} & \cdots & \tilde{\beta}_{k,1} \\ & 0 & \ddots & \vdots \\ & & \ddots & \tilde{\beta}_{k,k} \\ & & & 0 \end{pmatrix} = R_k.$$

Hence, we have constructed the orthogonal matrix

$$Q_k = G_k G_{k-1} \ldots G_1 G_0 \in \mathbb{R}^{(k+2)\times(k+2)},$$

which fulfils

$$Q_k \underline{e}^0 = G_k \ldots G_0 \begin{pmatrix} 1 \\ 0 \\ 0 \\ \vdots \\ 0 \\ 0 \end{pmatrix} = G_k \ldots G_1 \begin{pmatrix} a_0 \\ -b_0 \\ 0 \\ \vdots \\ 0 \\ 0 \end{pmatrix}$$

$$= G_k \ldots G_2 \begin{pmatrix} a_0 \\ a_1(-b_0) \\ (-b_0)(-b_1) \\ \vdots \\ 0 \\ 0 \end{pmatrix} = \begin{pmatrix} a_0 \\ a_1(-b_0) \\ a_2(-b_0)(-b_1) \\ \vdots \\ a_k(-b_0)\cdots(-b_{k-1}) \\ (-b_0)\cdots(-b_k) \end{pmatrix} \in \mathbb{R}^{k+2}.$$

From this, we find

$$\varrho_{k+1} = \|\underline{e}^0\|_2 |(Q_k \underline{e}^0)_{k+1}| = \|\underline{e}^0\|_2 \prod_{j=0}^{k} b_j.$$

With the definition of

$$b_j = \frac{\beta_{j,j+1}}{\sqrt{\beta_{j,j+1}^2 + \beta_{j,j}^2}} = \frac{\|\hat{\underline{v}}^j\|_2}{\sqrt{\|\hat{\underline{v}}^j\|_2^2 + (A\underline{v}^j, \underline{v}^j)^2}},$$

we conclude $b_j < 1$ when assuming $(A\underline{v}^j, \underline{v}^j) \neq 0$. Hence, the error is monotonic decreasing. In the case of the break down situation in the method of Arnoldi (Algorithm C.3), i.e. $\|\hat{\underline{v}}^k\|_2 = 0$, we find $b_k = 0$, and, therefore, $\varrho_{k+1} = \|\underline{r}^{k+1}\|_2 = 0$. In particular, $\underline{x}^{k+1} = \underline{x}$ is the solution of the linear system $A\underline{x} = \underline{f}$.

Summarising the above, we obtain the Generalised Method of the Minimal Residual (GMRES) as described in Algorithm C.4, see [96].

Algorithm C.4

1. Compute for an arbitrary given initial solution $\underline{x}^0 \in \mathbb{R}^N$

$$\underline{r}^0 = A\underline{x}^0 - \underline{f}, \quad \varrho_0 = \|\underline{r}^0\|_2, \quad \underline{v}^0 = \frac{1}{\varrho_0}\underline{r}^0, \quad p_0 = \varrho_0.$$

2. Iterate for $k = 0, \ldots, N-2$

$$\underline{w}^k = A\underline{v}^k, \quad \hat{\underline{v}}^{k+1} = \underline{w}^k - \sum_{\ell=0}^{k} \beta_{k\ell}\,\underline{v}^\ell, \quad \beta_{k\ell} = (\underline{w}^k, \underline{v}^\ell), \quad \beta_{kk+1} = \|\hat{\underline{v}}^{k+1}\|_2.$$

   Go to 3. if $\beta_{kk+1} = 0$ is satisfied.
   Otherwise compute
$$\underline{v}^{k+1} = \frac{1}{\beta_{kk+1}}\hat{\underline{v}}^{k+1}.$$

   For $\ell = 0, \ldots, k-1$ compute

$$\tilde{\beta}_{k\ell} = a_\ell\beta_{k\ell} + b_\ell\beta_{k\ell+1}, \quad \tilde{\beta}_{k\ell+1} = -b_\ell\beta_{k\ell} + a_\ell\beta_{k\ell+1}$$

   and

$$a_k = \frac{\beta_{kk}}{\sqrt{\beta_{kk}^2 + \beta_{kk+1}^2}}, \quad b_k = \frac{\beta_{kk+1}}{\sqrt{\beta_{kk}^2 + \beta_{kk+1}^2}}, \quad \tilde{\beta}_{kk} = \sqrt{\beta_{kk}^2 + \beta_{kk+1}^2}$$

   as well as

$$p_{k+1} = -b_k p_k, \quad p_k = a_k p_k, \quad \varrho_{k+1} = |p_{k+1}|.$$

   Stop, if $\varrho_{k+1} < \varepsilon\varrho_0$ is satisfied with some prescribed accuracy ε.

3. Compute the approximate solution, i.e. for $\ell = k, k-1, \ldots, 0$

$$\alpha_\ell = \frac{1}{\beta_{\ell\ell}}\left(p_\ell - \sum_{j=\ell+1}^{k} \beta_{\ell j}\alpha_j\right)$$

   and

$$\underline{x}^{k+1} = \underline{x}^0 - \sum_{\ell=0}^{k} \alpha_\ell\underline{v}^\ell.$$

References

1. B. Alpert, G. Beylkin, R. Coifman, and V. Rokhlin. Wavelet-like bases for the fast solution of second-kind integral equations. *SIAM J. Sci. Comput.*, 14(1):159–184, 1993.
2. S. Amini and A. T. J. Profit. Analysis of a diagonal form of the fast multipole algorithm for scattering theory. *BIT*, 39(4):585–602, 1999.
3. S. Amini and A. T. J. Profit. Multi-level fast multipole Galerkin method for the boundary integral solution of the exterior Helmholtz equation. In *Current trends in scientific computing (Xi'an, 2002)*, volume 329 of *Contemp. Math.*, pages 13–19. Amer. Math. Soc., Providence, RI, 2003.
4. D. N. Arnold and W. L. Wendland. On the asymptotic convergence of collocation methods. *Math. Comp.*, 41:349–381, 1983.
5. D. N. Arnold and W. L. Wendland. The convergence of spline collocation for strongly elliptic equations on curves. *Numer. Math.*, 47:317–341, 1985.
6. K. E. Atkinson. *The Numerical Solution of Integral Equations of the Second Kind.* Cambridge University Press, 1997.
7. M. Bebendorf. Approximation of boundary element matrices. *Numer. Math.*, 86(4):565–589, 2000.
8. M. Bebendorf and S. Rjasanow. Matrix compression for the radiation heat transfer in exhaust pipes. In W.L. Wendland, A.-M. Sändig, and W. Schiehlen, editors, *Multifield Problems. State of the Art*, pages 183–192. Springer, 2000.
9. M. Bebendorf and S. Rjasanow. Adaptive low–rank approximation of collocation matrices. *Computing*, 70(1):1–24, 2003.
10. M. Bebendorf and S. Rjasanow. Numerical simulation of exhaust systems in car industry - Efficient calculation of radiation heat transfer. In W. Jäger and H.-J. Krebs, editors, *Mathematics. Key Technology for the Future*, pages 55–62. Springer, 2003.
11. G. Beylkin, R. Coifman, and V. Rokhlin. Fast wavelet transforms and numerical algorithms. I. *Comm. Pure Appl. Math.*, 44(2):141–183, 1991.
12. M. Bonnet. *Boundary Integral Equation Methods for Solids and Fluids.* John Wiley & Sons, Chichester, 1999.
13. H. Brakhage and P. Werner. Über das Dirichletsche Aussenraumproblem für die Helmholtzsche Schwingungsgleichung. *Arch. Math.*, 16:325–329, 1965.
14. J. H. Bramble and J. E. Pasciak. A preconditioning technique for indefinite systems resulting from mixed approximations of elliptic problems. *Math. Comp.*, 50:1–17, 1988.

15. C. A. Brebbia, J. C. F. Telles, and L. C. Wrobel. *Boundary Element Techniques: Theory and Applications in Engineering.* Springer, Berlin, 1984.

16. A. Buchau, S. Kurz, O. Rain, V. Rischmüller, S. Rjasanow, and W. M. Rucker. Comparison between different approaches for fast and efficient 3D BEM computations. *IEEE Transaction on Magnetics*, 39(2):1107–1110, 2003.

17. A. Buffa, R. Hiptmair, T. von Petersdorff, and C. Schwab. Boundary element methods for Maxwell transmission problems in Lipschitz domains. *Numer. Math*, 95(3):459–485, 2003.

18. A. Buffa and S. Sauter. On the acoustic single layer potential: Stabilisation and Fourier analysis. *SIAM, J. Sci. Comp.*, 28(5):1974–1999, 2006.

19. A. J. Burton and G. F. Miller. The application of integral equation methods to the numerical solution of some exterior boundary-value problems. *Proc. Roy. Soc. London. Ser. A*, 323:201–210, 1971.

20. J. Carrier, L. Greengard, and V. Rokhlin. A fast adaptive multipole algorithm for particle simulations. *SIAM J. Sci. Statist. Comput.*, 9(4):669–686, 1988.

21. G. Chen and J. Zhou. *Boundary Element Methods.* Academic Press, New York, 1992.

22. A. H.-D. Cheng and D. T. Cheng. Heritage and early history of the boundary element method. *Engrg. Anal. Boundary Elements*, 29:268–302, 2005.

23. D. Colton and R. Kress. *Inverse acoustic and electromagnetic scattering theory.* Springer, Berlin, 1992.

24. M. Costabel. Boundary integral operators on Lipschitz domains: Elementary results. *SIAM J. Math. Anal.*, 19:613–626, 1988.

25. M. Costabel. Some historical remarks on the positivity of boundary integral operators. In M. Schanz and O. Steinbach, editors, *Boundary Element Analysis: Mathematical Aspects and Applications*. Lecture Notes is Applied and Computational Mechanics, vol. 29, Springer, Heidelberg, pp. 1–27, 2007.

26. M. Costabel and W. L. Wendland. Strong ellipticity of boundary integral operators. *Crelle's J. Reine Angew. Math.*, 372:34–63, 1986.

27. W. Dahmen, S. Prössdorf, and R. Schneider. Wavelet approximation methods for pseudodifferential equations. II. Matrix compression and fast solution. *Adv. Comput. Math.*, 1(3-4):259–335, 1993.

28. W. Dahmen, S. Prössdorf, and R. Schneider. Wavelet approximation methods for pseudodifferential equations. I. Stability and convergence. *Math. Z.*, 215(4):583–620, 1994.

29. R. Dautray and J. L. Lions. *Mathematical Analysis and Numerical Methods for Science and Technology. Volume 4: Integral Equations and Numerical Methods.* Springer, Berlin, 1990.

30. A. Douglis and L. Nirenberg. Interior estimates for elliptic systems of partial differential equations. *Commun. Pure Appl. Math.*, 8:503–538, 1955.

31. H. Forster, T. Schrefl, R. Dittrich, W. Scholz, and J. Fidler. Fast Boundary Methods for Magnetostatic Interactions in Micromagnetics. *IEEE Transaction on Magnetics*, 39(5):2513–2515, 2003.

32. M. Ganesh and O. Steinbach. Nonlinear boundary integral equations for harmonic problems. *J. Int. Equations Appl.*, 11:437–459, 1999.

33. C. F. Gauß. *Allgemeine Theorie des Erdmagnetismus.*, volume 5 of *Werke.* Dieterich, Göttingen, 1838, 1867.

34. C. F. Gauß. *Atlas des Erdmagnetismus*, volume 5 of *Werke.* Dieterich, Göttingen, 1840, 1867.

35. I. M. Gel'fand and G. E. Shilov. *Generalized functions. Vol. I: Properties and operations.* Translated by Eugene Saletan. Academic Press, New York, 1964.

36. S. A. Goreinov, E. E. Tyrtyshnikov, and N. L. Zamarashkin. A theory of pseudoskeleton approximations. *Linear Algebra Appl.*, 261:1–21, 1997.

37. L. Greengard and V. Rokhlin. A fast algorithm for particle simulations. *J. Comput. Phys.*, 73(2):325–348, 1987.

38. L. Greengard and V. Rokhlin. The rapid evaluation of potential fields in three dimensions. In *Vortex methods (Los Angeles, CA, 1987)*, pages 121–141. Springer, Berlin, 1988.

39. L. Greengard and V. Rokhlin. A new version of the fast multipole method for the Laplace equation in three dimensions. In *Acta numerica, 1997*, pages 229–269. Cambridge Univ. Press, Cambridge, 1997.

40. W. Hackbusch, editor. *Boundary Elements: Implementation and Analysis of Advanced Algorithms*, volume 54 of *Notes on Numerical Fluid Mechanics*, Vieweg, Braunschweig, 1996.

41. W. Hackbusch. A sparse matrix arithmetic based on $\mathcal{H}$-matrices. I. Introduction to $\mathcal{H}$-matrices. *Computing*, 62(2):89–108, 1999.

42. W. Hackbusch and B. N. Khoromskij. A sparse $\mathcal{H}$–matrix arithmetic. II. Application to multi–dimensional problems. *Computing*, 64:21–47, 2000.

43. W. Hackbusch, C. Lage, and S. A. Sauter. On the efficient realization of sparse matrix techniques for integral equations with focus on panel clustering, cubature and software design aspects. In *Boundary element topics (Stuttgart, 1995)*, pages 51–75. Springer, Berlin, 1997.

44. W. Hackbusch and Z. P. Nowak. On the fast matrix multiplication in the boundary element method by panel clustering. *Numer. Math.*, 54(4):463–491, 1989.

45. W. Hackbusch and S. A. Sauter. On the efficient use of the Galerkin method to solve Fredholm integral equations. *Appl. Math.*, 38:301–322, 1993.

46. W. Hackbusch and S. A. Sauter. On the efficient use of the Galerkin method to solve Fredholm integral equations. *Appl. Math.*, 38(4-5):301–322, 1993. Proceedings of ISNA '92—International Symposium on Numerical Analysis, Part I (Prague, 1992).

47. H. Han. The boundary integro–differential equations of three–dimensional Neumann problem in linear elasticity. *Numer. Math.*, 68:269–281, 1994.

48. H. Harbrecht and R. Schneider. Biorthogonal wavelet bases for the boundary element method. *Math. Nachr.*, 269/270:167–188, 2004.

49. H. Harbrecht and R. Schneider. Wavelet Galerkin schemes for boundary integral equations – implementation and quadrature. *SIAM J. Sci. Comput.*, 27:1347–1370, 2006.

50. F. Hartmann. *Introduction to Boundary Elements.* Springer, Berlin, 1989.

51. K. Hayami and S. A. Sauter. A panel clustering method for 3-D elastostatics using spherical harmonics. In *Integral methods in science and engineering (Houghton, MI, 1998)*, volume 418 of *Chapman & Hall/CRC Res. Notes Math.*, pages 179–184. Chapman & Hall/CRC, Boca Raton, FL, 2000.

52. D. Hilbert. *Grundzüge einer allgemeinen Theorie linearer Integralgleichungen.* Teubner, Leipzig, 1912.

53. L. Hörmander. *The Analysis of Linear Partial Differential Operators I.* Springer, Berlin, 1983.

54. G. C. Hsiao, P. Kopp, and W. L. Wendland. A Galerkin collocation method for some integral equations of the first kind. *Computing*, 25:89–130, 1980.

55. G. C. Hsiao and R. MacCamy. Solution of boundary value problems by integral equations of the first kind. *SIAM Rev.*, 15:687–705, 1973.

56. G. C. Hsiao, E. P. Stephan, and W. L. Wendland. On the integral equation method for the plane mixed boundary value problem of the Laplacian. *Math. Meth. Appl. Sci.*, 1:265–321, 1979.

57. G. C. Hsiao and W. L. Wendland. A finite element method for some integral equations of the first kind. *J. Math. Anal. Appl.*, 58:449–481, 1977.

58. G. C. Hsiao and W. L. Wendland. The Aubin–Nitsche lemma for integral equations. *J. Int. Equat.*, 3:299–315, 1981.

59. M. A. Jaswon and G. T. Symm. *Integral Equation Methods in Potential Theory and Elastostatics*. Academic Press, London, 1977.

60. B. N. Khoromskij and G. Wittum. *Numerical Solution of Elliptic Differential Equations by Reduction to the Interface*. Lecture Notes in Computational Science and Engineering, 36. Springer, Berlin, 2004.

61. R. Kress. *Linear Integral Equations*. Springer, Heidelberg, 1999.

62. V. D. Kupradze. *Three–dimensional problems of the mathematical theory of elasticity and thermoelasticity*. North–Holland, Amsterdam, 1979.

63. S. Kurz, O. Rain, V. Rischmüller, and S. Rjasanow. Periodic and Anti-Periodic Symmetries in the Boundary Element Method. In *Proceedings of 10th international IGTE Symposium, Graz, Austria*, pages 375–380, 2002.

64. S. Kurz, O. Rain, and S. Rjasanow. The Adaptive Cross Approximation Technique for the 3D Boundary Element Method. *IEEE Transaction on Magnetics*, 38(2):421–424, 2002.

65. S. Kurz, O. Rain, and S. Rjasanow. Application of the Adaptive Cross Approximation Technique for the Coupled BE-FE Solution of Electromagnetic Problems. *Comput. Mech.*, 32(4–6):423–429, 2003.

66. P. K. Kythe. *Fundamental Solutions for Differential Operators and Applications*. Birkhäuser, Boston, 1996.

67. C. Lage and C. Schwab. Wavelet Galerkin algorithms for boundary integral equations. *SIAM J. Sci. Comput.*, 20(6):2195–2222 (electronic), 1999.

68. U. Lamp, T. Schleicher, E. P. Stephan, and W. L. Wendland. Galerkin collocation for an improved boundary element method for a plane mixed boundary value problem. *Computing*, 33:269–296, 1984.

69. U. Langer and D. Pusch. Data-sparse algebraic multigrid methods for large scale boundary element equations. *Appl. Numer. Math.*, 54(3-4):406–424, 2005.

70. V. G. Mazya. Boundary integral equations. In V. G. Mazya and S. M. Nikolskii, editors, *Analysis IV*, volume 27 of *Encyclopaedia of Mathematical Sciences*, pages 127–233. Springer, Heidelberg, 1991.

71. W. McLean. *Strongly Elliptic Systems and Boundary Integral Equations*. Cambridge University Press, 2000.

72. S. G. Michlin. *Integral Equations*. Pergamon Press, London, 1957.

73. C. M. Miranda. *Partial Differential Equations of Elliptic Type*. Springer, Berlin, 1970.

74. N. I. Muskhelishvili. *Singular Integral Equations*. Noordhoff, Groningen, 1953.

75. T. Nakata, N. Takahashi, and K. Fujiwara. Summary of results for benchmark problem 10 (steel plates around a coil). *COMPEL*, 11:335–344, 1992.

76. J. C. Nédélec. Curved finite element methods for the solution of singular integral equations on surfaces in $\mathbb{R}^3$. *Comp. Meth. Appl. Mech. Engrg.*, 8:61–80, 1976.

77. J. C. Nédélec. Integral equations with non integrable kernels. *Int. Eq. Operator Th.*, 5:562–572, 1982.

78. J. C. Nédélec. *Acoustic and Electromagnetic Equations*. Springer, New York, 2001.

79. J. C Nédélec and J. Planchard. Une methode variationelle d'elements finis pour la resolution numerique d'un probleme exterieur dans $\mathbb{R}^3$. *R.A.I.R.O.*, 7:105–129, 1973.

80. C. Neumann. *Untersuchungen über das Logarithmische und Newtonsche Potential*. Teubner, Leipzig, 1877.

81. E. J. Nyström. Über die praktische Auflösung von linearen Integralgleichungen mit Anwendungen auf Randwertaufgaben der Potentialtheorie. *Soc. Sci. Fenn. Comment. Phys. Math.*, 4:1–52, 1928.

82. E. J. Nyström. Über die praktische Auflösung von linearen Integralgleichungen mit Anwendungen auf Randwertaufgaben. *Acta Math.*, 54:185–204, 1930.

83. G. Of, O. Steinbach, and W. L. Wendland. Applications of a fast multipole Galerkin boundary element method in linear elastostatics. *Comput. Vis. Sci.*, 8(3-4):201–209, 2005.

84. G. Of, O. Steinbach, and W. L. Wendland. The fast multipole method for the symmetric boundary integral formulation. *IMA J. Numer. Anal.*, 26(2):272–296, 2006.

85. N. Ortner. Fundamentallösungen und Existenz von schwachen Lösungen linearer partieller Differentialgleichungen mit konstanten Koeffizienten. *Ann. Acad. Sci. Fenn. Ser. A I Math.*, 4:3–30, 1979.

86. N. Ortner and P. Wagner. A survey on explicit representation formulae for fundamental solutions of linear partial differential operators. *Acta Appl. Math.*, 47:101–124, 1997.

87. J. Ostrowski, Z. Andjelić, M. Bebendorf, B. Crânganu-Cre₊tu, and J. Smajić. Fast BEM-Solution of Laplace Problems with H-Matrices and ACA. *IEEE Trans. on Magnetics*, 42(4):627–630, 2006.

88. J. Plemelj. *Potentialtheoretische Untersuchungen*. Teubner, Leipzig, 1911.

89. A. Pomp. *The Boundary–Domain Integral Method for Elliptic Systems. With an Application to Shells*, volume 1683 of *Lecture Notes in Mathematics*. Springer, Berlin, 1998.

90. B. Reidinger and O. Steinbach. A symmetric boundary element method for the Stokes problem in multiple connected domains. *Math. Methods Appl. Sci.*, 26:77–93, 2003.

91. F. J. Rizzo. An integral equation approach to boundary value problems of classical elastostatics. *Quart. Appl. Math.*, 25:83–95, 1967.

92. D. Rodger, N. Allen, H. C. Lai, and Leonard P. J. Calculation of transient 3D eddy currents in nonlinear media = verification using a rotational test rig. *IEEE Transaction on Magnetics*, 30(5(2)):2988–2991, 1994.

93. V. Rokhlin. Rapid solution of integral equations of classical potential theory. *J. Comput. Phys.*, 60(2):187–207, 1985.

94. V. Rokhlin and M. A. Stalzer. Scalability of the fast multipole method for the Helmholtz equation. In *Proceedings of the Eighth SIAM Conference on Parallel Processing for Scientific Computing (Minneapolis, MN, 1997)*, page 8 pp. (electronic), Philadelphia, PA, 1997. SIAM.

95. K Ruotsalainen and W. L. Wendland. On the boundary element method for some nonlinear boundary value problems. *Numer. Math.*, 53:299–314, 1988.

96. Y. Saad and M. H. Schultz. GMRES: a generalized minimal residual algorithm for solving nonsymmetric linear systems. *SIAM J. Sci. Statist. Comput.*, 7(3):856–869, 1986.

97. S. A. Sauter and A. Krapp. On the effect of numerical integration in the Galerkin boundary element method. *Numer. Math.*, 74:337–360, 1996.

98. S. A. Sauter and C. Lage. Transformation of hypersingular integrals and black–box cubature. *Math. Comp.*, 70:223–250, 2001.

99. S. A. Sauter and C. Schwab. *Randelementmethoden. Analyse, Numerik und Implementierung schneller Algorithmen.* B. G. Teubner, Stuttgart, Leipzig, Wiesbaden, 2004.

100. G. Schmidt. On spline collocation methods for boundary integral equations in the plane. *Math. Meth. Appl. Sci.*, 7:74–89, 1985.

101. G. Schmidt. Spline collocation for singular integro–differential equations over (0,1). *Numer. Math.*, 50:337–352, 1987.

102. C. Schwab and W. L. Wendland. On numerical cubatures of singular surface integrals in boundary element methods. *Numer. Math.*, 62:343–369, 1992.

103. S. Sirtori. General stress analysis method by means of integral equations and boundary elements. *Meccanica*, 14:210–218, 1979.

104. I. H. Sloan. Error analysis of boundary integral methods. *Acta Numerica*, 92:287–339, 1992.

105. O. Steinbach. *Numerische Näherungsverfahren für elliptische Randwertprobleme. Finite Elemente und Randelemente.* B. G. Teubner, Stuttgart, Leipzig, Wiesbaden, 2003.

106. O. Steinbach. A robust boundary element method for nearly incompressible elasticity. *Numer. Math.*, 95:553–562, 2003.

107. O. Steinbach. *Stability Estimates for Hybrid Coupled Domain Decomposition Methods.* Springer Lecture Notes in Mathematics, 1809. Springer, Heidelberg, 2003.

108. O. Steinbach and W. L. Wendland. On C. Neumann's method for second order elliptic systems in domains with non–smooth boundaries. *J. Math. Anal. Appl.*, 262:733–748, 2001.

109. E. P. Stephan. A boundary integral equations for mixed boundary value problems in $\mathbb{R}^3$. *Math. Nachr.*, 131:167–199, 1987.

110. M. Stolper. Computing and compression of the boundary element matrices for the Helmholtz equation. *J. Numer. Math.*, 12(1):55–75, 2005.

111. M. Stolper and S. Rjasanow. A compression method for the Helmholtz equation. In *Numerical mathematics and advanced applications*, pages 786–795. Springer, Berlin, 2004.

112. G. Verchota. Layer potentials and regularity for the Dirichlet problem for Laplace's equation in Lipschitz domains. *J. Funct. Anal.*, 59:572–611, 1984.

113. O. von Estorff, S. Rjasanow, M. Stolper, and O. Zaleski. Two efficient methods for a multifrequency solution of the Helmholtz equation. *Comput. Vis. Sci.*, 8(2–4):159–167, 2005.

114. T. von Petersdorff. Boundary integral equations for mixed Dirichlet, Neumann and transmission problems. *Math. Meth. Appl. Sci.*, 11:185–213, 1989.

115. T. von Petersdorff and C. Schwab. Wavelet approximations for first kind boundary integral equations on polygons. *Numer. Math.*, 74(4):479–516, 1996.

116. T. von Petersdorff, C. Schwab, and R. Schneider. Multiwavelets for second-kind integral equations. *SIAM J. Numer. Anal.*, 34(6):2212–2227, 1997.

117. W. L. Wendland, editor. *Boundary Element Topics*, Heidelberg, 1997. Springer.

118. K. Yosida. *Functional Analysis*. Springer, Berlin, Heidelberg, New York, 1980.

119. K. Zhao, M. N. Vouvakis, and J.-F. Lee. The Adaptive Cross Approximation Algorithm for Accelerated Method of Moments Computations of EMC Problems. *IEEE Transactions on electromagnetic compatibility*, 47(4):763–773, 2005.

Index